The book is in the honor of Prof. Ilya Gertsbakh and Prof. Igor Ushakov, who have made a great pioneering contribution to all areas of reliability and, in particular, to the topics of signatures and multi-state systems.

Preface

This book covers the recent developments in modern reliability theory, mainly in such important areas as signatures and multi-state systems and their influence on statistical inference. Research in these areas is growing rapidly due to many successful applications in very diverse problems. As the result, many industries have benefited from adopting the corresponding methods.

These methods have attracted increasing attention in recent years for solving many complex problems which were inspired by nature and technology. New methods have been successfully applied to solving many complex problems where traditional problem-solving methods have failed.

This book presents new theoretical issues that were not previously presented in the literature, as well as the solutions of important practical problems and case studies illustrating the application methodology.

The book provides an overview of the recent developments in the theory of signatures and demonstrates their role in the study of dynamic reliability and nonparametric inference for lifetime distribution of monotone systems. New properties of system signatures (D-spectra) and component importance D-spectra have been investigated. It was demonstrated how component Birnbaum importance measures can be expressed via these spectra and how bounds on lifetime variances for coherent and mixed systems can be found by using signatures. In addition, it was pointed out on the connection between several aspects of probability-signature and structure-signature.

Concerning multi-state system (MSS) reliability, the book introduces a special transform for a discrete-states continuous-time Markov process, so-called L_Z-transform and demonstrates the benefits of its applications. In MSS context, there issues such as practical availability modeling, a case-study for supermarket refrigerating system, finding optimal reserve structure for power generating system, determination of vital activities in reliability program, optimal incomplete maintenance, optimal multi-objective reliability allocation, importance analysis based on multiple-valued logic methods, and optimal replacement and protection strategy were also considered. A separate chapter is devoted to the novel issue of continuous-state system reliability. Absorbing controllable Markov processes were considered as the models of aging and degradation for some technical and/or biological objects, as well as a semi-Markov model of MSS operation reliability.

The book aims to be repository for modern theoretical methods and their applications in real-world reliability analysis and optimization. Recent advances in statistical inference are presented in this volume by reliability analysis of redundant systems with unimodal hazard rate functions, nonparametric estimation of marginal temporal functionals, frailty models in survival analysis and reliability, goodness of fit tests for reliability modeling and nonparametric estimators of the transition probabilities for three-state Markov model.

All chapters are written by leading experts in the corresponding areas. This book will be useful to postgraduate and doctoral students, researchers, reliability practitioners, engineers and industrial managers with interest in reliability theory and its applications.

We wish to thank all the authors for their insights and excellent contributions to this book. We would like to acknowledge the assistance of all involved in the review process of the book, without whose support this book could not have been successfully completed. We want to thank all who participated in the reviewing process: Prof. Somnath Datta, University of Louisville, USA, Prof. Ilya Gertsbakh, Ben Gurion University of the Negev, Israel, Dr. Gregory Gurevich, SCE–Shamoon College of Engineering, Israel, Prof. Alex Karagrigoriou, University of Cyprus, Cyprus, Prof. Ron S. Kenett, KPA Ltd., Israel, Dr. Edward Korczak, Telecommunications Research Institute, Poland, Prof. Michael Peht, University of Maryland, USA, Prof. Dmitrii Silvestrov, Stockholm University, Sweden, Dr. Armen Stepanyants, Institute of Control Science, RAS, Russia, Prof. Guram Tsitsiashvili, Institute for Applied Mathematics, Eastern Branch of RAS, Russia, Dr. Valentina Victorova, Institute of Control Science, RAS, Russia, Prof. Ilia Vonta, National Technical University of Athens, Greece, Prof. David Zucker, Hebrew University of Jerusalem, Israel.

We would like to express our sincere appreciation to Prof. Ilya Gertsbakh from Ben-Gurion University, Israel, for his great impact on book preparation.

We would like also to thank the SCE–Shamoon College of Engineering (Israel), and its president, Prof. Jehuda Haddad and the SCE Industrial Engineering and Management Department and its dean Prof. Zohar Laslo for their support and ever present help at all stages of the work.

Our special thanks to Boris Frenkel from Kivitek Ltd. (Israel) for his technical assistance and help.

It was indeed our pleasure working with the Springer Senior Editorial Assistant, Ms. Claire Protherough.

Israel, May 2011   Anatoly Lisnianski
                                                                        Ilia Frenkel

Contents

Contributors

N. Balakrishnan Department of Mathematics and Statistics, McMaster University, Hamilton, Canada, e-mail: bala@univmail.cis.mcmaster.ca

Emil Bashkansky Department of Industrial Engineering and Management, Ort Braude College, Karmiel, Israel, e-mail: ebashkan@braude.ac.il

Roberta Bazzo Dipartimento di Energia, Politecnico di Milano, Milan, Italy, e-mail: roberta.bazzo@alice.it

Somnath Datta Department of Bioinformatics and Biostatistics, University of Louisville, Louisville, KY, USA, e-mail: somnath.datta@louisville.edu

Jacobo deUña-Álvarez Department of Statistics and OR, Facultad de CC Económicas y Empresariales, Universidad de Vigo, Vigo, Spain, e-mail: jacobo@uvigo.es

Yi Ding Department of Electrical Engineering, Centre for Electric Technology, Technical University of Denmark, Lyngby, Denmark, e-mail: yding@elektro.dtu.dk

Shuki Dror Department of Industrial Engineering and Management, Ort Braude College, Karmiel, Israel, e-mail: dror@braude.ac.il

Dmitry Efrosinin Institut for Stochastic, Johannes Kepler University of Linz, Linz, Austria, e-mail: dmitry.efrosinin@jku.at

A. Nicole Ferguson Department of Bioinformatics and Biostatistics, University of Louisville, Louisville, KY, USA, e-mail: nicole.ferguson@louisville.edu

Ilia Frenkel Industrial Engineering and Management Department, Center for Reliability and Risk Management, Sami Shamoon College of Engineering, Beer Sheva, Israel, e-mail: iliaf@sce.ac.il

Tamar Gadrich Department of Industrial Engineering and Management, Ort Braude College, Karmiel, Israel, e-mail: tamarg@braude.ac.il

Ilya Gertsbakh Department of Mathematics, Ben Gurion University of the Negev, Beer Sheva, Israel, e-mail: elyager@bezeqint.net

Lalit Goel Division of Power Engineering, School of Electrical and Electronic Engineering College of Engineering, Nanyang Technological University, Singapore, Singapore, e-mail: elkgoel@ntu.edu.sg

Franciszek Grabski Department of Mathematics and Physics, Polish Naval University, Gdynia, Poland, e-mail: franciszekgr@onet.eu

Waltraud Kahle Institute of Mathematical Stochastics, Otto-von-Guericke-University, Magdeburg, Germany, e-mail: waltraud.kahle@ovgu.de

Alex Karagrigoriou Department of Mathematics and Statisrics, University of Cyprus, Nicosia, Cyprus, e-mail: alex@ucy.ac.cy

Lev Khvatskin Industrial Engineering and Management Department, Center for Reliability and Risk Management, Sami Shamoon College of Engineering, Beer Sheva, Israel, e-mail: khvat@sce.ac.il

Dong Seong Kim Department of Electrical and Computer Engineering, Duke University, Durham, NC, 27708, USA, e-mail: dk76@duke.edu

Gregory Levitin, The Israel Electric Corporation Ltd, Haifa, Israel, e-mail: levitin@iec.co.il

Anatoly Lisnianski The Israel Electric Corporation Ltd, Haifa, Israel, e-mail: anatoly-l@iec.co.il e-mail: lisnians@zahav.net.il

Poh Chiang Loh Division of Power Engineering, School of Electrical and Electronic Engineering College of Engineering, Nanyang Technological University, Singapore, Singapore, e-mail: epcloh@ntu.edu.sg

Jorge Navarro Facultad de Matemáticas, Universidad de Murcia, Murcia, Spain, e-mail: jorgenav@um.es

Szu Hui Ng Department of Industrial and Systems Engineering, National University of Singapore, Kent Ridge Crescent, Singapore, 119260, Singapore, e-mail: isensh@nus.edu.sg

Mikhail Nikulin IMB, Université Victor Segalen Bordeaux 2, Bordeaux 2, France, e-mail: mikhail.nikouline@u-bordeaux2.fr

Rui Peng Department of Industrial and Systems Engineering, National University of Singapore, Kent Ridge Crescent, Singapore, 119260, Singapore, e-mail: g0700981@nus.edu.sg

Tomasz Rychlik Institute of Mathematics, Polish Academy of Sciences, Chopina 12, Torun, 87100, Poland, e-mail: trychlik@impan.gov.pl

Vladimir Rykov Department of Applied Mathematics and Computer Modeling, Russian State University of Oil and Gas, Moscow, Russia, e-mail: vladimir_rykov@mail.ru

Noureddine Saaidia IMB, Université Victor Segalen Bordeaux 2, Bordeaux 2, France, e-mail: mikhail.nikouline@u-bordeaux2.f; Université Badji Mokhtar, Annaba, Algérie, e-mail: saaidianoureddine@yahoo.fr

Francisco J. Samaniego Department of Statistics, University of California, Davis, USA, e-mail: fjsamaniego@ucdavis.edu

Yoseph Shpungin Software Engineering Department, Sami Shamoon College of Engineering, Beer Sheva, Israel, e-mail: yosefs@sce.ac.il

Fabio Spizzichino Department of Mathematics, Università La Sapienza, Rome, Italy, e-mail: fabio.spizzichino@uniroma1.it

Yuanzhang Sun School of Electrical Engineering, Wuhan University, Wuhan, P.R. China, e-mail: yzsun@whu.edu.cn

Ramzan Tahir IMB, Université Victor Segalen Bordeaux 2, Bordeaux 2, France, e-mail: ramzantahir7@gmail.com

Kishor S. Trivedi Department of Electrical and Computer Engineering, Duke University, Durham, NC, 27708, USA, e-mail: kst@ee.duke.edu

Kobi Tsuri Department of Industrial Engineering and Management, Ort Braude College, Karmiel, Israel, e-mail: zuria10@walla.com

Filia Vonta Department of Mathematics, School of Applied Mathematical and Physical Sciences, National Technical University of Athens, Athens, Greece, e-mail: vonta@math.ntua.gr

Peng Wang Division of Power Engineering, School of Electrical and Electronic Engineering College of Engineering, Nanyang Technological University, Singapore, Singapore, e-mail: epwang@ntu.edu.sg

Qiuwei Wu Department of Electrical Engineering, Centre for Electric Technology, Technical University of Denmark, Lyngby, Denmark, e-mail: qw@elektro.dtu.dk

Min Xie Department of Industrial and Systems Engineering, National University of Singapore, Kent Ridge Crescent, Singapore, 119260, Singapore, e-mail: mxie@nus.edu.sg

Xiaoyan Yin Department of Electrical and Computer Engineering, Duke University, Durham, NC, 27708, USA, e-mail: xy15@duke.edu

Elena Zaitseva Department of Informatics, Faculty of Management Science and Informatics, University of Žilina, Žilina, Slovakia, e-mail: elena.zaitseva@fri.uniza.sk

E. Zio Chair "Systems Science and Energetic Challenge"European Foundation for New Energy – EDF, Ecole Centrale Paris- Supelec, Paris, France, e-mail: enrico.zio@ecp.fr e-mail: enrico.zio@polimi.it

Chapter 1
Signature Representation and Preservation Results for Engineered Systems and Applications to Statistical Inference

N. Balakrishnan, Jorge Navarro and Francisco J. Samaniego

Abstract The aim of this article is to provide an overview of some of the recent developments relating to the theory of signatures and their role in the study of dynamic reliability, systems with shared components and nonparametric inference for a component lifetime distribution. Some new results and interpretations are also presented in the process.

Keywords Coherent system · Mixed system · Signature · k-out-of-n system · Exchangeability · Order statistics · Stochastic ordering · Hazard rate ordering · Likelihood ratio ordering · Dynamic reliability · Dynamic signature · Systems with shared components · Burned-in systems · D-spectrum · Nonparametric inference · Parametric inference · Best linear unbiased estimator · Proportional hazard rate model

1.1 A Brief Overview of Signature Theory

Most work on reliability theory focuses on the study of coherent systems; see, for example, (Barlow and Proschan 1975). A reliability system is said to be a *coherent system* if

N. Balakrishnan (✉)
Department of Mathematics and Statistics, McMaster University,
Hamilton L8S 4K1, Canada
e-mail: bala@univmail.cis.mcmaster.ca

J. Navarro
Universidad de Murcia, Murcia, Spain
e-mail: jorgenav@um.es

F. J. Samaniego
Department of Statistics, University of California, Davis, USA
e-mail: fjsamaniego@ucdavis.edu

A. Lisnianski and I. Frenkel (eds.), *Recent Advances in System Reliability*,
Springer Series in Reliability Engineering, DOI: 10.1007/978-1-4471-2207-4_1,
© Springer-Verlag London Limited 2012

- it is monotone in its components (i.e., replacing a failed component by a working one cannot make the system worse), and
- every component is relevant (i.e., every component influences the functioning or failure of the system).

Suppose a coherent system has n components whose lifetimes $X_1,..., X_n$ are independent and identically distributed (i.i.d.) continuous random variables with distribution function F. Let $X_{1:n} < \cdots < X_{n:n}$ be the order statistics obtained by arranging the component lifetimes X_i's in increasing order of magnitude. Then, the system lifetime T will coincide with an order statistic $X_{i:n}$ for some $i \in \{1,..., n\}$, which led to the following notion of *system signature* in a natural way. Let s_i, for $i = 1,..., n$, be such that $P(T = X_{i:n}) = s_i$. Then, the system signature is simply the vector $\mathbf{s} = (s_1,..., s_n)$, as introduced by Samaniego (1985). Signatures are most useful in the comparison of system designs as amply demonstrated in the books by Samaniego (2007) and Gertsbakh and Shpungin (2010). The system signature is a pure distribution-free measure of a system's design, and it is important to recognize that it dissociates the quality or reliability of the components from that of the system. The comparison of systems otherwise becomes very difficult since a series system with 4 good components can outperform a parallel system with 4 poor components. Thus, signature vectors enable us to compare the performance characteristics of different systems in a complete nonparametric way without reference to the lifetime distribution of the components.

The signature vector facilitates the following representation for the system reliability.

Theorem 1 (Samaniego 1985) *Let $X_1,..., X_n \sim F$ be the i.i.d. component lifetimes of a coherent system of order n, and let T be the system lifetime and $\mathbf{s}$ its signature. Then,*

$$\overline{F}_T(t) = \Pr(T > t) = \sum_{i=1}^{n} s_i \Pr(X_{i:n} > t)$$

$$= \sum_{i=1}^{n} s_i \sum_{j=0}^{i-1} n\binom{n}{j} \{F(t)\}^j \{1 - F(t)\}^{n-j}, \quad for \ t > 0. \tag{1.1}$$

From representation (1.1), we also readily see, for example, that

$$E(T) = \sum_{i=1}^{n} s_i E(X_{i:n}).$$

Signatures also become useful in determining the reliability of one coherent system relative to another as demonstrated in the following representation result.

Theorem 2 (Hollander and Samaniego 2008) *Let T_1 and T_2 represent the lifetimes of coherent systems of orders n and m, with respective signatures $s_1 = (s_{1,1}, \ldots, s_{1,n})$ and $s_2 = (s_{2,1}, \ldots, s_{2,m})$ and ordered component lifetimes $\{X_{1:n}, \ldots, X_{n:n}\}$ and $\{Y_{1:m}, \ldots, Y_{m:m}\}$ drawn from independent i.i.d. samples from a common continuous distribution F. Then,*

$$\Pr(T_1 \leq T_2) = \sum_{i=1}^{n} \sum_{j=1}^{m} s_{1,i} s_{2,j} \Pr(X_{i:n} \leq Y_{j:m}). \tag{1.2}$$

Since the precedence probability $\Pr(X_{i:n} \leq Y_{j:m})$ is known to be

$$\Pr(X_{i:n} \leq Y_{j:m}) = \frac{1}{\binom{m+n}{n}} \sum_{\ell=i}^{n} \binom{j+\ell-1}{\ell} \binom{m+n-j-\ell}{n-\ell},$$

readily representation (1.2) shows that the relative reliability of two systems can be readily determined from their respective signatures. More on such representation results and also preservation results in terms of signatures are presented in the subsequent sections.

The above representations can be stated in a more general way through mixed systems. A *mixed system*, as defined by Boland and Samaniego (2004) is a stochastic (ST) mixture of coherent systems; so, any probability vector **s** in the simplex

$$\left\{ \mathbf{s} \in [0, 1]^n : \sum_{i=1}^{n} s_i = 1 \right\}$$

is the signature of a mixed system. Although mixed systems are not physical systems and are realized in practice only by using a randomization device which chooses a coherent system according to a fixed discrete probability distribution, it facilitates the study of reliability characteristics of systems in a general framework.

With this brief overview of signature theory, we are ready to proceed to the main discussions of this paper. The rest of the paper is organized as follows. In Sect. 1.2, we describe some general representation and preservation results based on system signatures. In Sect. 1.3, we discuss the concept of dynamic reliability and the notion of dynamic signature and associated representation and preservation results. Next, in Sect. 1.4, we consider the situation of two systems sharing some components and introduce the idea of joint signatures and then describe some distributional results and properties associated with them. Finally, in Sect. 1.5, we describe both nonparametric and parametric methods of inference for component lifetime distributions based on system lifetime data under the assumption that the system signature is known.

Fig. 1.1 Coherent system
with lifetime
$T = \min [X_1, \max (X_2, X_3)]$

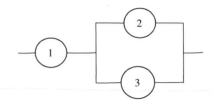

1.2 Signature-Based Representation and Preservation Results

We refer the readers to the book by Barlow and Proschan (1975) for basic details
on the coherent system reliability theory. There, it is stated (see page 12) that if
T is the lifetime of a coherent system with component lifetimes $X_1,\ldots, X_n$ and
minimal path sets $P_1,\ldots, P_k$, then

$$T = \max_{j=1,\ldots,k} X_{P_j}, \tag{1.3}$$

where $X_{P_j} = \min_{i \in P_j} X_i$ is the lifetime of the series system with components in P_j,
for $j = 1,\ldots, k$. A set $P \subseteq \{1,\ldots,n\}$ is a *path set* of a coherent system if the system
works when all the components in P work. A path set P is a *minimal path set* if it
does not contain other path sets (see Barlow and Proschan 1975, p. 9 or Gertsbakh
and Shpungin 2010, p. 38). For example, the minimal path sets of the series–parallel
system $T = \min[X_1, \max(X_2, X_3)]$ (see Fig. 1.1) are $P_1 = \{1, 2\}$ and $P_2 = \{1, 3\}$.
A k-out-of-n system is a system which works if at least k of its n component works.
Thus, the minimal path sets of a k-out-of-n, for $k = 1,\ldots, n$, system are all k-element
subsets of $\{1,\ldots, n\}$. It is evident that the lifetimes of the k-out-of-n systems are the
ordered component lifetimes $X_{1:n} < \cdots < X_{n:n}$, respectively.

From (1.3), it is easy to prove by means of inclusion–exclusion formula that the
reliability function $\overline{F}_T(t) = \Pr(T > t)$ of the system can be expressed as

$$\overline{F}_T(t) = \sum_{j=1}^{k} \overline{F}_{P_j}(t) - \sum_{i<j} \overline{F}_{P_i \cup P_j}(t) + \cdots + (-1)^{k+1} \overline{F}_{P_1 \cup \cdots \cup P_k}(t), \tag{1.4}$$

where $\overline{F}_P$ stands for the reliability function of the series system lifetime
$X_P = \min_{i \in P} X_i$ for $P \subseteq \{1,\ldots,n\}$. This representation can be traced back to
(Agrawal and Barlow 1984). Note that the system reliability function is a linear
combination of reliability functions of series systems with positive and negative
coefficients (that sum to one). Such representations are called *generalized mixtures*
since mixtures usually involve only positive coefficients. It may also be noted that
representation (1.4) holds for general coherent systems (without any assumption
about the components). For example, the reliability function of the system
depicted in Fig. 1.1 can be expressed as

$$\overline{F}_T(t) = \overline{F}_{\{1,2\}}(t) + \overline{F}_{\{1,3\}}(t) - \overline{F}_{\{1,2,3\}}(t). \tag{1.5}$$

If the random vector $(X_1,\ldots, X_n)$ has exchangeable component lifetimes (i.e., $(X_1,\ldots, X_n)$ is equal in distribution to $[X_{\sigma(1)},\ldots, X_{\sigma(n)}]$ for any permutation σ of the set $\{1,\ldots, n\}$), then $\overline{F}_P = \overline{F}_Q$ whenever P and Q have the same cardinality. In this case, representation (1.4) can be simplified to

$$\overline{F}_T(t) = \sum_{j=1}^{n} a_j \overline{F}_{1:j}(t), \tag{1.6}$$

where $\overline{F}_{1:j}(t) = \Pr(X_{1:j} > t)$ is the reliability function of the series system lifetime $X_{1:j} = \min (X_1,\ldots, X_j)$ and $a_1,\ldots, a_n$ are some coefficients that depend only on the structure of the system. Note that some coefficients can be negative and so here again we have a generalized mixture representation. The vector $\mathbf{a} = (a_1,\ldots, a_n)$ has been termed the *minimal signature* in Navarro et al. (2007). An analogous representation can also be presented by using the reliability functions of parallel systems, which has been termed the *maximal signature* in Navarro et al. (2007). For example, if the system depicted in Fig. 1.1 has exchangeable components, then representation (1.5) simplifies to

$$\overline{F}_T(t) = 2\overline{F}_{1:2}(t) - \overline{F}_{1:3}(t), \tag{1.7}$$

which gives the minimal signature of T to be $(0, 2, -1)$. The minimal signatures of all coherent systems with 1–5 components have been tabulated in Navarro and Rubio (2010a, b).

In particular, if the component lifetimes are i.i.d., then representation (1.6) can be simplified to

$$\overline{F}_T(t) = \sum_{j=1}^{n} a_j \overline{F}^{\,j}(t) = p\big(\overline{F}(t)\big), \tag{1.8}$$

where $\overline{F}(t) = \Pr(X_1 > t)$ is the common reliability function of the component lifetimes and $p(x) = \sum_{j=1}^{n} a_j x^j$ is the *reliability polynomial* of the system. This representation for the i.i.d case was obtained in (Birnbaum et al. 1966, Esary and Proschan 1963, Satyarananaya and Prabhakar 1978) and the coefficients in this polynomial are called *dominations*. For example, if the components of the system in Fig. 1 are independent, then representation (1.7) simplifies to

$$\overline{F}_T(t) = 2\overline{F}^2(t) - \overline{F}^3(t) = p\big(\overline{F}(t)\big),$$

where the reliability polynomial is $p(x) = 2x^2 - x^3$.

Evidently, the minimal signature of the series system $X_{1:n}$ is $(0, 0,\ldots,0, 1)$. The lifetime of the $(n-1)$-out-of-n system is $X_{2:n}$ and its minimal path sets are $P_i = \{1,\ldots, n\}-\{i\}$, for $i = 1,\ldots, n$. Hence, from (1.3), its minimal signature is $(0, 0,\ldots, 0, n, 1-n)$. In general, the minimal path sets of $X_{i:n}$ are all $(n-i + 1)$-element subsets of $\{1,\ldots, n\}$. It is then easy to see from (1.3) that its minimal signature $\mathbf{a} = (a_1,\ldots, a_n)$ satisfies $a_1 = \cdots = a_{n-i} = 0$ and $a_{n-i+1} = \binom{n}{n-i+1}$.

The complete set of coefficients can be obtained from the expressions given in David and Nagaraja (2003), p. 46. Therefore, if the component lifetimes are exchangeable, then the vector $\overline{\mathbf{F}}_{OS} = (\overline{F}_{1:n}, \ldots, \overline{F}_{n:n})$ of reliability functions of lifetimes of k-out-of-n systems (order statistics) can be obtained from the vector $\overline{\mathbf{F}}_{SER} = (\overline{F}_{1:1}, \ldots, \overline{F}_{1:n})$ of reliability functions of lifetimes of series systems as $\overline{\mathbf{F}}_{OS} = \overline{\mathbf{F}}_{SER}A_n$, where A_n is a non-singular $n \times n$ triangular matrix. Conversely, $\overline{\mathbf{F}}_{SER}$ can also be obtained from $\overline{\mathbf{F}}_{OS}$ as $\overline{\mathbf{F}}_{SER} = \overline{\mathbf{F}}_{OS}A_n^{-1}$, where A_n^{-1} is the inverse matrix of A_n. Hence, in this exchangeable case, upon replacing $\overline{F}_{1:1}, \ldots, \overline{F}_{1:n}$ by $\overline{F}_{1:n}, \ldots, \overline{F}_{n:n}$ in (1.6), we obtain

$$\overline{F}_T(t) = \sum_{j=1}^{n} s_j \overline{F}_{j:n}(t), \tag{1.9}$$

where the coefficients $s_1, \ldots, s_n$ depend only on the structure of the system. Hence, these coefficients should be the same as those in (1.1) for the i.i.d. continuous case. This shows that the coefficients in (1.9) are non-negative and consequently (1.9) is indeed a mixture representation. In fact, these coefficients are such that $s_i = \Pr(T = X_{i:n})$ for $i = 1, \ldots, n$, whenever the random vector $(X_1, \ldots, X_n)$ has a joint absolutely continuous distribution (see Navarro and Rychlik 2007). However, this is not necessarily the case when $(X_1, \ldots, X_n)$ has an arbitrary exchangeable joint distribution. For example, if we consider the series system $T = X_{1:2} = \min(X_1, X_2)$, evidently

$$\overline{F}_{1:2}(t) = 1 \cdot \overline{F}_{1:2}(t) + 0 \cdot \overline{F}_{2:2}(t),$$

and so $(1, 2)$ is its signature vector. However, if X_1 and X_2 are i.i.d. with a common Bernoulli distribution with parameter $p \in (0, 1)$ (i.e., $\Pr(X_i = 1) = p$ and $\Pr(X_i = 0) = 1 - p$ for $i = 1, 2$), then $\Pr(T = X_{2:2}) = p^2 + (1 - p)^2 \neq 0$. Consequently, representation (1.9) obtained in Navarro et al. 2008a, b) extends representation (1.1) for coherent systems having arbitrary exchangeable components, but by using the signature vector obtained in the i.i.d. continuous case. The coefficients s_i can be obtained from the domination coefficients a_i by using the matrix A_n or through the general expressions presented in (Boland et al. 2003).

Finally, we show how a system with n exchangeable components can also be represented as a mixture of ordered lifetimes from m similar components. This property will enable us to compare systems of different sizes. Let T be the lifetime of a coherent system with component lifetimes $X_1, \ldots, X_n$, and let $(X_1, \ldots, X_m)$ be an exchangeable random vector (with $m \geq n$) comprising component lifetimes. Now, recall from (1.6) that the reliability function of the system is a linear combination of $\overline{F}_{1:1}, \ldots, \overline{F}_{1:n}$, and consequently it is also a linear combination of $\overline{F}_{1:1}, \ldots, \overline{F}_{1:m}$ $(m \geq n)$. Then, by using the fact that $\overline{F}_{1:1}, \ldots, \overline{F}_{1:m}$ can be obtained from $\overline{F}_{1:m}, \ldots, \overline{F}_{m:m}$ (using a matrix A_m), we readily have

$$\overline{F}_T(t) = \sum_{j=1}^{m} s_j^{(m)} \overline{F}_{j:m}(t), \tag{1.10}$$

where the coefficients $s_1^{(m)}, \ldots, s_m^{(m)}$ depend only on the structure of the system. These coefficients can be computed either by using the triangle rule of order statistics or by using the general formulas presented in Samaniego (2007), Navarro et al. (2008a, b). In fact, it can be proved that if $(X_1, \ldots, X_m)$ has an absolutely continuous joint distribution, then the coefficients $s_j^{(m)}$ are such that $s_j^{(m)} = \Pr(T = X_{j:m})$. Thus, $s_j^{(m)}$ are non-negative and (1.10) shows that T is equal in distribution to a mixture of k-out-of-m systems with component lifetimes $X_1, \ldots, X_m$. The vector $\mathbf{s}^{(m)} = [s_1^{(m)}, \ldots, s_m^{(m)}]$ is called *signature of order m* for the system. Of course, the signature of order $m = n$ coincides with the usual signature presented in (1.1).

The mixture representation results can be used to carry out ST comparisons of systems by using signatures. The first result in this direction was obtained by Kochar et al. (1999) for coherent systems with i.i.d. components, which is stated below in Theorem 3. We refer the readers to the book by Shaked and Shanthikumar (2007) for definitions and various properties of the ST, hazard rate (HR), mean residual life (MRL) and likelihood ratio (LR) orders that are pertinent to subsequent discussions.

Theorem 3 (Kochar et al. 1999) *If $T_1 = \phi_1(X_1, \ldots, X_n)$ and $T_2 = \phi_2(X_1, \ldots, X_n)$ are the lifetimes of two coherent systems with respective signatures $\mathbf{s}_1 = (s_{1,1}, \ldots, s_{1,n})$ and $\mathbf{s}_2 = (s_{2,1}, \ldots, s_{2,n})$ and $X_1, \ldots, X_n$ are i.i.d. with a common continuous distribution F, then the following properties hold*:

- *If $\mathbf{s}_1 \leq_{ST} \mathbf{s}_2$, then $T_1 \leq_{ST} T_2$;*
- *If $\mathbf{s}_1 \leq_{HR} \mathbf{s}_2$, then $T_1 \leq_{HR} T_2$;*
- *If F is absolutely continuous and $\mathbf{s}_1 \leq_{LR} \mathbf{s}_2$, then $T_1 \leq_{LR} T_2$.*

These preservation results were extended to the exchangeable case by Navarro et al. (2008a, b). Moreover, their results, presented below in Theorem 4, can also be used to compare systems of different sizes by using the concept of signature of order m described above.

Theorem 4 (Navarro et al. 2008a, b) *If $T_1 = \phi_1(Y_1, \ldots, Y_{n_1})$ and $T_2 = \phi_2(Z_1, \ldots, Z_{n_2})$ are the lifetimes of two coherent systems with respective signatures of order n [for $n \geq \max(n_1, n_2)$] $\mathbf{s}_1 = (s_{1,1}, \ldots, s_{1,n})$ and $\mathbf{s}_2 = (s_{2,1}, \ldots, s_{2,n})$, $\{Y_1, \ldots, Y_{n_1}\}$ and $\{Z_1, \ldots, Z_{n_2}\}$ are contained in $(X_1, \ldots, X_n)$ with $(X_1, \ldots, X_n)$ having an exchangeable joint distribution, then the following properties hold*:

- *If $\mathbf{s}_1 \leq_{ST} \mathbf{s}_2$, then $T_1 \leq_{ST} T_2$;*
- *If $\mathbf{s}_1 \leq_{HR} \mathbf{s}_2$ and*

$$X_{1:n} \leq_{HR} \cdots \leq_{HR} X_{n:n}, \tag{1.11}$$

then $T_1 \leq_{HR} T_2$;
- *If $\mathbf{s}_1 \leq_{HR} \mathbf{s}_2$ and*

$$X_{1:n} \leq _{\text{MRL}} \cdots \leq _{\text{MRL}} X_{n:n}, \qquad (1.12)$$

then $T_1 \leq _{\text{MRL}} T_2;$
- *If* $(X_1,\ldots, X_n)$ *has an absolutely continuous joint distribution,* $\mathbf{s}_1 \leq _{\text{LR}} \mathbf{s}_2$ *and*

$$X_{1:n} \leq _{\text{LR}} \cdots \leq _{\text{LR}} X_{n:n}, \qquad (1.13)$$

- *then* $T_1 \leq _{\text{LR}} T_2.$

These properties are proved by using the representation (1.10) and the mixture preservation results in Shaked and Shanthikumar (2007). It should be noted that in Theorem 4 we need the ordering properties (1.11), (1.12) and (1.13) for the order statistics which need not hold for exchangeable distributions (see Navarro and Shaked 2006). However, since these ordering properties hold in the i.i.d. case, they can be dropped from the statement of the theorem in this case. Also, observe that the condition $\mathbf{p} \leq _{\text{HR}} \mathbf{q}$ is required for the MRL ordering property.

1.3 Dynamic Reliability with Representation and Preservation Results

A system that is working at time t has a profile at that point in time which would differ from its profile at time 0 when the system was new. This may be due to the aging of its components which, if not for the exponential lifetime distribution, will typically result in poorer performance than when the components were new. But there may also be a change in the system itself which, while still working, may be operating with one or more failed components. In this section, we will describe some recent work aimed at characterizing the lifetime characteristics of working used systems at a particular inspection time t at which some information about the system and its components may have become available. There are many different formulations possible for this problem, and here we shall discuss three of them. For a detailed discussion on these problems and their solutions, one may refer to the recent works of Navarro et al. (2008a, b) and Samaniego et al. (2009).

Before embarking on our intended survey, it is useful to make some remarks on a slightly broader applicability of system signatures than has been typical in the existing literature on this topic. The typical definition of the signature of a system begins with the assumption that the system is *coherent*, as described, for example, in Sect. 1.1. However, if several components have failed by time t during which the system has been in service, the used system may in fact no longer be coherent. The monotonicity of the original system will of course be inherited by the used system, but it is no longer true that every component is necessarily relevant. This is apparent from the following simple example. Suppose the 4-component coherent system depicted in Fig. 1.2 is put into service.

Fig. 1.2 The 4-component parallel-series coherent system with lifetime $T = \max [\min (X_1, X_2), \min (X_3, X_4)]$

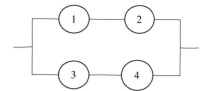

Suppose, at time t, it is noted that the system is still working, but that Component 1 has failed. It is clear that Component 2 is now irrelevant to the functioning of the used system, and that the system now behaves exactly as a 2-component series system with Components 3 and 4. Thus, at time t, such a used system is no longer coherent, but is still a monotone 3-component system. The natural question that arises then is whether the notion of system signature is applicable to a used incoherent system, and fortunately the answer to this question is *affirmative*. In the above example, by denoting the failure times of the three working components at time t as Y_1, Y_2 and Y_3, and their ordered values by $Y_{1:3}$, $Y_{2:3}$ and $Y_{3:3}$, one may easily compute the signature vector $\mathbf{s}$, with $s_i = \Pr (T = Y_{i:3})$ for $i = 1, 2, 3$, as $\mathbf{s} = \left(\frac{2}{3}, \frac{1}{3}, 0\right)$. Indeed, it is straightforward to show that the representation and preservation theorems for system signatures described in Sect. 1.2 apply in the broader context of monotone systems. In what follows, we will use the standard definition and notation for system signatures without making specific reference to whether the system in question is coherent or simply monotone.

We now turn our attention to the ST behavior of working used systems. We focus, first, on the case in which the system is inspected at time t and is simply noted to be working. This is simply equivalent to studying the system residual lifetime $T{-}t$, given that it is known that $T > t$. Navarro et al. (2008a, b) established that the following representation of the conditional residual reliability of a system, given $T > t$, holds:

$$\Pr(T - t > x | T > t) = \sum_{i=1}^{n} p_i(t) \Pr(X_{i:n} - t > x | X_{i:n} > t), \qquad (1.14)$$

where $p_i(t) = s_i \Pr(X_{i:n} > t | T > t)$, with $\mathbf{s}$ being the signature of the system when new and $X_{i:n}$ being the ith order statistic (for $i = 1,\ldots, n$) of the lifetimes of the n i.i.d components of the original system. It should be noted that the vector $\mathbf{p}(t) = [p_1(t),\ldots, p_n(t)]$ depends on the system design (through s) as well as the common component lifetime distribution F. In this way, unlike the signature of a new system, this signature vector is a not a distribution-free measure. In this case, the following preservation theorem has been established in Navarro et al. (2008a, b).

Theorem 5 (Navarro et al. 2008a, b) *Let $p_1(t)$ and $p_2(t)$ be the vectors of the coefficients in representation (1.14) of two coherent systems, both based on n components with i.i.d. lifetimes distributed according to a common continuous distribution F, and let T_1 and T_2 be their respective lifetimes. Then, the following properties hold:*

- If $p_1(t) \leq_{ST} p_2(t)$, then $(T_1 - t | T_1 > t) \leq_{ST} (T_2 - t | T_2 > t)$;
- If $p_1(t) \leq_{HR} p_2(t)$, then $(T_1 - t | T_1 > t) \leq_{HR} (T_2 - t | T_2 > t)$;
- If F is absolutely continuous and $p_1(t) \leq_{LR} p_2(t)$, then

$$(T_1 - t | T_1 > t) \leq_{LR} (T_2 - t | T_2 > t).$$

Navarro et al. (2008a, b) also discussed the behavior of a working used system at time t, but with the information that at least i components have failed by time t. They obtained the following representation theorem in this case.

Theorem 6 (Navarro et al. 2008a, b) *If T is the lifetime of a coherent system with n i.i.d components having a common continuous distribution function F and $i \in \{1, 2, \ldots, n-1\}$ such that $P(T > t, X_{i:n} < t) > 0$, then there exist coefficients $p_1(t, i), \ldots, p_n(t, i)$ (that depend on F) such that $\sum_{j=1}^{n} p_j(t, i) = 1$ and*

$$\Pr(T - t > x | T > t, X_{i:n} < t) = \sum_{j=1}^{n} p_j(t, i) \Pr(X_{j:n} - t > x | X_{j:n} > t) \qquad (1.15)$$

for all $x \geq 0$.

Some coefficients in (1.15) can be negative and it is therefore a generalized mixture representation. One may refer to (Navarro et al. 2008a, b) for some comments on the interpretation and computation of the coefficients $p(t, i)$, as well as a preservation result for the ST order.

A detailed study of *dynamic system signatures* has been made by Samaniego et al. (2009). While the coefficients in representations (1.14) and (1.15) depend on both the system design and the underlying component distribution F, it is of natural interest to obtain the signature of a used system working at the inspection time t that is distribution-free and is therefore a measure on the design, just as the signature of a new system is. This becomes possible under a different form of conditioning than those considered above. The following concept of the dynamic signature of a used system working at time t, given that exactly i components of the system have failed by time t, has been formulated in Samaniego et al. (2009).

Theorem 7 (Samaniego et al. 2009) *Let $\mathbf{s}$ be the signature of a coherent system based on n components with i.i.d lifetimes having a common continuous distribution F. Let T be the system lifetime and let $X_{k:n}$ (for $k = 1, \ldots, n$) be the kth ordered component lifetime. Moreover, let $E_i = \{X_{i:n} \leq t < X_{i+1:n}\}$ for $i \in \{0, \ldots, n-1\}$, and assume that $\Pr(E_i \cap \{T > t\}) > 0$. Then, the dynamic signature of the system, given $E_i \cap \{T > t\}$, is the $(n-i)$-dimensional vector $\mathbf{s}_{n-i}(n-i)$ with its elements as*

$$s_{n-i,k}(n-i) = \Pr(T = X_{k:n} | E_i \cap \{T > t\}) = \frac{s_k}{\sum_{j=i+1}^{n} s_j}, \quad \text{for } k = i+1, \ldots, n. \qquad (1.16)$$

Remark The ratio on the RHS of (1.16) is in fact the conditional probability mass function of an integer-valued random variable Y whose distribution function is given by the cumulative signature, which is called *D-spectrum* (see Gertsbakh and Shpungin 2010). In their setting, Y stands for the random number of sequentially destroyed components needed to cause the failure of the system.

From Theorem 7, the following representation result is readily deduced (see Samaniego et al. 2009):

$$\Pr(T > t + x | E_i \cap \{T > t\}) = \sum_{j=1}^{n-i} s_{n-i,j}(n-i)\overline{G}_{j:n-i|t}(x), \quad \text{for } x > 0, \quad (1.17)$$

where $\overline{G}_{j:n-i|t}(x)$ is the reliability function of the jth order statistic from a random sample of size $n-i$ from the population with conditional reliability function $\overline{G}(x|t) = \overline{F}(x+t)/\overline{F}(t)$ and $s_{n-i,j}(n-i)$ is the jth element of the signature vector $\mathbf{s}_{n-i}(n-i)$ of the working used system with exactly i failed components. A similar representation, but of order n, is given by

$$\Pr(T > t + x | E_i \cap \{T > t\}) = \sum_{j=1}^{n-i} s_{n,j}(n-i)\overline{G}_{j:n|t}(x), \quad \text{for } x > 0, \quad (1.18)$$

where $\overline{G}_{j:n|t}(x)$ is the reliability function of the jth order statistic from a random sample of size n from the population with conditional reliability function $\overline{G}(x|t) = \overline{F}(x+t)/\overline{F}(t)$. The vector $\mathbf{s}_n(n-i)$ with elements $s_{n,j}(n-i)$ in representation (1.18) is called the *dynamic signature of order n*. Under an i.i.d. assumption ($\sim F$) on component lifetimes, Samaniego (2007, p.32) proved that, for any coherent system of size k, and for $n > k$, there is a coherent system of size n with the same lifetime distribution. In the notation used here, $s_k(k)$ is the signature of the system of size k and $s_n(k)$ is the signature of the system of size n that is equivalent to it. Using this vector, the following preservation theorem has been established in Samaniego et al. (2009).

Theorem 8 (Samaniego et al. 2009) *Let $\mathbf{s}_n^{(1)}(n-i)$ and $\mathbf{s}_n^{(2)}(n-j)$ be the dynamic signatures of order n of two coherent systems, both based on n components with i.i.d. lifetimes having a common continuous distribution F. Let T_1 and T_2 be their respective lifetimes and suppose both systems are working at the inspection time t and have exactly i and j failed components, respectively, by time t. Then, the following properties hold:*

- *If $\mathbf{s}_n^{(1)}(n-i) \leq_{ST} \mathbf{s}_n^{(2)}(n-j)$, then $(T_1 - t | T_1 > t, E_i) \leq_{ST} (T_2 - t | T_2 > t, E_j)$;*
- *If $\mathbf{s}_n^{(1)}(n-i) \leq_{HR} \mathbf{s}_n^{(2)}(n-j)$, then $(T_1 - t | T_1 > t, E_i) \leq_{HR} (T_2 - t | T_2 > t, E_j)$;*
- *If F is absolutely continuous and $\mathbf{s}_n^{(1)}(n-i) \leq_{LR} \mathbf{s}_n^{(2)}(n-j)$, then $(T_1 - t | T_1 > t, E_i) \leq_{LR} (T_2 - t | T_2 > t, E_j)$.*

Signature-based necessary and sufficient conditions for various orderings of the residual lifetimes of the systems compared in the above theorem have also been given by Samaniego et al. (2009).

The representation of the residual reliability of a used system working at time t, given $X_{i:n} \leq t < X_{i+1:n}$, facilitates a novel study of certain well-known notions of aging. In the definitions below, drawn from Samaniego et al. (2009), the notions of *conditional New Better than Used* (*i*-NBU) lifetime distributions and *Uniformly NBU* (UNBU) lifetime distributions are introduced.

Definition 1 Consider a coherent system based on n components with i.i.d. lifetimes $X_1, \ldots, X_n \sim F$, where F is a continuous distribution with support $(0, \infty)$. Let T be the system's lifetime, and let $E_i = \{X_{i:n} \leq t < X_{i+1:n}\}$, where $X_{0:n} \equiv 0$. For fixed $i \in \{0, \ldots, n-1\}$, T is conditionally NBU, given i failed components, (denoted by i-NBU) if for all $t > 0$, either

- $\Pr(E_i \cap \{T > t\}) = 0$ or
- $\Pr(E_i \cap \{T > t\}) > 0$ and

$$\Pr(T > x) \geq \Pr(T > x + t | E_i \cap \{T > t\}) \quad \text{for all} \quad x > 0. \tag{1.19}$$

Definition 2 An n component coherent system is said to be UNBU if it is i-NBU for $i \in \{0, 1, \ldots, n-1\}$.

Sufficient conditions on the common component lifetime distribution F and on the system's dynamic signatures to ensure that the system is UNBU have also been given by Samaniego et al. (2009), and their result is as follows.

Theorem 9 (Samaniego et al. 2009) *Let* $\mathbf{s}_n(n-i)$, $i = 0, \ldots, n-1$, *be the dynamic signatures and T be the lifetime of a coherent system based on n components whose lifetimes are i.i.d. with common continuous distribution F. Assume that F is NBU and that*

$$\mathbf{s}_n(n) \geq_{ST} \mathbf{s}_n(n - i) \quad \text{for } i = 1, \ldots, n - 1. \tag{1.20}$$

Then, the system is UNBU.

An interesting application of these dynamic signatures to evaluation of burned-in systems has also been discussed by Samaniego et al. (2009). The engineering practice of *burn-in* is widely used as a vehicle for weeding out poor systems or poor components before a product is deployed or released for sale. The testing of new computer software for bugs that might be detected and removed constitutes a prototypical example. Using a *performance per unit cost* criterion, the options of fielding a new system or fielding a system burned into the ith component failure (that is, to $X_{i:n}$) have been compared in Samaniego et al. (2009). For an n component system with i.i.d. component lifetimes ($\sim F$), three modeling scenarios have been investigated: F is exponential, F is increasing failure rate (IFR) Weibull (i.e., with shape parameter larger than 1) and F is increasing failure rate (DFR) Weibull (i.e., with shape parameter less than 1). Conditions are identified in which a system burned into the ith component failure, for some given $i \in \{1, \ldots, n-1\}$, will provide better performance per unit cost than a new system. The answers obtained are shown to depend critically on the relationship between the fixed cost A of building the system and the cost C of each of its components. We refer the readers to Samaniego et al. (2009) for further details.

1.4 Joint Signatures and Systems with Shared Components

Let us consider two coherent systems with lifetimes $T_1 = \phi_1(Y_1, \ldots, Y_{n_1})$ and $T_2 = \phi_2(Z_1, \ldots, Z_{n_2})$, where $\{Y_1, \ldots, Y_{n_1}\}$ and $\{Z_1, \ldots, Z_{n_2}\}$ are the respective sets of component lifetimes. We shall assume that $\{Y_1, \ldots, Y_{n_1}\}$ and $\{Z_1, \ldots, Z_{n_2}\}$ are contained in $\{X_1, \ldots, X_n\}$, where $X_1, \ldots, X_n$ are i.i.d. with a common distribution function F. Under this setup, note that T_1 and T_2 may share some components and thus can be dependent. The dependence can be represented by their joint distribution function

$$G(t_1, t_2) = \Pr(T_1 \leq t_1, T_2 \leq t_2).$$

Two cases of special interest are (a) when $T_1 = X_{i:n}$ for a fixed $i \in \{1, \ldots, n\}$ and $T_1 < T_2$, and (b) when $T_1 = X_i$ for a fixed $i \in \{1, \ldots, n\}$ and $X_i < T_2$.

The joint distribution function G of these two coherent systems can also be represented in terms of the distribution functions $F_{1:n}, \ldots, F_{n:n}$ of the associated k-out-of-n system lifetimes. This result, due to Navarro et al. (2010), is as follows.

Theorem 10 (Navarro et al. 2010) *The joint distribution function G of T_1 and T_2 can be written as*

$$G(t_1, t_2) = \sum_{i=1}^{n} \sum_{j=0}^{n} s_{i,j} F_{i:n}(t_1) F_{j:n}(t_2) \text{ for } t_1 \leq t_2 \qquad (1.21)$$

and

$$G(t_1, t_2) = \sum_{i=0}^{n} \sum_{j=1}^{n} s_{i,j}^* F_{i:n}(t_1) F_{j:n}(t_2) \text{ for } t_1 > t_2, \qquad (1.22)$$

where $F_{0:n} = 1$ (by convention) and $\{s_{i,j}\}$ and $\{s_{i,j}^*\}$ are collections of coefficients (which do not depend on F) such that $\sum_{i=1}^{n} \sum_{j=0}^{n} s_{i,j} = \sum_{i=0}^{n} \sum_{j=1}^{n} s_{i,j}^* = 1$.

The proof given in Navarro et al. (2010) is based on the minimal cut set representation obtained in Barlow and Proschan (1975), p. 12 and representation (1.10). Observe once again that both expressions (1.21) and (1.22) are generalized mixture representations. Moreover, $\Pr(T_1 = T_2)$ can be positive and consequently G can have a singular part. For this reason, it is not possible to obtain a common mixture representation based on $F_{i:n}(t_1)$ and $F_{j:n}(t_2)$ for all t_1 and t_2. The vector of matrices $\mathbf{S} = (S, S^*)$, where $S = (s_{i,j})$ and $S^* = (s_{i,j}^*)$, with the coefficients in representations (1.21) and (1.22), has been termed the *joint signature* of the systems by Navarro et al. (2010). Their procedure for computing these coefficients is illustrated in the following example. Incidentally, this example also shows that some coefficients $s_{i,j}$ and $s^*_{i,j}$ can be negative.

Example Let us consider the coherent system depited in Fig. 1.1 with lifetime $T_1 = \min [X_1, \max (X_2, X_3)]$ and $T_2 = X_{3:3}$, where X_1, X_2, X_3 are i.i.d. variables with a continuous distribution function. First of all, note that $T_1 < T_2$. The joint distribution function G of (T_1, T_2) can be written, for $t_1 \leq t_2$, as

$$
\begin{aligned}
G(t_1, t_2) &= \Pr(\min(X_1, \max(X_2, X_3)) \leq t_1, \max(X_1, X_2, X_3) \leq t_2) \\
&= \Pr(\{X_1 \leq t_1\} \cup \{\max(X_2, X_3) \leq t_2\}, \max(X_1, X_2, X_3) \leq t_2) \\
&= \Pr(X_1 \leq t_1, \max(X_1, X_2, X_3) \leq t_2) \\
&\quad + \Pr(\max(X_2, X_3) \leq t_1, \max(X_1, X_2, X_3) \leq t_2) \\
&\quad - \Pr(X_1 \leq t_1, \max(X_2, X_3) \leq t_1, \max(X_1, X_2, X_3) \leq t_2) \\
&= \Pr(X_1 \leq t_1, X_2 \leq t_2, X_3 \leq t_2) + \Pr(X_2 \leq t_1, X_3 \leq t_1, X_1 \leq t_2) \\
&\quad - \Pr(X_1 \leq t_1, X_2 \leq t_1, X_3 \leq t_1) \\
&= F(t_1)F^2(t_2) + F^2(t_1)F(t_2) - F^3(t_1) \\
&= F_{1:1}(t_1)F_{2:2}(t_2) + F_{2:2}(t_1)F_{1:1}(t_2) - F_{3:3}(t_1).
\end{aligned}
$$

Then, since the signatures of order 3 of $X_{1:1}$, $X_{2:2}$ and $X_{3:3}$ are $\left(\frac{1}{3}, \frac{1}{3}, \frac{1}{3}\right)$, $\left(0, \frac{1}{3}, \frac{2}{3}\right)$ and $(0, 0, 1)$, respectively (see Navarro et al. 2008a, b), we obtain

$$
\begin{aligned}
G(t_1, t_2) &= \frac{1}{9}F_{1:3}(t_1)F_{2:3}(t_2) + \frac{2}{9}F_{1:3}(t_1)F_{3:3}(t_2) \\
&\quad + \frac{1}{9}F_{2:3}(t_1)F_{1:3}(t_2) + \frac{2}{9}F_{2:3}(t_1)F_{2:3}(t_2) + \frac{3}{9}F_{2:3}(t_1)F_{3:3}(t_2) \\
&\quad - F_{3:3}(t_1) + \frac{2}{9}F_{3:3}(t_1)F_{1:3}(t_2) + \frac{3}{9}F_{3:3}(t_1)F_{2:3}(t_2) + \frac{4}{9}F_{3:3}(t_1)F_{3:3}(t_2)
\end{aligned}
$$

for $t_1 \leq t_2$. Similarly, for $t_1 > t_2$, we obtain

$$
\begin{aligned}
G(t_1, t_2) &= \Pr(T_1 < t_1, T_2 < t_2) \\
&= \Pr(T_2 < t_2) \\
&= F_{3:3}(t_2).
\end{aligned}
$$

Therefore, the joint signature in this case is determined by

$$
S = \begin{pmatrix} 0 & 0 & 1/9 & 2/9 \\ 0 & 1/9 & 2/9 & 3/9 \\ -1 & 2/9 & 3/9 & 4/9 \end{pmatrix} \text{ and } S^* = \begin{pmatrix} 0 & 0 & 1 \\ 0 & 0 & 0 \\ 0 & 0 & 0 \\ 0 & 0 & 0 \end{pmatrix}.
$$

It is of interest in this case to note that $\Pr(T_1 = T_2) = 0$ and so the joint distribution function G is absolutely continuous.

A similar representation holds for the joint reliability function of lifetimes of two coherent systems sharing some components. Also, representations in terms of series and parallel systems can be obtained similarly (see Navarro et al. 2010). Finally, some preservation results can be obtained for the lower orthant and upper orthant bivariate orders based on some matrix-ordering properties for the joint signature of these systems. Interested readers may refer to Navarro et al. (2010) for further details.

1.5 Statistical Inference from System Lifetime Data

The first problem we will treat in this section is the problem of estimating the component lifetime distribution from a random sample of failure times of systems whose components have i.i.d. lifetimes with common distribution F. This problem is of some interest and importance in engineering reliability. Since the behavior of the components may differ under laboratory and field conditions, solving the above problem may be the only approach available for accurately estimating F. While the problem of estimating F from observed system lifetimes is treated in the reliability literature under varied assumptions, the estimator described below has the advantage of being applicable to systems of arbitrary size and design and of being fully nonparametric, that is, free of the assumption that F has a known parametric form. The solution described below has a number of desirable properties: it is the non-parametric maximum likelihood estimator (NPMLE) of F and is a consistent, asymptotically normal and nonparametrically efficient estimator of F. More details on properties of this estimator may be found in Bhattacharya and Samaniego (2010).

In what follows, we restrict our attention to the class of coherent systems and ST mixtures of them (i.e., mixed systems), and we tacitly assume again that the components of the systems considered here have i.i.d. lifetimes with a common continuous distribution F. Suppose the mixed system of interest has a fixed, known design with signature vector $\mathbf{s}$. Then, under these conditions, it is known (see Sect. 1.1) that the reliability function $\overline{F}_T$ of the system lifetime T is given by

$$\overline{F}_T(t) = \sum_{i=1}^{n} s_i \sum_{j=n-i+1}^{n} \binom{n}{j} (\overline{F}(t))^j (F(t))^{n-j}. \tag{1.23}$$

Much of what is done in the sequel will utilize the particular form of the relationship in (1.23) given in (1.8) and based on the domination coefficients (or, equivalently, in the minimal signature).

Suppose that a random sample of system failure times $T_1, \ldots, T_N \sim_{i.i.d.} F_T$ is available. The empirical reliability function of the sample system lifetimes $\widehat{\overline{F}}_{T,N}(t)$ is, of course, the NPMLE of the system reliability function $\overline{F}_T(t)$ and is a consistent, asymptotically normal and nonparametrically efficient estimator of $\overline{F}_T(t)$. Now, all that remains to be done is to solve the inverse problem, that is, to find the estimator $\widehat{\overline{F}}_N(t)$ of $\overline{F}(t)$ which solves, for all $t > 0$, the equation

$$\widehat{\overline{F}}_{T,N}(t) = \sum_{i=1}^{n} a_i \left(\widehat{\overline{F}}_N(t) \right)^i.$$

For any $t > 0$, the asymptotic distribution of $\widehat{\overline{F}}_{T,N}(t)$ may be expressed as

$$\sqrt{N} \left(\widehat{\overline{F}}_{T,N}(t) - \overline{F}_T(t) \right) \to_D U \sim N(0, F_T(t)\overline{F}_T(t)). \qquad (1.24)$$

As noted above, for a given fixed t, the estimators $\widehat{\overline{F}}_N(t)$ and $\widehat{\overline{F}}_{T,N}(t)$ are explicitly related via the equation

$$\widehat{\overline{F}}_{T,N}(t) = p\left(\widehat{\overline{F}}_N(t) \right), \qquad (1.25)$$

where p is the reliability polynomial defined earlier in (1.8). Since the reliability polynomial p is a strictly increasing function when its argument is in the interval $[0, 1]$, we may obtain a well-defined estimator of $\widehat{\overline{F}}_N(t)$ by inverting the relationship in (1.25). The estimator of interest, which is, by the invariance property of maximum likelihood estimation, the NPMLE of $\overline{F}(t)$, may be expressed as

$$\widehat{\overline{F}}_N(t) = p^{-1} \left(\widehat{\overline{F}}_{T,N}(t) \right).$$

Since the asymptotic distribution of $\widehat{\overline{F}}_{T,N}(t)$ is known and is as given in (1.24), and since p is a smooth one-to-one function, we may apply the δ-method to obtain the asymptotic distribution of $\widehat{\overline{F}}_N(t)$. In the standard presentation of the δ-method, the exact expression for the asymptotic variance of the transformed variable $p^{-1}\left(\widehat{\overline{F}}_{T,N}(t) \right)$ involves the derivative of the function p^{-1}. Indeed, we may write

$$\sqrt{N} \left(\widehat{\overline{F}}_N(t) - \overline{F}(t) \right) \to_D U \sim N\left(0, \left[\frac{d}{dy} p^{-1}(y)|_{y=\overline{F}_T(t)} \right]^2 F_T(t)\overline{F}_T(t) \right). \qquad (1.26)$$

However, since p is a polynomial of (potentially high) degree n, one is not generally able to obtain the asymptotic variance of $\widehat{\overline{F}}_N(t)$ in (1.26) in closed form. It has been shown by Bhattacharya and Samaniego (2010) that the asymptotic result in (1.26) may be written in a somewhat more useful form as

$$\sqrt{N} \left(\widehat{\overline{F}}_N(t) - \overline{F}(t) \right) \to_D W \sim N\left(0, \left(\sum_{i=1}^{n-1} i a_i \left[p^{-1}(\overline{F}_T(t)) \right]^{i-1} \right)^{-2} F_T(t)\overline{F}_T(t) \right).$$

$$(1.27)$$

Since $y = p(x)$ is a known function, its inverse $p^{-1}(y)$ may be obtained numerically at any given y (say, by interval halving). Suppose that the random sample of N system lifetimes gives rise to the set of ordered failure times $t_{1:N}, \ldots, t_{N:N}$. The reliability function $\overline{F}_T(t)$ is estimated by the empirical distribution $\widehat{\overline{F}}_{T,N}(t)$ of the observed system failure times, with

$$
\widehat{\overline{F}}_{T,N}(t) = \begin{cases} 1 & \text{for} \quad t < t_{1:N}, \\ \frac{N-i}{N} & \text{for} \quad t_{i:N} \leq t < t_{i+1:N}, \quad i = 1, \ldots, N-1, \\ 0 & \text{for} \quad t \geq t_{N:N}. \end{cases}
$$

The estimator $\widehat{\overline{F}}_N(t)$ of $\overline{F}(t)$ is simply the step function with jumps at times $t_{1:N}, \ldots, t_{N:N}$ and values given by

$$
\widehat{\overline{F}}_N(t) = \begin{cases} 1 & \text{for} \quad t < t_{1:N}, \\ p^{-1}\left(\frac{N-i}{N}\right) & \text{for} \quad t_{i:N} \leq t < t_{i+1:N}, \quad i = 1, \ldots, N-1, \\ 0 & \text{for} \quad t \geq t_{N:N}. \end{cases}
$$

The estimator $\widehat{\overline{F}}_N(t)$ is, asymptotically, the optimal estimator of $\overline{F}$ in the nonparametric setting explained above. Bhattacharya and Samaniego (2010) carried out Monte Carlo simulations, based on samples of size 50 and 100, to evaluate the performance of $\widehat{\overline{F}}_N$ for the well-known five-component bridge system. Five parametric models—exponential, Weibull, lognormal, gamma and Pareto—as component distributions F. These results provide support to the claim that $\widehat{\overline{F}}_N$ does indeed perform well, over a reasonably broad class of possible models for F, even for moderate sample sizes like $N = 50$ and certainly for sample sizes of 100 or more. Furthermore, by using (1.27), it is possible to obtain, numerically, approximate but quite reliable confidence intervals for $\overline{F}(t)$ when the sample size N is sufficiently large. Bhattacharya and Samaniego (2010) have also discussed some possible extensions of the above inversion method to non-i.i.d. settings. They have noted that, even in the case of independent but non-identically distributed (INID) component lifetimes, the estimation of component distributions is not generally possible, as the component distributions are not identifiable from data on system failure times. However, in the special case in which, for each i, the ith component of the system has reliability function $h_i(p)$, where h_i is strictly increasing for $p \in [0, 1]$, with $h_i(0) = 0$ and $h_i(1) = 1$, the inversion technique described above leads to the identification of the NPMLEs of the n component reliability functions and to the determination of their asymptotic distributions.

Another development on nonparametric inference for component lifetime distribution, different in nature compared to the one detailed above, is due to Balakrishnan et al. (2011b). This exact method utilizes the order statistics $T_{1:N} < \cdots < T_{N:N}$ obtained from the lifetimes $T_1, \ldots, T_N$ of N identical systems under test. Once again, by assuming that the components in each system are i.i.d. with a continuous distribution F, and that the signature of the system is specified, these authors have used order statistics $T_{i:N}$ to develop exact nonparametric confidence intervals for the qth quantile of the component lifetime distribution F, denoted by ξ_q. Specifically, for determining an upper $100(1-\alpha)\%$ confidence limit for ξ_q, they proposed to find the minimal index of U (for $U = 1, 2,..., N$) that satisfies

$$\Pr(T_{U:N} \geq \xi_q) \geq 1 - \alpha,$$

i.e.,

$$U(\xi_q) := \inf\{U : \Pr(T_{U:N} \geq \xi_q) \geq 1 - \alpha\}.$$

Clearly,

$$\Pr(T_{U:N} \geq \xi_q) = \sum_{\ell=0}^{U-1} \binom{N}{\ell} \{F_T(\xi_q)\}^\ell \{1 - F_T(\xi_q)\}^{N-\ell}$$

$$= \sum_{\ell=0}^{U-1} \binom{N}{\ell} \{\rho(q)\}^\ell \{1 - \rho(q)\}^{N-\ell}, \qquad (1.28)$$

where $\rho(x) = 1 - \sum_{i=1}^{n} a_i(1-x)^i = 1 - p(1-x)$, with $p(x)$ being the reliability polynomial defined earlier in (1.8). Then, for a given value of α, a one-sided (upper) $100(1-\alpha)\%$ confidence interval for ξ_q is given by $(0, T_{U:N})$, where U is the smallest integer such that

$$\sum_{\ell=0}^{U-1} \binom{N}{\ell} \left\{1 - \sum_{i=1}^{n} a_i(1-q)^i\right\}^\ell \left\{\sum_{i=1}^{n} a_i(1-q)^i\right\}^{N-\ell} \geq 1 - \alpha.$$

Although this is an exact result, for a small sample size N, the exact confidence coefficient $1-\alpha$ may not be achieved due to the discreteness of the probabilities involved, and so the value of U needs to be chosen so that the probability value is the closest to $1-\alpha$. In an analogous manner, Balakrishnan et al. (2011b) have also discussed the construction of lower confidence limits ($T_{L:N}$, say) and two-sided confidence intervals for the quantile ξ_q of the component lifetime distribution. They have presented some tables of the values of U and L for different choices of N, q and α for some selected 4-component coherent systems. These authors have also discussed the determination of nonparametric tolerance limits and intervals for the component lifetime distribution along the same lines.

Alternatively, one could also develop optimal parametric methods of inference in this context. Let us assume that the component lifetimes belong to the scale family of distributions with density

$$f_X(x; \sigma) = \frac{1}{\sigma} f^*\left(\frac{x}{\sigma}\right), \quad x > 0, \ \sigma > 0,$$

where $f^*(\cdot)$ is the standard member of the scale family. Then, the distribution of T, the system lifetime, also belongs to the scale family of distributions with pdf

$$f_T(t; \sigma) = \frac{1}{\sigma} f^* \left(\frac{t}{\sigma} \right) \sum_{i=1}^{n} a_i i \left\{ \bar{F}^* \left(\frac{t}{\sigma} \right) \right\}^{i-1}, \quad t > 0, \ \sigma > 0,$$

where $\bar{F}^*(\cdot)$ is the reliability function corresponding to $f^*(\cdot)$. This readily reveals that, for a given Type-II right censored sample $T = (T_{1:N}, \ldots, T_{r:N})'$ from the system lifetimes, $T_{1:N}^* < T_{2:N}^* < \cdots < T_{r:N}^*$, where $T_{j:N}^* = \frac{T_{j:N}}{\sigma}$, form a Type-II right censored sample from the standard system lifetime distribution with pdf

$$f_{T^*}^*(t) = f^*(t) \sum_{i=1}^{n} a_i i \{ \bar{F}^*(t) \}^{i-1}, \quad t > 0. \tag{1.29}$$

This fact facilitates the derivation of the best linear unbiased estimator of σ based on the given Type-II censored sample T (see Balakrishnan et al. 2011a). Specifically, with $\mathbf{m}^* = (\mu_{1:N}^*, \mu_{2:N}^*, \ldots, \mu_{r:N}^*)'$ and $\mathbf{S}^* = ((\sigma_{i,j:N}^*))$, $i,j = 1, \ldots, r$, denoting the mean vector and variance–covariance matrix of $T^* = (T_{1:N}^*, T_{2:N}^*, \ldots, T_{r:N}^*)'$, respectively, the BLUE of σ is given by (see Balakrishnan et al. 2011a)

$$\tilde{\sigma} = \left(\frac{\mathbf{m}^{*\prime} \mathbf{S}^{*-1}}{\mathbf{m}^{*\prime} \mathbf{S}^{*-1} \mathbf{m}^*} \right) T;$$

further, its variance equals

$$\mathrm{Var}(\tilde{\sigma}) = \frac{\sigma^2}{\mathbf{m}^{*\prime} \mathbf{S}^{*-1} \mathbf{m}^*}.$$

Balakrishnan et al. (2011a) have described the computational issues involved in the above expressions and also presented some numerical results for the exponential case some selected choices of r and N and for some known reliability systems.

In a similar vein, Balakrishnan et al. (2011a) also developed best linear unbiased estimators for a general location-scale family of distributions. Suppose the component lifetimes or the log-transformed component lifetimes belong to the location-scale family of distributions with pdf

$$f_X(x; \mu, \sigma) = \frac{1}{\sigma} f^* \left(\frac{x - \mu}{\sigma} \right), \quad -\infty < \mu < \infty, \ \sigma > 0,$$

where $f^*(\cdot)$ is the standard member of the location-scale family. As above, let $T = (T_{1:N} < \cdots < T_{r:N})'$ be a Type-II right censored sample of system lifetimes obtained from N systems under test. Then, it can be shown that $T^* = (T_{1:N}^* \leq \cdots \leq T_{r:N}^*)'$, where $T_{j:N}^* = (T_{j:N} - \mu)/\sigma$, forms a Type-II right censored sample from the standard system lifetime distribution with pdf $f_{T^*}^*(t)$ presented in (1.29). Using now the facts that $E(T) = \mu \mathbf{1} + \sigma \mathbf{m}^*$ and $Cov(T) = \sigma^2 \mathbf{S}^*$, where $\mathbf{m}^*$ and $\mathbf{S}^*$ are the mean vector and variance–covariance matrix of T^*, as before, and $\mathbf{1}$ is a vector of $\mathbf{1}$'s, Balakrishnan et al. (2011a) derived the BLUEs of μ and σ as

$$\tilde{\mu} = \mathbf{m}^{*'}\mathbf{G}T \quad \text{and} \quad \tilde{\sigma} = -\mathbf{1}'\mathbf{G}T, \tag{1.30}$$

where

$$\mathbf{G} = \frac{\mathbf{S}^{*-1}(\mathbf{m}^*\mathbf{1}' - \mathbf{1}\mathbf{m}^{*'})\mathbf{S}^{*-1}}{\mathbf{D}},$$

and

$$\mathbf{D} = (\mathbf{m}^{*'}\mathbf{S}^{*-1}\mathbf{m}^*)(\mathbf{1}'\mathbf{S}^{*-1}\mathbf{1}) - (\mathbf{m}^{*'}\mathbf{S}^{*-1}\mathbf{1})^2.$$

Furthermore, the variances and covariance of the BLUEs in (1.30) are given by

$$\text{Var}(\tilde{\mu}) = \sigma^2 \left(\frac{\mathbf{m}^{*'}\mathbf{S}^{*-1}\mathbf{m}^*}{\mathbf{D}} \right),$$

$$\text{Var}(\tilde{\sigma}) = \sigma^2 \left(\frac{\mathbf{1}^{*'}\mathbf{S}^{*-1}\mathbf{1}}{\mathbf{D}} \right), \tag{1.31}$$

$$\text{Cov}(\tilde{\mu}, \tilde{\sigma}) = -\sigma^2 \left(\frac{\mathbf{m}^{*'}\mathbf{S}^{*-1}\mathbf{1}}{\mathbf{D}} \right).$$

Balakrishnan et al. (2011a) have described the computational issues involved in the above expressions and illustrated the results by applying them to the extreme value distribution (which is the log-transformed form of Weibull distribution) for some known reliability systems.

A final mention should be made of the work of Ng et al. (2011) wherein parametric estimators for $\overline{F}$ have been developed in the general case when the component reliability function belongs to the proportional HR model, i.e., when $\overline{F}(t) = \overline{G}^\alpha(t)$ for a known baseline reliability function $\overline{G}$ and an unknown positive parameter α. This model includes several parametric models such as exponential, Weibull (with a fixed shape parameter) and Pareto. For this case, they have studied the methods of moments, maximum likelihood and least-squares for the estimation of the proportionality parameter α, with the assumption of a known signature for the coherent system under test. Then, by focusing on the case when the lifetime distribution of the components is exponential, they have carried out a Monte Carlo simulation study to compare the performance of all these methods of estimation.

Acknowledgments Our sincere thanks go to the Editors of this volume for extending an invitation to prepare this article. We also appreciate their patience and support during the course of this work. We are especially grateful to Professor Gertbakh for several helpful suggestions during the preparation of this paper. We also thank the support from the Natural Sciences and Engineering Research Council of Canada (to NB), Ministerio de Ciencia y Tecnolog under grant MTM2009-08311 and Fundación Séneca (C.A.R.M.) under grant 08627/PI/08 (to JN), and the U.S. Army Research Office under grant W911NF-08-1-0077 (to FJS) are all gratefully acknowledged.

References

Agrawal A, Barlow RE (1984) A survey of network reliability and domination theory. Op Res 12:611–631

Balakrishnan N, Ng HKT, Navarro J (2011a) Linear inference for Type-II censored lifetime data of reliability systems with known signatures. IEEE Trans Reliab 60:426–440

Balakrishnan N, Ng HKT, Navarro J (2011b) Exact nonparametric inference for component lifetime distribution based on lifetime data from systems with known signatures. J Nonparametric Stat doi:10.1080/10485252.2011.559547

Barlow RE, Proschan F (1975) Statistical theory of reliability and life testing. Holt Rinehart and Winston, New York

Bhattacharya D, Samaniego FJ (2010) On estimating component characteristics from system failure-time data. Naval Res Logist 57:380–389

Birnbaum ZW, Esary JD, Marshall AW (1966) A stochastic characterization of wear-out for components and systems. Annals Math Stat 37:816–825

Boland PJ, Samaniego FL (2004) The signature of a coherent system and its applications in reliability. In: Soyer R, Mazzuchi T, Singpurwalla ND (eds) Mathematical reliability: an expository perspective. Kluwer, Boston, pp 1–29

Boland PJ, Samaniego FJ, Vestrup EM (2003) Linking dominations and signatures in network reliability theory. In: Lindquist B, Doksum KA (eds) Mathematical and statistical methods in reliability. World Scientific, Singapore, pp 89–103

David HA, Nagaraja HN (2003) Order statistics. Wiley, Hoboken

Esary JD, Proschan F (1963) Relationship between system failure rate and component failure rates. Technometrics 5:183–189

Gertsbakh IB, Shpungin Y (2010) Models of network reliability. CRC Press, Boca Raton

Hollander M, Samaniego FJ (2008) On comparing the reliability of arbitrary systems via stochastic precedence. In: Bedford T, Quigley J, Walls L, Alkali B, Daneshkhah A, Hardman G (eds) Advances in mathematical modeling for reliability. IOS Press, Amsterdam, pp 129–137

Kochar S, Mukerjee H, Samaniego FJ (1999) The "signature" of a coherent system and its application to comparison among systems. Naval Res Logist 46:507–523

Navarro J, Rubio R (2010a) Computation of signatures of coherent systems with five components. Commun Stat Simul Comput 39:68–84

Navarro J, Rubio R (2010b) Comparisons of coherent systems using stochastic precedence. Test 19:469–486

Navarro J, Rychlik T (2007) Reliability and expectation bounds for coherent systems with exchangeable components. J Multivar Anal 98:102–113

Navarro J, Shaked M (2006) Hazard rate ordering of order statistics and systems. J Appl Probab 43:391–408

Navarro J, Ruiz JM, Sandoval CJ (2007) Properties of coherent systems with dependent components. Commun Stat Theory Methods 36:175–191

Navarro J, Balakrishnan N, Samaniego FJ (2008a) Mixture representations of residual lifetimes of used systems. J Appl Probab 45:1097–1112

Navarro J, Samaniego FJ, Balakrishnan N, Bhattacharya D (2008b) On the application and extension of system signatures in engineering reliability. Naval Res Logist 55:313–327

Navarro J, Samaniego FJ, Balakrishnan N (2010) Joint signature of coherent systems with shared components. J Appl Probab 47:235–253

Ng HKT, Navarro J, Balakrishnan N (2011) Parametric inference for system lifetime data with signatures available under a proportional hazard rate model. Metrika (to appear)

Samaniego FJ (1985) On closure of the IFR class under formation of coherent systems. IEEE Trans Reliab R-34:69–72

Samaniego FJ (2007) System signatures and their applications in engineering reliability. Springer, New York

Samaniego FJ, Balakrishnan N, Navarro J (2009) Dynamic signatures and their use in comparing the reliability of new and used systems. Naval Res Logist 56:577–591

Satyarananaya A, Prabhakar A (1978) New topological formula and rapid algorithm for reliability analysis of complex networks. IEEE Trans Reliab TR-30:82–100

Shaked M, Shanthikumar JG (2007) Stochastic orders. Springer, New York

Chapter 2
Using D-Spectra in Network Monte Carlo: Estimation of System Reliability and Component Importance

Ilya B. Gertsbakh and Yoseph Shpungin

Abstract We present a combinatorial definition of network-type system destruction spectrum (signature) and component importance D-spectra, and demonstrate how Birnbaum importance measures (BIMs) can be expressed via these spectra. We demonstrate an efficient algorithm which calculates simultaneously system reliability spectrum and BIM spectra for all its components. Use of BIMs in optimal system design is discussed.

Keywords D-spectrum · Signature · System reliability · Birnbaum importance measure · Importance D-spectrum · Network design

2.1 Introduction

It is a well-known fact that there exists a wide gap between theoretical analysis of reliability problems and the ability to compute reliability parameters for large or even moderate networks. Our purpose is to describe a Monte Carlo approach which allows efficient estimation of various network reliability parameters and may also be used in the optimal network design.

This approach is based on using the so-called *D-spectrum* for network reliability and the *importance D-spectrum* for computing network component

I. B. Gertsbakh (✉)
Department of Mathematics, Ben Gurion University of the Negev,
Beer Sheva, Israel
e-mail: elyager@bezeqint.net

Y. Shpungin
Software Engineering Department, Sami Shamoon College of Engineering,
Beer Sheva, Israel
e-mail: yosefs@sce.ac.il

A. Lisnianski and I. Frenkel (eds.), *Recent Advances in System Reliability*,
Springer Series in Reliability Engineering, DOI: 10.1007/978-1-4471-2207-4_2,
© Springer-Verlag London Limited 2012

importance measures. These notions were introduced and investigated by Elperin et al. (1991), Gertsbakh and Shpungin (2004, 2008, 2009). Examples of simulated D-spectra for a family of complete graphs and the dodecahedron networks were presented in Elperin et al. (1991) and named network *internal distributions*. Numerically, they are identical to so-called *signatures* introduced six years earlier by Samaniego (1985).

The paper is organized as follows. In Sect. 2.2, we give the basic notions and definitions of D-spectra. In Sect. 2.3, we present an efficient algorithm which calculates simultaneously system reliability D-spectrum and BIM D-spectra for all system components. In Sect. 2.4, we describe the use of BIMs D-spectra in network design.

2.2 Basic Notions and Definitions

2.2.1 Network and Its Reliability

By network $\mathbf{N} = (V, E, T)$ we denote an undirected graph with a node-set V, $|V| = n$, an edge-set E, $|E| = m$, and a set $T \subseteq V$ of special nodes called *terminals*. Each element a (node or edge) is associated with a probability p_a of being *up* and probability $q_a = 1 - p_a$ of being *down*. We postulate that element failures are mutually independent events. In a given network $\mathbf{N}$ the state of $\mathbf{N}$ is induced by all its elements which are in the *up* state.

In this paper we deal with *terminal connectivity* operational criterion. By this criterion the network state is *UP* if any pair of terminals is connected by the elements in the *up* state. Otherwise, the state is *DOWN*. The terminal connectivity has the property of being monotone: each subset of the *DOWN* state is a *DOWN* state, and each superset of the *UP* state is an *UP* state.

We define the network reliability $R(N)$ as the probability that the network is in *UP* state.

2.2.2 Birnbaum Importance Measure

Let us introduce the Birnbaum Importance Measure (BIM) (Barlow and Proschan 1975) of system component j, $j = 1, 2, \ldots, k$. If system reliability is a function $R = \Psi(p_1, p_2, \ldots, p_k)$ of component reliability p_j, then BIM of component j is defined as

$$BIM_j = \frac{\partial \Psi(p_1, \ldots, p_k)}{\partial p_j} = \Psi(p_1, \ldots, p_{j-1}, 1, p_{j+1}, \ldots, p_k)$$
$$- \Psi(p_1, \ldots, p_{j-1}, 0, p_{j+1}, \ldots, p_k). \quad (2.1)$$

Fig. 2.1 Network with two terminals

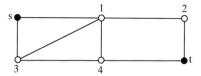

BIM has a transparent probabilistic meaning: it is the gain in system reliability received from replacing a *down* component j by an absolutely reliable one. BIM_j gives the approximation to the system reliability increment δR resulted from component j reliability increment by δp_j.

The first expression on the right-hand side of (2.1) is the reliability of a system in which component j is permanently *up* and the second term is the reliability of the system in which component j is permanently *down*.

The use of BIM in practice was limited since usually the reliability function $\Psi(\cdot)$ is not available in explicit form. It turns out however that for the case of equal component reliability $p_j \equiv p$, there is a surprising connection between the network spectrum and the reliability function which allows to estimate the component BIMs without knowing the analytic form of the reliability function (Gertsbakh and Shpungin 2008).

2.3 Network Reliability and BIM D-Spectrum

2.3.1 Reliability Destruction Spectrum (D-Spectrum)

Definition 1 Let π be a permutation of all unreliable elements (edges or nodes) $e_1, \ldots, e_k : \pi = (e_{i_1}, e_{i_2}, \ldots, e_{i_k})$. Start with a network with all elements being *up* and "erase" the elements in the order they appear in π, from left to right. Stop at the first element e_{i_r} when the network becomes *DOWN*. The ordinal number r of this element is called the *anchor* of permutation π and denoted $r(\pi)$.#

Example 1 Consider the network shown in Fig. 2.1. In this network with two terminals, the edges are reliable and nodes 1,2,3,4 fail with probability $q = 1 - p$. Consider an arbitrary permutation π of node numbers, e.g. $\pi = (3, 2, 4, 1)$. Let us begin the following destruction process. We start with a network with all elements in the *up* state, and erase one node after another in the order prescribed by π, from left to right. Erasing a node means erasing all edges incident to this node. The network becomes *DOWN* after erasing the third node, i.e. node 4. So we have $r(\pi) = 4$.

Note that the anchor value for given π depends only on the network structure and its *DOWN* definition. It is completely separated from the stochastic mechanism which governs the node or edge failures in a real network destruction process.

Definition 2 Let x_i be the total number of permutations such that their anchor equals i. The set

$$D = \left\{ d_1 = \frac{x_1}{k!}, d_2 = \frac{x_2}{k!}, \ldots, d_k = \frac{x_k}{k!} \right\} \tag{2.2}$$

is called the *D-spectrum of the network*.#

For example, it is easy to check that for the network in Fig. 2.1, the D-spectrum is $\left\{ 0, \frac{1}{2}, \frac{1}{2}, 0 \right\}$.

Remark "*D*" in Definition 2 refers to the "destruction" process of erasing network elements from left to right in the permutation π. D-spectrum is a distribution of the anchor value, and obviously $\sum_{i=1}^{k} d_i = 1$. Numerically, the D-spectrum coincides with the so-called *signature* introduced first in (Samaniego 1985). It was proved there (see also Samaniego 2007) that if system components fail independently and their lifetimes X_i have identical continuous distribution function $F(t)$, then the system lifetime distribution $F_s(t) = \sum_{i=1}^{k} d_i \cdot F_{i:k}(t)$ where $F_{i:k}$ is the cumulative distribution function of the ith order statistics in random sample $X_1, X_2, \ldots, X_k$.#

Definition 3 Let $y_b = \sum_{i=1}^{b} d_i, b = 1, 2, \ldots, k$. Then the set $(y_1, y_2, \ldots, y_k)$ is called the *cumulative D-spectrum*.#

Theorem 1 *If all $p_i \equiv p$, then network static reliability $R(N)$ can be expressed in the following form:*

$$R(\mathbf{N}) = 1 - \sum_{i=1}^{k} y_i \frac{k! q^i p^{k-i}}{i!(k-i)!}. \# \tag{2.3}$$

We omit the proof of this statement which can be found in Gertsbakh and Shpungin (2009), Chap. 8. This theorem is equivalent to the following interesting combinatorial fact: $y_i \cdot k!/(i!(k-i)!) = C(i)$, where $C(i)$ is the total number of cut sets of size i in the system. A statement equivalent to Theorem 1 is presented in Samaniego (2007), Sect. 6.1.

2.3.2 BIM Spectrum

We use the abbreviation BIM for the Birnbaum Importance Measure.

Definition 4 Denote by $Z_{i,j}$ the number of permutations satisfying the following two conditions:

- If the first i elements in the permutation are *down*, then the network is *DOWN*;
- Element j (node or edge) is among the first i elements of the permutation.

Table 2.1 $BIM\diamondsuit S$ for network

i	$z_{i,1}$	$z_{i,2}$	$z_{i,3}$	$z_{i,4}$
1	0	0	0	0
2	1/3	1/6	1/6	1/3
3	3/4	3/4	3/4	3/4
4	1	1	1	1

The collection of $z_{i,j} = Z_{i,j}/k!$ values, $i = 1, 2, \ldots, k; j = 1, 2, \ldots, k$, is called BIM-spectrum of the network and denoted $BIM\diamondsuit S$. The set of $z_{i,j}$ values for fixed j and $i = 1, \ldots, k$ is called the BIM_j D-spectrum, or the importance spectrum of component j.

Example 1 (continued) Let us turn to the network in Fig. 2.1 and calculate one of the $Z_{i,j}$ values, say $Z_{2,2}$. The permutations which satisfy the conditions of the above definition, are the following: $(2, 4, 1, 3)$, $(2, 4, 3, 1)$, $(4, 2, 1, 3)$, $(4, 2, 3, 1)$. So we have $Z_{2,2} = 4$. The Table 2.1 presents the $BIM\diamondsuit S$ for our network.

The columns in this table are the BIM_j spectra.

The proof of the following theorem is presented in Gertsbakh and Shpungin (2009), Chap. 10.

Theorem 2 BIM_j, $j = 1, 2, \ldots, k$, *equals*

$$BIM_j = \sum_{i=1}^{k} \frac{k!(z_{i,j}q^{i-1}p^{k-i} - (y_i - z_{i,j})q^i p^{k-i-1})}{i!(k-i)!} . \# \qquad (2.4)$$

Note that $y_k - z_{k,j} = 0$, *which means that in the second term of the numerator of* (2.4) *one can assume that index i changes from 1 to* $k-1$.

The exact calculation of network D-spectrum and BIM-spectra is a formidable task. We suggest estimating the spectra using Monte Carlo approach. Below we present an algorithm which simultaneously estimates the D-spectrum and the BIM-spectra for all network components.

The core of Algorithm 1 is Operator 3 which finds out the anchor of the random permutation of network elements simulated by Operator 2. A single permutation and its anchor allow obtaining one observation replica for estimating the D-spectrum and all BIM-spectra. A substantial feature of this operator is that only a *single* spanning tree is constructed to find out the anchor for given π. More details about this algorithm can be found in Gertsbakh and Shpungin (2004, 2009).

The algorithm is adjusted to the case of n unreliable edges in the network.

Algorithm 1 Computing BIM Spectra and D-Spectrum

1. `Initialize` all a_i and $b_{i,j}$ `to be zero,` $i = 1, \ldots, n; j = 1, \ldots, n$.
2. `Simulate` permutation $\pi \in \Pi_E$. (Π_E `is the set of all edge permutations.`)

Fig. 2.2 Hypercube H_4 with
3 terminals

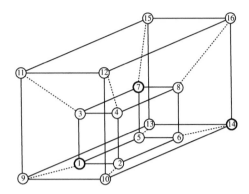

Table 2.2 Network H_4: cumulative reliability spectrum

i	y_i	i	y_i	i	y_i	i	y_i
1	0	9	0.011709	17	0.404952	25	0.994868
2	0	10	0.020265	18	0.547724	26	0.998277
3	0	11	0.035539	19	0.687263	27	0.999590
4	0.000079	12	0.053547	20	0.802654	28	0.999947
5	0.000426	13	0.083682	21	0.885786	29	1.0
6	0.001297	14	0.128297	22	0.939116	30	1.0
7	0.003053	15	0.192645	23	0.970280	31	1.0
8	0.006217	16	0.284164	24	0.986978	32	1.0

3. **Find out** the minimal index of the edge $r = r(\pi)$ such that the first r edges in π create network *DOWN* state (r is the anchor of π.)
4. **Put** $a_r := a_r + 1$.
5. **Find** all j such that e_j occupies one of the first r positions in π, and for each such j put $b_{r,j} := b_{r,j} + 1$.
6. **Put** $r := r + 1$. **If** $r \le k$, **GOTO** 4.
7. **Repeat** 2–6 M times.
8. **Estimate** $y_i, z_{i,j}$ via $\hat{y}_i = \frac{a_i}{M}$, $\hat{z}_{i,j} = \frac{b_{i,j}}{M}$.

Example 2 Consider a three-terminal hypercube H_4 with 16 reliable nodes and 32 unreliable edges. This hypercube is shown on Fig. 2.2. Let its terminals be nodes 1, 7, and 14. Edges will be denoted by (k,s) where k and s are node numbers. Table 2.2 presents the values of spectrum for this network. Table 2.3 presents hypercube reliability R for different values of p. The values of R may be computed via the spectrum.

It follows from the simulated BIMs spectra (not presented here) that in our network there are several groups of edges having approximately equal estimated BIMs within each group, while the groups can be partially ranked according to their importance measures. The first group consists of edges (1, 5) and (5, 7)

Table 2.3 Network H_4: reliability for different values of p

p	R	p	R	p	R
0.1	0.000529	0.4	0.344084	0.7	0.966237
0.15	0.003855	0.45	0.498603	0.75	0.985145
0.2	0.015373	0.5	0.647305	0.8	0.994341
0.25	0.046276	0.55	0.775587	0.85	0.998327
0.3	0.107647	0.6	0.869351	0.9	0.999694
0.35	0.207249	0.65	0.930455	0.95	0.999977

Fig. 2.3 BIMs spectra for various edges

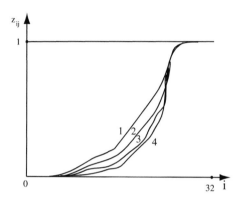

connecting terminals 1 and 7. This group dominates all other edges by its BIMs for all values of p. The second group consists of two edges (6, 14) and (13, 14). Their BIMs dominate the remaining edge BIMs also for all p values. The third group of approximately identical BIMs constitute four edges $\{(1, 3), (3, 7), (5, 6), (5, 13)\}$. The third group, however, does not dominate another group of edges uniformly for all p values. A similar behavior exhibits the fourth group of the rest edges.

Figure 2.3 presents the graphs of accumulated BIM-spectra for edges representing each of the four groups of edges.

2.4 BIM-Based Heuristic for Network Optimal Design

In this section we consider network optimal design problem in the following form. Suppose that s, $1 \leq s \leq k$, elements (nodes or edges) can be simultaneously replaced by more reliable ones having the *up* probability $p^* > p$. The problem is to choose the s candidates for the reinforcement in order to maximize the network reliability.

The solution of this problem is not trivial and involves the notion of *joint importance measure*. We will not investigate here this issue but instead provide a well-working heuristic method. This method uses the estimated network element importance measures.

Table 2.4 Network reliability after four edges replacement

Edges	$p = 0.7$	$p = 0.8$	$p = 0.9$	$p = 0.95$
(1,5), (5,7), (6,14), (13,14)	0.904522	0.93793	0.968138	0.982422
(1,3), (3,7), (5,6), (5,13)	0.893605	0.917477	0.940232	0.952468
(1,2), (1,9), (7,8), (7,15)	0.899953	0.923532	0.942001	0.948896
(2,6), (6,8), (9,13), (10,14)	0.884582	0.898797	0.911858	0.917419

The background for the proposed method is the following theorem (Gertbakh and Shpungin 2008, 2009).

Theorem 3 *Suppose we are given the BIM$\lozenge$S for our network. Let us fix two indices α and $\beta \neq \alpha$, and the corresponding $z_{i,\alpha}$ and $z_{i,\beta}$ values from the BIM$\lozenge$S.*

- If for all $i, i = 1, 2, \ldots, k$, $z_{i,\alpha} \geq z_{i,\beta}$, then $BIM_\alpha \geq BIM_\beta$ for all p values.
- Suppose that the first sentence does not take place, and let s be the maximal index such that $z_{s,\alpha} \neq z_{s,\beta}$. Without loss of generality, assume that $z_{s,\alpha} > z_{s,\beta}$.

Then there exists p_0 such that for $p \geq p_0$ $BIM_\alpha > BIM_\beta$.

The following heuristic algorithm gives rather good, although not necessary optimal solution for the problem of choosing s elements for replacement. The advantage of this algorithm is its simplicity and computational efficiency.

Algorithm 2 Heuristic Replacement
1. **Estimate** the BIM's values for all elements.
2. **Range** the elements by their BIM's values, from the ``best'' to the ``worst''.
3. **Take** the first ``best'' s elements and **replace** them by more reliable ones.

Example 3 Consider the network H_4 in Fig. 2.2, with three terminals: 1, 7, and 14. Suppose that all edges *up* probabilities equal 0.6. The appropriate network computed reliability is $R = 0.86917$. Let us replace four edges by more reliable ones in accordance with the above heuristic algorithm. For the first, we try the edges with the highest BIMs, from the first group (such are two edges) and from the second (also two edges). After this we take edges from the third, fourth, and fifth groups. We see from Table 2.4 that the replacing by the edges with the highest BIMs is better than all other replacings. Replacing by the edges from the second group is slightly worse than from the third group, for the probabilities $p = 0.7, 0.8, 0.9$, and slightly better for $p = 0.95$.

2.5 Conclusions

For the case of networks with equally reliable components (nodes, or edges or both), we have developed a new efficient computation scheme for evaluating the BIMs. Its main idea is based on calculating, via a Monte Carlo scheme, network combinatorial invariant called spectrum and on the connection between the spectrum and the component BIMs.

We have proposed a new efficient approach to one of the problems in optimal network design.

The techniques developed in this approach can be viewed as a first step toward developing other efficient algorithms for optimal network reliability design.

References

Barlow RE, Proschan F (1975) Statistical theory of reliability and life testing. Holt, Rinehart and Winston, New York

Elperin T, Gertsbakh I, Lomonosov M (1991) Estimation of network reliability using graph evolution models. IEEE Trans Reliab R-40:572–581

Gertsbakh I, Shpungin Y (2004) Combinatorial approaches to Monte Carlo estimation of network lifetime distribution. Appl Stoch Models Bus Ind 20:49–57

Gertsbakh I, Shpungin Y (2008) Network reliability importance measures: combinatorics and Monte Carlo based computations. WSEAS Trans Comp 4(7):216–227

Gertsbakh I, Shpungin Y (2009) Models of network reliability: analysis, combinatorics, and Monte Carlo. CRC Press, Boca Raton

Samaniego FJ (1985) On closure of the IFR under formation of coherent systems. IEEE Trans Reliab 34:69–72

Samaniego FJ (2007) System signatures and their applications in engineering reliability. Springer, New York

Chapter 3
Signatures and Symmetry Properties of Coherent Systems

Fabio Spizzichino and Jorge Navarro

Abstract The theory of signatures originally emerged as a useful and elegant tool in the analysis of coherent systems with components having independent, identically distributed lifetimes. Recently this theory has been extended to the case of exchangeable lifetimes. Here, we discuss some related concepts that arise when the lifetimes are not exchangeable.

Keywords Average systems · Projected systems · Weak-exchangeability · Signature-exchangeability · Order-statistics-exchangeability · Well-designed systems

3.1 Introduction

The concept of *signature* is an extremely simple, elegant, and powerful concept that emerges in a completely natural way in the analysis of coherent systems. Despite that, only relatively recently it appeared in the reliability literature, having been introduced by Samaniego (1985). Independent from Samaniego's original study, the same concept was also later rediscovered by Elperin et al. (1991) and presented under the different term of *internal distribution*. Later on, the term *Destruction-spectrum* (or *D-spectrum*) was also alternatively used. In the work by

F. Spizzichino (✉)
Department of Mathematics, Università La Sapienza,
Piazzale A. MORO 2, 00185 Rome, Italy
e-mail: fabio.spizzichino@uniroma1.it

J. Navarro
Facultad de Matemáticas, Universidad de Murcia, Murcia, Spain
e-mail: jorgenav@um.es

A. Lisnianski and I. Frenkel (eds.), *Recent Advances in System Reliability*,
Springer Series in Reliability Engineering, DOI: 10.1007/978-1-4471-2207-4_3,
© Springer-Verlag London Limited 2012

Samaniego, signature came out in the study about the classical problem of the closure of the IFR class of distributions under the formation of coherent systems. In the work by Elperin et al. (1991) it manifested in the analysis of network reliability. Even though, already from Samaniego (1985), it emerged as an evident fact that signature could have a large importance in the analysis of coherent systems, some years elapsed until the potentiality of signature-based analysis could be actually pointed out.

But after a few years, and especially more recently, more and more papers have appeared on the subject and signature progressively revealed to be a useful tool in the analysis of several different problems in reliability.

In particular, the use of signatures turned out also to be very useful in the comparison of systems (Block et al. 2007a, b; Kochar et al. 1999; Navarro and Shaked 2006), in the analysis of direct and indirect majority systems (Boland 2001) and of consecutive k-out-of-n systems (Boland and Samaniego 2004a; Navarro and Eryilmaz 2007).

A different notion, also useful for computational purposes in system reliability, is the one of *domination,* introduced by Satyanarayana and Prabhakar (1978). The link between domination and signature has been discussed in Boland et al. (2003).

Nowadays a number of papers have appeared on the topic of signature and two different books have recently been published (Gertsbakh and Shpungin 2010; Samaniego 2007). An interesting discussion paper is Boland and Samaniego (2004b).

Coming to the motivation for this paper we point out that, in the beginning, signature was exclusively considered in the analysis of systems with i.i.d. components' lifetimes. There are several motivations for this special choice that any reader can likely understand basing on her/his own experience; a useful discussion can be found in Samaniego (2007).

Recently, it has been pointed out that, in any case, arguments about signatures can be extended, in a natural way, to the case of exchangeability (Navarro and Rychlik 2007; Navarro et al. 2008). This is an important extension for several reasons. In particular, it is to be taken into account that one often deals with components that are conditionally i.i.d. rather than just i.i.d.

Furthermore, it has been shown (Spizzichino 2008; Navarro et al. 2010; Marichal and Mathonet 2011) the interest, and the possibility, of considering the concept of signature even in cases of non-exchangeability. A main feature emerges at this point, however: two different concepts of signature arise in the case of non-exchangeability. We are led to introduce in fact two different concepts that do coincide in the special case of i.i.d. or exchangeable components' lifetimes.

In the more general case, one concept of signature is related with the study of symmetry properties of the system (Spizzichino 2008), while the other concept has a role in the approximation of system reliability. These two concepts suggest to consider the notions of *average system* (Navarro et al. 2010; Spizzichino 2008) and *projected system* (Navarro et al. 2010).

In this note we consider some possible developments along this direction and present the related considerations and examples.

The notions of average system and projected system, in their turn, are related with a concept of weak-exchangeability pointed out already in Navarro et al. (2008). Here we distinguish between two different notions of weak-exchangeability, namely signature-exchangeability and order-statistics-exchangeability. As we shall discuss, both the notions are related with the idea of signature and can be of interest in the analysis of symmetry properties of a coherent system with non-exchangeable components.

More precisely, the contents of the paper is as follows. In Sect. 3.2 we briefly recall the concept of signature and compare the case of exchangeable lifetimes with the non-exchangeable case. In Sect. 3.3 we illustrate the concepts of average system, projected system, and weak-exchangeability. Related to the above discussion, Sect. 3.4 presents some examples that can also constitute possible hints for further research. Finally, we devote Sect. 3.5 to a brief discussion and some concluding remarks.

3.2 Signatures in the Exchangeable Case and in the Non-Exchangeable Case

In order to recall concepts related to signatures we consider, as a natural setting, a (two-states) *reliability system* S formed with n (two-states) *components* $C_1, \ldots, C_n$.

It is assumed that $C_1, \ldots, C_n$ are new when installed and that S starts working at time 0. We denote by T_S, or simply T, the *lifetime* of S and by

$$R_S(t) := P\{T_S > t\}$$

the *reliability function* (or survival function) of T_S.

$T_1, \ldots, T_n$ are the *lifetimes* of $C_1, \ldots, C_n$; $\overline{F}(t_1, \ldots, t_n)$ denotes the *joint survival function* of $\mathbf{T} \equiv (T_1, \ldots, T_n)$ and $\overline{G}_k(t)$ denotes the *marginal survival function* of T_k, $k = 1, \ldots, n$, i.e.

$$\overline{F}(t_1, \ldots, t_n) := P\{T_1 > t_1, \ldots, T_n > t_n\},$$

$$\overline{G}_k(t) := P\{T_k > t\} = \overline{F}(0, \ldots, 0, t, 0, \ldots, 0),$$

t being the kth coordinate of the vector $(0, \ldots, 0, t, 0, \ldots, 0)$.

$T_{1:n}, \ldots, T_{n:n}$ will denote the *order statistics* of $T_1, \ldots, T_n$ and, for $k = 1, \ldots, n$, $R_{k:n}(t)$ will denote the marginal reliability function of $T_{k:n}$.

We also denote by φ the *structure function* of S and (by letting, as usual, 1 for *working* and 0 for *failed*) we look at φ as a boolean function $\varphi : \{0, 1\}^n \to \{0, 1\}$.

Of course the function $R_S(t)$ is determined by the pair $(\varphi, \overline{F})$. Concerning $(\varphi, \overline{F})$, we make the following basic assumption.

Assumption 1

$$P\{T_S = T_{k:n}, \text{ for one and only one } k\} = 1. \tag{3.1}$$

The most natural cases when Assumption 1 is verified are obviously those when $P\{T_i = T_j\} = 0$ for $i \neq j$ and S is a *coherent system* (Barlow and Proschan 1975); but other cases are also possible; in particular one can think of the mixed systems considered in Boland and Samaniego (2004b) and Navarro et al. (2008).

In view of Assumption 1 we can consider the $\{1, \ldots, n\}$-valued random variable J, associated to S and defined by

$$J = k \Leftrightarrow T_s = T_{k:n}. \tag{3.2}$$

We can consider furthermore the partition $\{E_1, \ldots, E_n\}$, where

$$E_k := \{T_S = X_{k:n}\} \equiv \{J = k\}, \quad k = 1, 2, \ldots, n. \tag{3.3}$$

It is immediately seen that, under Assumption 1, a central role is played by the random variables $T_{1:n}, \ldots, T_{n:n}$ and by the events $E_1, \ldots, E_n$. In fact, as for the computation of $R_S(t)$, we can use the rule of total probabilities and write

$$R_S(t) = \sum_{k=1}^{n} P(T_S = T_{k:n}) \cdot P(T_{k:n} > t | T_S = T_{k:n}), \quad t > 0,$$

or

$$R_S(t) = \sum_{k=1}^{n} P(E_k) \cdot R_{k:n}(t | E_k), \quad t > 0, \tag{3.4}$$

where the terms with $P(E_k) = 0$ are of course to be ignored.

In the cases when $T_1, \ldots, T_n$ are i.i.d. with a continuous distribution G, the quantities $P(E_k)$, $(k = 1, \ldots, n)$ are the coordinates s_k of the *signature* of $(S, \overline{F})$, as originally defined in the paper (Samaniego 1985). Then $\mathbf{s} \equiv (s_1, s_2, \ldots, s_n)$ is a probability distribution over $\{1, 2, \ldots, n\}$.

For our purposes, we maintain the notation $s_k = P(E_k)$, $k = 1, 2, \ldots, n$ also for the general case when $T_1, \ldots, T_n$ are not necessarily i.i.d. and present the following definition extracted from Navarro et al. (2010).

Definition 1 We will call $\mathbf{s} \equiv (s_1, s_2, \ldots, s_n)$ the probability-signature of $(S, \overline{F})$.

Concerning the concept of signature, in the following remark we summarize some topics contained, in particular, in Kochar et al. (1999), Navarro et al. (2008, 2010), Samaniego (1985), Spizzichino (2008).

Remark 1 In the special case when S is coherent and $T_1, \ldots, T_n$ are exchangeable (or, in particular, i.i.d.) and with an absolutely continuous joint distribution we have, for any k such that $s_k > 0$,

$$R_{k:n}(t|E_k) = R_{k:n}(t), \tag{3.6}$$

so that (3.4) becomes

$$R_S(t) = \sum_{k=1}^{n} s_k \cdot R_{k:n}(t), \quad t > 0. \tag{3.7}$$

The reliability functions $R_{k:n}(t)$ $(k = 1, \ldots, n)$, which depend (of course) on the joint distribution of $T_1, \ldots, T_n$, are not influenced by the particular structure of the system S. On the other side, as a relevant aspect of Eq. (3.7), it is remarkable that under the mentioned assumptions the vector $\mathbf{s}$ only depends on the structure of S and is not affected by the joint distribution of $T_1, \ldots, T_n$. More precisely, the following property holds for $\mathbf{s}$ in the i.i.d. and in the exchangeable cases.

Denote by P_n the set of the permutations of $\{1, \ldots, n\}$ and, for $\pi \in \mathrm{P}_n$, briefly write $\pi \equiv (\pi(1), \ldots, \pi(n))$ to mean the permutation $\pi \equiv \begin{pmatrix} 1 & \cdots & n \\ \pi(1) & \cdots & \pi(n) \end{pmatrix}$.

Now consider the subsets of P_n defined by

$$A_k := \{\pi \in \mathrm{P}_n | \{T_{1:n} = T_{\pi(1)}, \ldots, T_{n:n} = T_{\pi(n)}\} \Rightarrow E_k\}.$$

$A_1, \ldots, A_n$ form a partition of P_n and we let

$$\rho_k := \frac{|A_k|}{n!}, \tag{3.8}$$

where $|A_k|$ denotes the cardinality of A_k.

The vector $\boldsymbol{\rho} \equiv (\rho_1, \ldots, \rho_n)$ is a combinatorial characteristic of the system S that is not influenced by the joint distribution of $T_1, \ldots, T_n$. When S is coherent we can say that $\boldsymbol{\rho}$ only depends on φ_S, the structure function of S. In Navarro et al. (2010) has been introduced the term *structure-signature* to designate $\boldsymbol{\rho}$.

It is easy to check the following: when S is coherent and $T_1, \ldots, T_n$ are exchangeable and satisfy (3.1), then

$$s_k = \rho_k, \quad k = 1, \ldots, n. \tag{3.9}$$

We can then say that, in such cases, the probability-signature coincides with the structural signature and, then, it is a purely combinatorial characteristic of the structure of the system. In particular, as a relevant fact, we note that in the case of i.i.d. lifetimes with an absolutely continuous survival function $\overline{F}$, the probability-signature $\mathbf{s}$ does not depend on $\overline{F}$. Thus, starting from Eq. (3.9), we can also write

$$R_S(t) = \sum_{k=1}^{n} \rho_k \cdot R_{k:n}(t), \quad t > 0. \tag{3.10}$$

The formula (3.10) is at the basis of the numerous applications of signature concerning different aspects of coherent system and network reliability that we partly mentioned in the Introduction and that are illustrated in the cited references.

3.3 Signatures and Partial Symmetry Properties of Coherent Systems

As widely demonstrated in the literature and briefly recalled above, the concept of structure-signature has then a very important role in the analysis of coherent systems with exchangeable (in particular i.i.d.) components.

More generally, we can say that such a concept is strictly related with the symmetry properties of a reliability system. For this reason, as it has been argued in Navarro et al. (2010) and Spizzichino (2008), the same concept can provide a deep insight into coherent systems even in the case of non-exchangeable components. In particular, an analysis based on signature-related properties can be of help in the approximation of system reliability and in the decision of where to allocate the components among the different positions of a coherent system.

On these topics we aim to add here some considerations, of potential interest for future research. Our arguments will, in particular, introduce the examples presented in the next section.

In what follows, we consider coherent systems and assume that the probability distribution of the components' lifetimes admit a joint probability density.

For a given joint density function $f(t_1, \ldots, t_n)$ and a given structure function $\phi(x_1, \ldots, x_n)$, the corresponding system will also be denoted by (φ, f). Denote by $\mathbf{s}(\varphi, f)$ the probability-signature of (φ, f).

For our aims, the following definition is useful. Let (φ', f') and (φ'', f'') be two systems where $\varphi', \varphi'' : \{0, 1\}^n \to \{0, 1\}$ are coherent structure functions and f', f'' joint probability densities over $\mathbf{R}_+^n$.

Definition 2 We say that (φ', f') and (φ'', f'') are reliability-equivalent, and write $(\varphi', f') \tilde{=}_R (\varphi'', f'')$, if and only if we have that $\mathbf{s}(\varphi', f') = \mathbf{s}(\varphi'', f'')$, and

$$P_{\varphi', f'}(T_{k:n} > t | E_k) = P_{\varphi'', f''}(T_{k:n} > t | E_k)$$

for all k such that $s_k(\varphi', f') > 0$ and $s_k(\varphi'', f'') > 0$.

Remark 2 In view of Eq. (3.4), it is obvious that if the relation $(\varphi', f') \tilde{=}_R (\varphi'', f'')$ holds, then the two coherent systems characterized by (φ', f') and (φ'', f'') respectively, have the same reliability function. Of course this is a sufficient but not a necessary condition for two systems to share the same reliability function.

Remark 3 We saw in the previous section that the probability-signature $\mathbf{s}$ coincides with the structure-signature ρ defined by Eq. (3.8) when $T_1, \ldots, T_n$ are exchangeable and satisfy (3.1). This identity is not only important from a very conceptual point of view, but also useful for computational purposes.

Concerning the computation of $\mathbf{s}$ for the non-exchangeable case, a useful formula has been provided in the recent paper (Marichal and Mathonet 2010).

It is now of basic importance for our study pointing out the different meanings that the permutations $\pi \in P_n$ may have in the analysis of a system.

Let us consider then a coherent system S formed with n components $C_1, \ldots, C_n$, with structure function φ, and let f be the joint density function of $T_1, \ldots, T_n$, the lifetimes of $C_1, \ldots, C_n$. We make the hypothesis that each component out of $\{C_1, \ldots, C_n\}$ is potentially able to work in any of the n positions of the system. In practice, this hypothesis is purely conceptual and only has a mathematical role. For many real systems, it is a rather curious assumption from an engineering point of view; however it can be perfectly justified in several cases as, for instance, in network reliability (Gertsbakh and Shpungin 2010).

It is also useful to add a further assumption, purely conceptual and innocuous as well: after the system has started to work, each component will go on working until its own failure, even if the system already failed at a certain instant of time. In view of this assumption, we conceive that it is possible to observe sequentially the failures of all the components and this observation then gives rise to a random permutation of the indexes $\{1, \ldots, n\}$; such a permutation describes the order according to which the components $C_1, \ldots, C_n$ fail.

Thus the observation of the performance of the system gives rise to a random permutation of $\{1, \ldots, n\}$, i.e. to a random element of P_n. The probability distribution of such a permutation is uniform over P_n, when $T_1, \ldots, T_n$ are exchangeable and satisfy (3.1), but it may be not at all uniform in more general cases.

In all the cases when such a distribution is actually uniform over P_n, the structure-signature and the probability-signature coincide, by definition.

In our analysis however, the elements of P_n can, not only be seen as random events in the experiment of observing the order of the subsequent failures of the components of the system, but we can also look at them as at transformations acting on the system.

More precisely we note that any $\pi \in P_n$ can be seen as an operator that acts on φ. In fact we can define, for $(x_1, \ldots, x_n) \in \{0, 1\}^n$,

$$\phi_\pi(x_1, \ldots, x_n) := \phi\left(x_{\pi(1)}, \ldots, x_{\pi(n)}\right).$$

For $\pi \in P_n$, we denote by S_π the reliability system that admits φ_π as its structure function. S_π is still a coherent system that is obtained by installing the component C_{π_h} in the position h of the system S ($h = 1, \ldots, n$).

Denoting by $\circ$ the composition between two elements of P_n, we can write

$$\varphi_{\pi' \circ \pi''}(x_1, \ldots, x_n) = \varphi\left(x_{\pi' \circ \pi''(1)}, \ldots, x_{\pi' \circ \pi''(n)}\right) = \varphi\left(x_{\pi'(\pi''(1))}, \ldots, x_{\pi'(\pi''(n))}\right)$$
$$= (\varphi_{\pi''})_{\pi'}(x_1, \ldots, x_n).$$

By the symbol π^{-1} we then denote the inverse of π with respect to the operation $\circ$, so that $S_{\pi^{-1}}$ is the system obtained by installing the component C_h in the position $\pi^{-1}(h)$ of the system S ($h = 1, \ldots, n$).

Similar to the above, any $\pi \in P_n$ can also be seen as an operator acting on $f(t_1, \ldots, t_n)$. In fact, we can consider the joint density function f_π of the random vector $\left(X_{\pi(1)}, \ldots, X_{\pi(n)}\right)$, i.e. we define for $(t_1, \ldots, t_n) \in R_+^n$,

$$f_\pi(t_1, \ldots, t_n) := f(t_{\pi^{-1}(1)}, \ldots, t_{\pi^{-1}(n)}).$$

At this point, it is useful to note (even though it is almost obvious) that the reliability function of the system S_π with components' lifetimes $T_1, \ldots, T_n$ coincides with the reliability function of the system S with components' lifetimes having joint density function f_π. Thus, we can in particular write $(\varphi_\pi, f) \cong (\varphi, f_\pi)$.

Remark 4 We mentioned above that the structure-signature $\boldsymbol{\rho} = \boldsymbol{\rho}^{(\phi)}$ of a coherent system S is determined by its structure function φ. By the very definition of $\boldsymbol{\rho}$ it trivially follows that all the structures φ_π admit the same vector of structure-signature $\boldsymbol{\rho}^{(\varphi)}$.

On the contrary it can happen that, for some non-exchangeable f and some $\pi \in P_n$, the probability-signature of (φ_π, f) is different from the one of (φ, f) (see Example 3 in the next section).

Remark 5 The probability-signature (seen as a probability distribution) of a coherent system has the same support of the structure-signature. It follows that all the probability-signatures of (φ_π, f) have the same support, for $\pi \in P_n$.

For $\pi \in P_n$, set now $\mathbf{T}_\pi = (T_{\pi(1)}, \ldots, T_{\pi(n)})$, so that $\mathbf{T}_\pi$ has a joint density f_π. Furthermore, let the random vector $\tilde{\mathbf{T}} \equiv (\tilde{T}_1, \ldots, \tilde{T}_n)$, be defined by $\tilde{\mathbf{T}} = \mathbf{T}_\pi$ with probability $1/n!$.

It is promptly seen that $\tilde{\mathbf{T}}$ is exchangeable and the following facts hold: the joint survival function of $\tilde{\mathbf{T}}$ is given by

$$\bar{F}_{\tilde{\mathbf{T}}}(t_1, \ldots, t_n) = \frac{1}{n!} \sum_{\pi \in P} \bar{F}(t_{\pi(1)}, \ldots, t_{\pi(n)}); \qquad (3.11)$$

its one-dimensional marginal survival function is $\bar{G}(t) = \frac{1}{n} \sum_{k=1}^{n} \bar{G}_k(t)$; the joint density function of $\tilde{\mathbf{T}}$ is given by

$$f_{\tilde{\mathbf{T}}}(t_1, \ldots, t_n) = \frac{1}{n!} \sum_{\pi \in P} f(t_{\pi(1)}, \ldots, t_{\pi(n)}); \qquad (3.12)$$

Remark 6 A simple property of $\bar{F}_{\tilde{\mathbf{T}}}$ is relevant for our discussion: the random vectors $(T_{1:n}, \ldots, T_{n:n})$ and $(\tilde{T}_{1:n}, \ldots, \tilde{T}_{n:n})$ share the same joint distribution (Spizzichino 2008). Then, in particular, they share the same marginal reliability functions $R_{k:n}(t)$, $k = 1, \ldots, n$. It is furthermore shown in Navarro and Spizzichino (2010a) that the connecting copula of $(T_{1:n}, \ldots, T_{n:n})$ depends on C, the connecting copula of $(T_1, \ldots, T_n)$, and on $(\bar{G}_1, \ldots, \bar{G}_n)$, only through $\tilde{C}(C; \bar{G}_1, \ldots, \bar{G}_n)$, the connecting copula of $(\tilde{T}_1, \ldots, \tilde{T}_n)$.

We remind that, for a vector of random variables $X_1, \ldots, X_n$ with a joint distribution function $F(x_1, \ldots, x_n)$ and marginal distribution functions $F_1, \ldots, F_n$ respectively, the connecting copula is the function $C : [0, 1]^n \to [0, 1]$ defined by

$$C(u_1, \ldots, u_n) = F(F_1^{-1}(u_1), \ldots, F_n^{-1}(u_n)).$$

C can be seen as the restriction to $[0, 1]^n$ of the n-dimensional distribution function of the vector $(F_1(X_1), \ldots, F_n(X_n))$. For more details, properties, meaning, and applications of the notions of copula and connecting copula (see Nelsen 2006).

Let S be a coherent system with structural signature $\boldsymbol{\rho} \equiv (\rho_1, \ldots, \rho_n)$, and non-exchangeable $T_1, \ldots, T_n$. The expression $\sum_{k=1}^{n} \rho_k \cdot R_{(k)}(t)$ appearing in Eq. (3.10) corresponds to the reliability function of the system $\overline{S}$ that has the same structure of S and is formed with components whose lifetimes are jointly distributed as $\tilde{\mathbf{T}}$. Such an expression can then be seen as the reliability function of the system S when its components $C_1, \ldots, C_n$ are installed at random in the different positions of S. $\overline{S}$ can be seen as the average of all the systems S_π. For this reason it can be called the "average system". For more detailed discussions about the interpretation of $\overline{S}$ and related considerations see Navarro et al. (2010) and Spizzichino (2008).

Remark 7 By the very definition of the average system, it is immediately clear that all the systems S_π admit the same average system.

In Navarro et al. (2010) has also been introduced the notion of *projected system*, as the (possibly fictitious) system that has a reliability function given by $\sum_{k=1}^{n} s_k \cdot R_{k:n}(t)$. The notion of projected system leads to useful approximations of system's reliability; a different notion, having properties of optimality in the approximation problem and that can be seen as dual to the one of projected system, is the *best symmetric approximation* considered in Marichal and Mathonet (2011).

The original system S, its average system, and its projected system do coincide when $T_1, \ldots, T_n$ are exchangeable. On the contrary they have, generally, different reliability functions when $T_1, \ldots, T_n$ are not exchangeable.

The possible comparisons among such reliability functions suggest us to consider the following properties of weak-exchangeability for a pair (φ, f).

Definition 3 The pair (φ, f) is signature-exchangeable (S-E) if the condition (3.9) holds.

Definition 4 The pair (φ, f) is order-statistics-exchangeable (OS-E) if the condition (3.6) holds for all k such that $\rho_k > 0$.

Both S-E and OS-E are then satisfied when S is coherent and $T_1, \ldots, T_n$ are exchangeable. However the study of these two conditions can also be of interest when coherency of S and/or exchangeability of lifetimes fail.

Under Assumption 1, both Definitions 3 and 4 are related with (partial) symmetry properties of the pair (φ, f). Condition (3.6) is related with the concept of *weak-exchangeability* introduced in Navarro et al. (2008); see also Navarro et al. (2010). The condition (3.9) says that the projected system coincides with the average system; in such a case the probability-signatures $\mathbf{s}_\pi$ of the systems S_π, $\pi \in P_n$, are equal to each other.

We now describe in detail the idea of (partial) symmetry properties of the pair (φ, f). To this purpose, we start with a simple remark.

Consider a system S of the type k-out-of-n, for some $1 \le k \le n$. In terms of the concept of signature, such cases are characterized by the condition that the structure-signature is a degenerate distribution:

$$\boldsymbol{\rho} = (0, \ldots, 0, 1, 0, \ldots, 0), \tag{3.13}$$

1 appearing as the kth coordinate of $\boldsymbol{\rho}$.

We also necessarily have that $\mathbf{s} = \boldsymbol{\rho}$ (see also Remark 5) and that S coincides with all the permuted systems S_π, with $\pi \in P_n$. We can say that φ is *completely symmetric* and it follows that the system S, its average system, and its projected system do coincide, although $T_1, \ldots, T_n$ are not exchangeable.

Generally we will have pairs (φ, f) where neither φ nor f are completely symmetric; i.e. neither the system is of the form k-out-of-n as in (3.13), nor the components' lifetimes are exchangeable.

In any case we can consider the following two subsets of P_n:

$$P_n^{(\varphi)} := \{\pi \in P_n | \phi_\pi = \varphi\},$$

$$P_n^{(f)} := \{\pi \in P_n | f_\pi = f\}.$$

P_n being a group (*the symmetric group*) with respect to the composition $\circ$, we can look at $P_n^{(\varphi)}$ and $P_n^{(f)}$ as subgroups of P_n.

Let us denote by $P_n^{(\varphi,f)}$ the subgroup of P_n generated by $P_n^{(f)} \cup P_n^{(\varphi)}$.

Obviously $P_n^{(\varphi,f)} = P_n$, when T is exchangeable or φ is of the form (3.13).

It is intuitive that the bigger the cardinality of $P_n^{(\varphi,f)}$, the more the structure (φ, f) is symmetric or, in other words, the closer are the reliability functions of S and $\overline{S}$. In this respect we can formally state the following result.

Proposition 1 *For* $\pi \in P_n^{(\varphi,f)}$, *all the pairs* (φ_π, f) *are reliability-equivalent to each other, in the sense of Definition 2.*

Proof We noted above that, for any $\pi \in P_n$, the two pairs (φ_π, f) and (φ, f_π) are obviously reliability-equivalent.

Let us now consider $\pi' \in P_n^{(f)}$, so that $f_{\pi'} = f$ and thus $(\varphi_{\pi'}, f) \tilde{=} (\varphi, f_{\pi'}) \tilde{=} (\varphi, f)$, also for $\pi'' \in P_n^{(\phi)}$, we obviously have $(\varphi_{\pi''}, f) \tilde{=} (\varphi, f)$.

Consider now $\pi \in P_n^{(\phi,f)}$. If $\pi \notin P_n^{(f)}$ and $\pi \notin P_n^{(\phi)}$, we can at least find $\pi' \in P_n^{(f)}$ and $\pi'' \in P_n^{(\varphi)}$ such that $\pi = \pi' \circ \pi''$.

Whence it follows $(\varphi_\pi, f) \tilde{=} (\varphi, f)$, since $(\varphi_\pi, f) = (\varphi_{\pi' \circ \pi''} = ((\varphi_{\pi''})_{\pi'}, f) \cong (\varphi_{\pi''}), f \cong (\varphi, f)$

Corollary 1 *If* $P_n^{(f)} \cup P_n^{(\varphi)} = P_n$ *then all the pairs* (φ_π, f) *are reliability-equivalent to each other, both* (3.6) *and* (3.9) *hold, and the average system coincides with the original system.*

As for the cardinality $|P_n^{(f)} \cup P_n^{(\varphi)}|$ of $P_n^{(f)} \cup P_n^{(\varphi)}$, we obviously have

$$\max(|P_n^{(\varphi)}|, |P_n^{(f)}|) \leq |P_n^{(f)} \cup P_n^{(\varphi)}| \leq |P_n^{(\varphi)}| \cdot |P_n^{(f)}|.$$

On the other hand it is interesting to note that the knowledge of the structure-signature ρ directly contains some information about $|P_n^{(\varphi)}|$. Some basic considerations of algebraic type, in particular, show that, for any k such that $\rho_k > 0$, the number $|A_k| = n! \cdot \rho_k$ must be a multiple of $|P_n^{(\varphi)}|$ (see Spizzichino 2008, Proposition 3).

3.4 Examples

Here, we present a few examples that point out some aspects related with the above arguments.

In particular these examples allow us to build up cases where

1. Both the properties of signature-exchangeability and order-statistics exchangeability hold even if neither the lifetimes are exchangeable nor the structure is of the form $\hat{k} : n$;
2. Signature-exchangeability holds but order-statistics exchangeability does not hold;
3. Signature-exchangeability does not hold but order-statistics exchangeability does hold;
4. Neither signature-exchangeability nor order-statistics exchangeability hold.

To this purpose, we first concentrate attention on the joint distribution of lifetimes considered in Example 4.1 of Navarro et al. (2010) and described as follows.

Let $n = 4$ and let the joint survival function of (T_1, T_2, T_3, T_4) be of the form

$$\overline{F}(t_1, t_2, t_3, t_4) = \overline{W}(t_1, t_2) \overline{W}(t_3, t_4) \tag{3.14}$$

where $\overline{W}$ is a bivariate (absolutely continuous) exchangeable survival function with one-dimensional marginal $\overline{H}$. Then T_1, T_2, T_3, T_4 are identically distributed according to $\overline{H}$, however they are not jointly exchangeable, except in the case $\overline{W}(x, y) = \overline{H}(x)\overline{H}(y)$.

Under this model we have $P_n^{(f)} = \{(1,2,3,4), (2,1,3,4), (2,1,4,3), (1,2,4,3)\}$.

Example 1 Consider the coherent system with structure defined by

$$S = \min(T_1, \max(T_2, T_3), T_4).$$

In Navarro et al. (2010), Example 4.1, it was shown that

1. The structure-signature of the system is $\rho \equiv \left(\frac{1}{2}, \frac{1}{2}, 0, 0\right)$,
2. The condition (3.9) holds,
3. The lifetime of the average system $\overline{S}$ coincides with that of the original system S.

Note that the properties 2. and 3. are not affected by the special dependence structure of $\overline{W}$. This is an example where the conditions (3.6) and (3.9) hold even if the components' lifetimes are not exchangeable. We note that there are actually remarkable symmetries in the system; in particular, besides (3.15), we have

$$P_n^{(\phi)} = \{(1, 2, 3, 4), (1, 3, 2, 4), (4, 2, 3, 1), (4, 3, 2, 1)\}.$$

This yields $P_n^{(\varphi, f)} = P_n$.

Example 2 Consider the system with structure defined by

$$S' = \min(T_1, \max(T_4, T_3), T_2),$$

i.e. $S' = S_\pi$, with $\pi = (1, 4, 3, 2)$, where S is the system considered in the example above.

Since S' is obtained from S by a permutation of the components, then the structure-signatures of the two systems are equal. Also, the probability-signature for S' is the same as for S.

In fact the event E_1 means $\{(T_{1:4} = T_1) \cup (T_{1:4} = T_2)\}$ and then

$$P(E_1) = \frac{1}{2}, P(E_2) = 1 - P(E_1) = \frac{1}{2},$$

i.e. the condition (3.9) still holds. In this case we have

$$P_n^{(\varphi)} = \{(1, 2, 3, 4), (1, 3, 2, 4), (4, 2, 3, 1), (4, 3, 2, 1)\} = P_n^{(f)}$$

$$P_n^{(\varphi, f)} = P_n^{(\varphi)} = P_n^{(f)} \subset P_n. \tag{3.15}$$

Note that $\varphi_{S'}$ has the same symmetry structure as φ_S and also the joint distribution of (T_1, T_2, T_3, T_4) remains the same as in Example 1. However, the symmetries of the components' lifetimes and those of the system combine in a way different from those in Example 1.

We may find special conditions of stochastic dependence between (T_1, T_2) and (T_3, T_4) under which condition (3.6) is not necessarily verified and, more in particular, it can happen that $R_{1:4}(t) \geq R_{1:4}(t|E_1)$ and $R_{2:4}(t) > R_{2:4}(t|E_2)$.

An interesting fact in such situations is that, since S and S' are obtained one from the other by permutation of the components the lifetimes of their average systems are stochastically equal. We can conclude then that S' may be not a *well-designed system*, in the sense defined in Navarro et al. (2010), i.e. the system is less reliable than its average system (obtained by a random design).

The next example shows how we could find cases where, still under the position (3.14), neither (3.9) nor (3.6) holds.

Example 3 Here, we take the coherent system with

$$S'' = \min(\max(T_1, T_2), \max(T_3, T_4)),$$

whose structure-signature is $\left(0, \frac{1}{3}, \frac{2}{3}, 0\right)$. The event E_2 happens if and only if the first and the second failing components are mates in one of the two parallel subsystems. The event E_3 happens if and only if the first and the second failing components belong to the two different parallel systems. Furthermore, taking into account the total probability formula

$$R_{2:4}(t) = R_{2:4}(t|E_2)P(E_2) + R_{2:4}(t|E_3)P(E_3),$$

we note that $R_{2:4}(t|E_2) = R_{2:4}(t)$ if and only if $R_{2:4}(t|E_2) = R_{2:4}(t|E_3)$.

We could then find conditions of stochastic dependence for W, under which $p_2 \neq \frac{1}{3} = s_2$ and $R_{2:4}(t|E_2) \neq R_{2:4}(t)$. Note that, also in this case, we have $P_n^{(\varphi)} = P_n^{(f)}$ and then $P_n^{(\varphi f)} = P_n^{(f)}$, whence $P_n^{(\varphi f)}$ only contains four different permutations.

Example 4 Finally, we present a simple example where condition (3.9) does not hold, whereas order-statistics exchangeability can hold in an approximate form.

Consider the coherent system made with three components C_1, C_2, C_3 and such that $S''' = \min(T_1, \max(T_2, T_3))$, and then the structure-signature is $\rho \equiv \left(\frac{1}{3}, \frac{2}{3}, 0\right)$.

We assume T_1, T_2, T_3 to be stochastically independent. Let furthermore T_2, T_3 be identically distributed with survival function $\overline{G}$ and let T_1 be distributed according to a different survival function $\overline{H}$. Let the component C_1 be more reliable than C_2 and C_3, e.g. assume

$$\overline{G}(t) = \exp\{-mt\}, \overline{H}(t) = \exp\{-t\} \tag{3.16}$$

with $m > 1$.

From now on, let us use the notation $P^{(m)}(E)$ to denote the probability of the event E corresponding to the choice of the parameter m in (3.16); similarly we shall use the notation $\rho^{(m)}, s^{(m)}, R_{k:3}^{(m)}, \ldots$

The event E_1 is obviously equivalent to $\{T_{(1)} = T_1\}$ and thus

$$s_1^{(m)} = P^{(m)}\left(T_{(1)} = T_1\right).$$

Now it is easy to check that $P^{(m)}(T_{(1)} = T_1) = P^{(m)}(T_1 > \min(T_2, T_3)) = \frac{1}{1+2m}$, and thus the vectors $\rho^{(m)}$ and $s^{(m)}$ are necessarily different.

On the other hand we can claim that the order-statistics exchangeability conditions $R_{1:3}^{(m)}(t|E_1) = R_{1:3}^{(m)}(t)$ and $R_{2:3}^{(m)}(t|E_2) = R_{2:3}^{(m)}(t)$ approximately hold. In fact, for very large m, the considered system tends to reduce to a series of two independent, similar, components; so that $\lim_{m\to\infty} \frac{R_{1:3}^{(m)}(t|E_1)}{R_{1:3}^{(m)}(t)} = 1$ and $\lim_{m\to\infty} \frac{R_{2:3}^{(m)}(t|E_2)}{R_{2:3}^{(m)}(t)} = 1$.

3.5 Concluding Remarks

In this paper we pointed out several aspects related with the two concepts of probability-signature and structure-signature and with comparisons between them.

In the exchangeable case, under condition (3.1), such two concepts coincide with each other and have a basic role in the computation of system reliability.

In the non-exchangeable case they do not necessarily coincide anymore and cannot be used, generally, for a direct computation of system reliability. However, they can have a useful role in the study of (partial) symmetry properties of a system and in the approximation of its reliability.

Our interest in this topic is motivated by the fact that the analysis of non-exchangeable cases deserves special care and appropriate methods.

In particular we note the following: it is a common opinion that the study of a system with i.i.d. components' lifetimes can be of basic interest even for the case of non-exchangeable components, in that it allows one to compare different systems designs. Our examples above can show that the situation is actually a bit more complicate than that.

In the effort to understand what can be the effect of non-exchangeability on the performance of the system, it can be natural to start by considering two different special cases:

1. independence (and non-identical distribution) between component's lifetimes;
2. identical marginal distribution of component's lifetimes (and different types of correlations between different subsets of components).

These two classes of special cases constitute basic situations of non-exchangeability and, when considered separately, allow us to understand the effects of components' asymmetry on the performance of a system.

Actually a vast literature has been dedicated to the case (a). Concerning this case we just mention that one might construct examples of non-well-designed systems (as shown by Example 4), by considering situations of heterogeneity among marginal distributions, rather than of heterogeneity in the form of dependence (as shown by Examples 1–3).

We also remark that a natural generalization of the case of independent, marginally heterogeneous components arises by replacing the assumption of independence (i.e. product copula) by an arbitrary exchangeable copula. As it has been discussed in Navarro and Spizzichino (2010b), the form under which heterogeneity among the components affect the performance of a coherent system may be affected by the specific choice of the exchangeable copula.

To the best of our knowledge, the case considered in the above Examples 1–3, of marginal identical distributions (and lack of symmetry in the correlations among different sub-families of components) has not been, on the contrary, considered extensively in the literature. This case can however be of real interest for several respects, in particular in the field of network reliability, where it is not so unrealistic to admit that all the arcs connecting pairs of nodes are equally reliable, and that possible factors of asymmetry are rather related with the topology of the network and with interconnections among the arcs.

Acknowledgments We thank Professor Ilya Gerstbakh for his helpful remarks on a previous version of the paper. J. Navarro has been partially supported by Ministerio de Ciencia y Tecnologia de Espãna under grant MTM2009-08311 and by Fundacion Seneca (C.A.R.M.) under grant 08627/PI/08. F. Spizzichino has been partially supported in the frame of the Research Project *Problemi di convergenza e ottimizzazione in alcuni modelli stocastici*, Universita' La Sapienza, Ateneo Federato per la Scienza e la Tecnologia.

References

Barlow RE, Proschan F (1975) Statistical theory of reliability and life testing: probability models. Holt, Rinehart and Wiston, New York

Block HW, Dugas MR, Samaniego FJ (2007a) Characterizations of the relative behavior of two systems via properties of their signature vectors. In: Balakrishnan N, Sarabia JM, Castillo E (eds) Advances in distribution theory, order statistics and inference. Birkhäuser, Boston, pp 279–289

Block HW, Dugas MR, Samaniego FJ (2007b) Signature-related results on system lifetimes. In: Nair V (ed) Advances in statistical modeling and inference. World Scientific, Singapore, pp 115–129

Boland PJ (2001) Signature of indirect majority systems. J Appl Probab 38(2):597–603

Boland PJ, Samaniego FJ (2004a) Stochastic ordering results for consecutive k-out-of-n systems. IEEE Trans Reliab 53(1):7–10

Boland PJ, Samaniego FJ (2004b) The signature of a coherent system and its applications in reliability. In: Soyer R, Mazzuchi T, Singpurwalla N (eds) Mathematical reliability: an expository perspective. Kluwer, Boston, pp 3–29

Boland PJ, Samaniego FJ, Vestrup EM (2003) Linking dominations and signatures in network reliability theory. In: Lindqvist BH, Doksum K (eds) Mathematical and statistical methods in reliability. World Scientific, Singapore, pp 89–103

Elperin T, Gertsbakh IB, Lomonosov M (1991) Estimation of network reliability using graph evolution models. IEEE Trans Reliab 40(5):572–581

Gertsbakh IB, Shpungin Y (2010) Models of network reliability. CRC Press, Boca Raton

Kochar S, Mukerjee H, Samaniego FJ (1999) The "signature" of a coherent system and its application to comparisons among systems. Naval Res Logist 46(5):507–523

Marichal JL, Mathonet P (2011) System signatures for dependent lifetimes: explicit expressions and interpretations. J Multivar Anal 102(5):931–936

Navarro J, Eryilmaz S (2007) Mean residual lifetimes of consecutive-k-out-of-n systems. J Appl Prob 44(1):82–98

Navarro J, Rychlik T (2007) Reliability and expectation bounds for coherent systems with exchangeable components. J Multivar Anal 98(1):102–113

Navarro J, Shaked M (2006) Hazard rate ordering of order statistics and systems. J. Appl Probab 43(2):391–408

Navarro J, Spizzichino F (2010a) On the relationships between copulas of order statistics and marginal distributions. Stat Prob Lett 80:473–479

Navarro J, Spizzichino F (2010b) Comparisons of series and parallel systems with components sharing the same copula. Appl Stoch Models Bus Ind 26(6):775–791

Navarro J, Samaniego FJ, Balakrishnan N, Bhattacharaya D (2008) On the application and extension of system signatures in engineering reliability. Nav Res Log 55(4):313–327

Navarro J, Spizzichino F, Balakrishnan N (2010) Applications of average and projected systems to the study of coherent systems. J Multivar Anal 101(6):1471–1482

Nelsen R (2006) An introduction to copulas. Springer, New York

Samaniego FJ (1985) On closure of the IFR class under formation of coherent systems. IEEE Trans Reliab 34(1):69–72

Samaniego FJ (2007) System signatures and their application in engineering reliability. Springer, Berlin

Satyanarayana A, Prabhakar A (1978) New topological formula and rapid algorithm for reliability analysis of complex networks. IEEE Trans Reliab 27(2):82–100

Spizzichino F (2008) The role of signature and symmetrization for systems with non-exchangeable components. In: Bedford T, Quigley J, Walls L, Alkali B, Daneshkhah A, Hardman G (eds) Mathematical modeling for reliability. IOS Press, Amsterdam, pp 138–148

Chapter 4
Multidimensional Spectra of Multistate Systems with Binary Components

Ilya Gertsbakh and Yoseph Shpungin

Abstract We consider coherent systems with $K > 1$ *UP* states and binary components. Introducing uniform measure on $n!$ permutations of component numbers, we define a system's combinatorial invariant, so-called multidimensional D-spectrum. By means of its marginal distributions we determine the probability that the system is in state $i \in [0, 1, 2, \ldots, K]$, where 0 denotes system *DOWN* state. Formulas establishing the connection between the cumulative D-spectra and the number of system failure sets are derived. We present an example of a four-dimensional cubic network with 16 nodes, 32 edges and four terminals and the probabilistic description of its disintegration into four isolated clusters in the process of random edge failures.

Keywords Multistate coherent systems · Multidimensional D-spectra · Signatures · State probabilities

4.1 Introduction

In this paper we will consider coherent systems with binary components and more than two states. The paper is devoted to the investigation of a *multidimensional* analogue of the D-spectrum defined for *binary* coherent systems (Elperin et al. 1991; Gertsbakh and Shpungin 2004, 2008, 2009; Samaniego 1985).

I. Gertsbakh (✉)
Department of Mathematics, Ben Gurion University of the Negev,
Beer Sheva, Israel
e-mail: elyager@bezeqint.net

Y. Shpungin
Software Engineering Department, Sami Shamoon College of Engineering,
Beer Sheva, Israel
e-mail: yosefs@sce.ac.il

Let us recall the definition of system D-spectrum for the case of two-state coherent system. Consider an n-component coherent system. Each component may be in two states—*up* and *down*. Components are numbered as $1, 2, 3, \ldots, n$. System state is described by a vector $x = (x_1, x_2, \ldots, x_n)$, where $x_i = 1$ if the ith component is *up*, and $x_i = 0$, otherwise. The whole system also can be in only two states, *DOWN* and *UP*. The system state is determined by the so-called structure function $\phi(x)$, which maps each state vector x into the set $S = \{0, 1\}$. $\phi(\cdot) = 1$ means that the system is *UP* and $\phi(\cdot) = 0$ means that the system is *DOWN*. As usual, it is assumed that $\phi(1, 1, \ldots, 1) = 1$ and $\phi(0, 0, \ldots, 0) = 0$ and that the structure function is monotone, i.e. if $x < y$ then $\phi(x) \leq \phi(y)$. ($x \leq y$ means that $x_i \leq y_i$ for all $i, i = 1, \ldots, n$. If, in addition, at least one component of x is smaller than the corresponding component of y we say that $x < y$.)

For a fundamental description of binary systems with binary components and their properties see (Barlow and Proschan 1975).

Let π be a permutation of numbers $1, 2, \ldots, n$: $\pi = (e_{i_1}, e_{i_2}, \ldots, e_{i_n})$. Imagine that we start with all components being *up* and turn them from *up* to *down* in the order they appear in π, from left to right. Stop at the first element e_{i_r} when the system becomes *DOWN*. The ordinal number r of this element is called the *anchor* of permutation π and denoted $r(\pi)$. #

Definition 4.1 *D-spectrum for a binary system* Associate with each permutation the probability $P(\pi) = 1/n!$. Denote by f_i the probability that the random permutation anchor is equal to $i, i = 1, \ldots, n$. The collection of probabilities $\{f_i, i = 1, 2, \ldots, n\}$, is called the *D-spectrum* of a two-state binary system. #

In D-spectrum, "D" stands for "destruction" taking place in the process of turning *down* the components in a permutation.

In simple words, the D-spectrum is the distribution of the anchor position. Obviously, $\sum_{i=1}^{n} f_i = 1$.

Definition 4.2 *Cumulative D-spectrum for a binary system* The cumulative distribution function (CDF) $F(k)$ of the anchor position is called the cumulative D-spectrum:

$$F(k) = \sum_{i=1}^{k} f_i, k = 1, 2, \ldots, n. \# \tag{4.1}$$

We will accompany all definitions by examples.

Example 4.1   Figure 4.1 presents a bridge-type network with five components (edges). By definition, the system is *UP* if there is a path from s to t. Otherwise, it is *DOWN*.

The permutation $\pi = \{1, 3, 5, 4, 2\}$ has anchor $r(\pi) = 3$. Analyzing all $5! = 120$ permutations it is easy to find out that for this system the D-spectrum is $\{0, 1/5, 3/5, 1/5, 0\}$. Obviously, $F(1) = 0, F(2) = 1/5, F(3) = 4/5, F(4) = F(5) = 1$. #

Let us make a historical remark. The D-spectrum introduced in Definition 4.1 coincides with the so-called coherent system *signature* first discovered by

Fig. 4.1 Bridge system

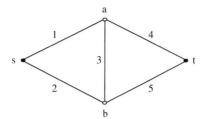

Samaniego (1985). The same object was introduced in (Elperin et al. 1991) under the name *Internal Distribution* (ID). This paper presents a collection of Monte Carlo simulated IDs for several complete graphs and dodecahedron-type networks. In (Gertsbakh and Shpungin 2004, 2008, 2009) the ID was termed *D-spectrum*. It is worth stressing that our definition of the D-spectrum is of purely combinatorial nature. After imposing a uniform measure on the set of all $n!$ permutations, the D-spectrum depends only on system properties reflected in the structure function and *does not* depend on any stochastic mechanism governing the system component failures. The fact that the system structure function determines the signature was noted also in (Navarro and Rubio 2010). It is shown there that the signature is determined by the minimal path sets of the system.

Now imagine that the system consisting of binary components can be in more than a single *UP* state. Number these states by integers: $1, 2, \ldots, K$, $1 < K$. Denote by zero the system *DOWN* state. It is convenient to introduce system structure function $\phi(x)$ which in this case maps the binary state vector x into a set of system states $S = \{0, 1, 2, \ldots, K\}$. Similar to the case of binary $\phi(\cdot)$, $\phi(1, 1, \ldots, 1) = K$, $\phi(0, 0, \ldots, 0) = 0$ and the structure function is monotone, i.e. if $x < y$ then $\phi(x) \leq \phi(y)$.

In addition we will postulate the following *regularity property* of the structure function. Suppose that $\phi(x) = r, r > 0$. Then for any y which is smaller than x and such that the Manhattan distance between x and y equals 1, $\phi(x) - \phi(y) \leq 1$. In simple words, the failure of a single component either leaves the system state unchanged or causes a change of system state by *one unit*.

Imagine again that all components are initially *up* and arranged in a row according to some permutation π and we start sequentially turning them *down*, from left to right. Then we will observe *more than one anchor*. Each one of them will signify a change in system state (i.e. the drop of $\phi(\cdot)$ by one unit) in the process of turning *down* the components.

Example 4.2   Assume that all edges of the bridge system in Fig. 4.1 have capacity 1. Define the system state as the *maximal flow* which can be delivered from s to t. Obviously, if all edges are *up*, this flow equals 2. If, for example edge 1 is *down* the flow will be 1. If, in addition, edge $2 = (s, b)$ gets *down*, the flow will be 0. If we define the system state as the maximal flow from s to t, we will have, therefore, two *UP* states, 2 and 1, and a *DOWN* state when the flow is 0. Consider now edge permutation $\pi = \{1, 5, 2, 4, 3\}$. Turning edges from *up* to *down*, we observe that the system changes its state at erasing the first component 1 (the flow changes from 2 to 1) and at erasing the third component in the permutation, i.e. the component 2—here the flow drops to 0. We see therefore that here we have a new situation when each permutation has *two* anchors. It is natural, therefore, to speak about a *two-dimensional* distribution of the anchor positions preserving the assumption that all permutations are equally probable. The system in this example is *regular* since elimination of an edge either leaves the flow from s to t unchanged, or causes its drop by one unit. #

Example 4.3   Consider a network $N = (V, E)$, where V is a node set, and E is the edge (link) set. Suppose that N has $|V| = 5$ nodes, network elements subject to failure are the edges. Suppose that initially the network is connected. Define network state as a function of the number of connected components in the network. If all nodes are connected, we define network state as $K = 4$. After an edge is erased, the network either remains connected or falls apart into two components (state 3). Continuing the process of edge removal, the number of components will increase to 3 (state 2), then to 4 (state 1) and finally 5 (which will be called state 0). This system, by definition is regular.

Imagine now a star-type system with a center a and four peripheral nodes b, c, d, e connected to node a by the links $(a, b), (a, c), (a, d), (a, e)$. Suppose that now the *nodes* (and not the links) are subject to failure. Node failure means erasing all links incident to it. Define system state as the number of isolated components in the network. (A single node is also a component). This system is *not regular*: failure of the central node a converts the initially connected network into a collection of five isolated components. #

This paper is organized as follows. Section 4.2 is devoted to the definition of multidimensional D-spectrum of a multistate system and marginal D-spectra. Section 4.3 presents several combinatorial properties of coherent multistate systems described via the marginal D-spectra and formulas for calculating system state probabilities. Section 4.4 presents an example of a four-dimensional cube network with 16 nodes and 32 edges. It has four special nodes called *terminals* and network's state is determined by the number of so-called *clusters*, i.e. isolated components containing terminals. We consider this network disintegration into four isolated clusters in the process of random elimination of its edges and present the probabilistic description of this process.

4.2 Multidimensional D-spectrum

Let us make the following assumptions.

1. The system has binary components and K *UP* states, $K > 1$, denoted as $K, K - 1, \ldots, 1$. $\varphi(1, 1, \ldots, 1) = K$, $\phi(0, 0, \ldots, 0) = 0$. System *DOWN* state is denoted by 0.

2. The system is regular, i.e. turning a single component from *up* to *down* either does not change the system state, or changes its state by exactly one unit.

Let us consider an arbitrary permutation of system component numbers

$$\pi = \{i_1, i_2, i_3, \ldots, i_n\}.$$

Imagine that all components are *up* and we start turning them *down*, moving from left to right along this permutation.

Definition 4.3 *Multidimensional anchor of* π Denote by $(r_1(\pi), r_2(\pi), \ldots, r_K(\pi))$ the positions in π of the components whose erasing causes the system state to change. Obviously, $r_i < r_{i+1}, i = 1, \ldots, K - 1$. Removal of the first r_1 components in the permutation causes transition from state K to state $K - 1$. Removal of the additional $r_2 - r_1$ components causes transition from state $K - 1$ to $K - 2$, and so on. When r_K components are turned *down*, the system enters *DOWN*.

$$A_K(\pi) = (r_1(\pi), r_2(\pi), \ldots, r_{K-1}(\pi), r_K(\pi))$$

is called the *K-dimensional anchor* of permutation π. #

The *s*th component of $A_K(\pi)$ is the exit point from state $K - s + 1$ into state $K - s, s = 1, \ldots, K$ and is called the *s*th anchor.

Example 4.4   Bridge network, unit flow capacities of edges (Fig. 4.1). The network has states 2, 1 and 0 ($K = 2$). Consider, for example $\pi = \{1, 3, 4, 5, 2\}$. Here $A_2(\pi) = (1, 4)$. Indeed, the flow drops by one after eliminating edge 1, and edge 5, positioned according to their anchors, on the first and fourth places in the permutation.

Consider all $5! = 120$ permutations and let us present all possible locations of the two anchors. There are 24 permutations of type $(1, 2)$, i.e. with the location of the first anchor on the first position and the second—on the second, like in permutation $\pi = \{1, 2, 3, 4, 5\}$. There are 56 permutations of type $(1, 3)$, with the first anchor on the first position and the second—on the third, like in $\pi = \{1, 4, 2, 3, 5\}$. There are 16 permutations of type $(1, 4)$, like in $\pi = \{1, 4, 3, 2, 5\}$. We counted 16 permutations of type $(2, 3)$, i.e. the first anchor being on the second position and the second—on the third, like in $\pi = \{3, 1, 2, 4, 5\}$. Finally, there are eight permutations of type $(2, 4)$, like in $\pi = \{3, 1, 4, 2, 5\}$. #

Definition 4.4 *Multidimensional D-spectrum*   Assign to each of the $n!$ permutations the probability $1/n!$. The collection of probabilities

$$P(r_1(\pi) = m_1, r_2(\pi) = m_2, \ldots, r_K(\pi) = m_K), 1 \leq m_1 < m_2 < \cdots < m_K \leq n \quad (4.2)$$

will be called the *multidimensional D-spectrum* of the system with K *UP* states. #

Now we are ready to define the main object of our interest, the marginal distributions of the anchors.

Definition 4.5 *Marginal D-spectrum*   The marginal distribution of the sth anchor obtained from the multidimensional D-spectrum (4.2) is called the marginal D-spectrum of the sth anchor, $s = 1, \ldots, K$. #

In simple words: the sth marginal D-spectrum, or simply, the sth spectrum is the distribution of the position of that system component in random permutation π whose removal (i.e. turning from *up* to *down*) causes the system to leave the state $K - s + 1$ and to enter state $K - s$.

We will use the notation $f^{(s)} = \{f_1^{(s)}, f_2^{(s)}, \ldots, f_n^{(s)}\}$ for the sth marginal D-spectrum, where $f_i^{(s)}$ is the probability that the sth anchor equals i.

The sth marginal D-spectrum is obtained by summing all probabilities of type (4.2) over all possible values of $m_1, \ldots, m_{s-1}, m_{s+1}, \ldots, m_K$ keeping the value of the sth anchor fixed. Let us illustrate this by an example.

Example 4.5   Marginal spectra for bridge system on Fig. 4.1. Bridge state is defined as the maximal flow it can deliver from s to t. In our example the states are 2, 1 or 0. So, $K = 2$. According to our notation, we assign to the system states the numbers equal to the delivered flow. The first anchor signifies exit of state 2 into state 1. The second anchor signifies the transition from state 1 into state 0.

From the data given in Example 2.1, $f_1^{(1)} = \sum_{x=2}^{5} P(r_1 = 1, r_2 = x)$ which gives $f_1^{(1)} = (24 + 56 + 16)/120 = 4/5$.

Similarly $f_2^{(1)} = (8 + 16)/120 = 1/5$, and $f_3^{(1)} = f_4^{(1)} = f_5^{(1)} = 0$. The first cumulative spectrum is $F^{(1)}(1) = 4/5$, $F^{(1)}(2) = 1 = F^{(1)}(3) = F^{(1)}(4) = F^{(1)}(5)$.

The second marginal D-spectrum is $\{0, 1/5, 3/5, 1/5, 0\}$, which coincides with our regular one-dimensional D-spectrum of the bridge system because the *DOWN* state means no s–t connection. #

Definition 4.6 *Cumulative marginal D-spectrum*   The discrete cumulative distribution function of the sth anchor

$$F^{(s)}(x) = \sum_{i=1}^{x} f_i^{(s)}, \quad x = 1, 2, \ldots, n \tag{4.3}$$

is called the sth *cumulative* D-spectrum. #

4.3 Marginal D-spectra and System Failure Sets

Let us return to a binary system with only states: *UP* and *DOWN*. Let Ω be the set of all binary n-dimensional vectors, $\Omega = \Omega_U \bigcup \Omega_D$, where $\Omega_U = \{x : \varphi(x) = 1\}$ and $\Omega_D = \{x : \varphi(x) = 0\}$. Obviously, $y = (1, 1, \ldots, 1) \in \Omega_U$ and $\mathbf{V} = (0, 0, \ldots, 0) \in \Omega_D$.

The component permutation π has now only a single anchor $r(\pi)$ designating the transition $UP \rightarrow DOWN$. Assigning measure $1/n!$ to each permutation π, we define the D-spectrum of the system as the discrete distribution

$$\{f_1, f_2, \ldots, f_n\}, \sum_{i=1}^{n} f_i = 1, \tag{4.4}$$

where f_i is the probability that the anchor's position is i (index s is omitted), and $F(x) = \sum_{i=1}^{x} f_i$ as the *cumulative* D-spectrum.

For further exposition we need the following fundamental combinatorial property of the D-spectrum of a binary system.

Theorem 1   *Let $C(x)$ be the number of system failure sets containing exactly x down components and $(n - x)$ up components. Then*

$$C(x) = F(x) \cdot \frac{n!}{x!(n - x)!} . \# \tag{4.5}$$

Before we turn to the proof of this theorem let us consider an example.

Example 4.6   Number of failure sets of size 3 for bridge system, Fig. 4.1. Failure of the bridge is defined as loss of connection between s and t. Bridge has D-spectrum $\{0, 1/5, 3/5, 1/5, 0\}$. $F(3) = 4/5$ and by (4.5) $C(3) = 8$.

Indeed, these are the eight failure sets of size $3:(1, 2, 3)$, $(1, 2, 4), (1, 2, 5)$, $(1, 3, 5), (2, 3, 4), (4, 5, 1), (4, 5, 2), (4, 5, 3).\#$

Theorem 1 establishes a purely combinatorial fact. Its proof which uses combinatorial arguments is given in (Gertsbakh and Shpungin 2009), Chap. 8. Let us present here an alternative proof (see also Samaniego 2007).

Proof Suppose all components of the system have i.i.d. continuous lifetimes $X_1, X_2, \ldots, X_n$, distributed according to CDF $V(t) = P(X_i \leq t)$. The system starts operating at $t = 0$. Denote by τ_{sys} system lifetime. By the well-known fact established in (Samaniego 1985),

$$F_{\text{sys}}(t_0) = P(\tau_{\text{sys}} \leq t_0) = \sum_{i=1}^{n} f_i \cdot V_{(i)}(t_0), \tag{4.6}$$

where $V_{(i)}(t_0)$ is the CDF of the ith *order statistic* of the sample $X_1, \ldots, X_n$:

$$V_{(i)}(t_0) = \sum_{j=i}^{n} \frac{n!}{j!(n - j)!} (V(t_0))^j (1 - V(t_0))^{n-j}. \tag{4.7}$$

If $\tau_{\text{sys}} \leq t_0$, at the instant t_0 the system is *DOWN*. Similarly, $V_j(t_0)$ is the probability that component j is down at t_0. Denote $1 - V(t_0) = p$, $V(t_0) = q$. Now denote by $P(DOWN)$ the probability that the system is *DOWN* at t_0. Combining (4.6) and (4.7) we obtain that

$$P(DOWN) = \sum_{i=1}^{n} f_i \cdot \sum_{j=i}^{n} \frac{n!}{j!(n-j)!} q^j p^{n-j}. \qquad (4.8)$$

Now change the order of summation in (4.8) and sample together the terms containing $(n!/i!(n-i)!)q^i p^{n-i}$. We arrive at the following formula:

$$P(DOWN) = \sum_{i=1}^{n} F(i) \cdot \frac{n!}{i!(n-i)!} q^i p^{n-i}. \qquad (4.9)$$

Now note that if all components are independent and are *up* with probability p, and *down* with probability q then system *DOWN* probability

$$P(DOWN) = \sum_{i=1}^{n} C(i) q^i p^{n-i}. \qquad (4.10)$$

Comparing (4.9) and (4.10) we arrive at the formula (4.5). #

Remark 1 Formula (4.5) allows the following interesting interpretation. The coefficient at $F(x)$ is the total number of sets of x components chosen from a set of n components. Suppose that all system components are *up*, we choose randomly a set of x components out of n and turn them *down*. There are, as we know, $C(x)$, such sets which cause the system to be *DOWN*. Then, according to (4.5), the probability that a *randomly* chosen set of x components will cause system failure is exactly $F(x)$. Equation (4.5) is useful in establishing the damage of a random "attack" on the network system, see e.g. (Levitin et al. 2011). #

Remark 2 Suppose that the system has independently failing components and the failure probabilities lie in the interval $[q_{min}, q_{max}]$. Then, due to the monotonicity property (Barlow and Proschan 1975), Theorem 1.2, p. 22, system failure probability $P(DOWN)$ lies in the interval $[Q_{min}, Q_{max}]$, where

$$Q_{min} = \sum_{i=1}^{n} C(i) q_{min}^i (1 - q_{min})^{n-i}, \ Q_{max} = \sum_{i=1}^{n} C(i) q_{max}^i (1 - q_{max})^{n-i}. \text{ # } \quad (4.11)$$

Remark 3 A formula dual to (4.5) is presented in (Samaniego 2007), p. 80. It follows from (6.3), (6.4) there that

$$\sum_{i=n-j+1}^{n} f_i \frac{n!}{j!(n-j)!}, \quad j = 1, \ldots, n,$$

equals to the number of system *path sets* with j *up* components and $(n-j)$ components *down*. It is worth noting that the cumulative D-spectrum can be expressed via the number of failure (cut) sets of given size or, equivalently, via the number of path sets of given size. Both these characteristics are combinatorial parameters of the system determined solely by its structure function. #

Now let us return to the regular multi state system with $K > 1$ *UP* states. Our goal is to find the probability $P_{sys}(r)$ that the system is in a particular state r, $0 \le r \le K$.

We will use a rather simple approach: the set of all $K + 1$ states of the multi state system will be divided into two sets: the *UP*-set and its complement which we will call the *DOWN*-set. Let us illustrate this approach by declaring that the *UP*-set consists only of the state K, and all other system states will be declared as the system *DOWN* state. Obviously for this case we can write a formula similar to (4.9):

$$P_{sys}(DOWN) = \sum_{\alpha=0}^{K-1} P_{sys}(\alpha) = \sum_{i=1}^{n} F^{(1)}(i)[n!/((n-i)!i!)]q^i p^{n-i}. \qquad (4.12)$$

Note that we write here $F^{(1)}(i)$ and not $F(i)$ because we take into account the CDF of the *first* marginal anchor of the system. All other anchors now are irrelevant. It remains to specify for this situation the exact meaning of the expression which is the multiple at $q^i p^{n-i}$.

Let us denote by $C(\alpha; i)$ the number of sets of components which have the following properties:

1. they have exactly i components *down* and $n - i$ components *up*;
2. the system is in state α, $\alpha = 0, 1, 2, \ldots, K$.

Finally, denote $\sum_{\alpha=0}^{m} C(\alpha; i) = C_{0,m}(i)$.
In this notation,

$$C_{0,K-1}(x) = F^{(1)}(x) \cdot \frac{n!}{x!(n-x)!}. \qquad (4.13)$$

Example 4.7   Let us consider the bridge with unit flow capacities of the edges. The system can be in three states: 2, 1, 0. Declare state 2 as *UP* and all other states as *DOWN*. From Example 4.5 we know that the first cumulative spectrum is $F^{(1)}(1) = 4/5$, $F^{(1)}(2) = 1$. By (4.13), $C_{0,1}(2) = 1 \cdot 5!/(2!3!) = 10$. There are, therefore ten failure sets with exactly two down edges which cause the system to be in state 0 or 1. The following eight failure sets (cut sets) with *down* components $(1, 3), (2, 3), (4, 3), (5, 3), (1, 4), (1, 5), (2, 4), (2, 5)$ correspond to state 1, and two failure sets with *down* components $(1, 2), (4, 5)$ cause the system to be in state 0. #

Now we return to the multi state system and move the "border" between *UP* and *DOWN* to include into *UP* the states K *and* $K - 1$, then $K \cup (K - 1) \cup (K - 2)$, etc. We will obtain the following generalization of (4.12):

$$\sum_{\alpha=0}^{K-r} P_{sys}(\alpha) = \sum_{i=1}^{n} F^{(r)}(i)[n!/((n-i)!i!)]q^i p^{n-i}. \qquad (4.14)$$

From it immediately follows a generalization of (4.13):

$$C_{0,K-r}(i) = F^{(r)}(i) \cdot \frac{n!}{i!(n-i)!}.$$ (4.15)

Finally we arrive at the following

Theorem 2 *The probability that the multi state system is in state* $K - \beta$, $\beta = 1, 2, \ldots, K$ *equals*

$$P_{\text{sys}}(K - \beta) = \sum_{i=1}^{n} [F^{(\beta)}(i) - F^{(\beta+1)}(i)] \cdot [n!/((n-i)!i!)]q^i p^{n-i},$$ (4.16)

where $F^{(K+1)}(i) \equiv 0$. #

4.4 Example: A Four-Dimensional Cube Network with Four Terminals

In this section we will consider the so-called four-dimensional cube network H_4, see Fig. 4.2.

This network has node set V with 16 nodes, four of which are special nodes called *terminals*. The edge set E has 32 edges, and the H_4 components subject to failure are edges. A connected subgraph of this network is called *cluster* if at least one of its nodes is a terminal. An isolate terminal is also defined as a cluster. Image that initially all edges of H_4 are present and afterwards are sequentially deleted from the network in random order. Initially there will be one cluster (all terminals connected to each other). Then in the process of edge deletion the network will fall apart into two clusters isolated from each other. Continuing the process, there will appear three clusters and finally the network will totally fall apart into four isolated clusters, each having one terminal. In addition, isolated nonterminal nodes or isolated subgraphs *without* terminals may appear, but they are not considered as clusters. Figure 4.3 presents a picture of a disintegrated network when it has three clusters.

To be consistent with the previously introduced system states, we make an agreement to assign system state number 3 if it has a single cluster, number 2—if it has two clusters, number 1—when there are three clusters and state 0 (*DOWN*)—for the case of 4 clusters meaning total network disintegration. Obviously, our system is regular, since elimination of a single edge either leaves the number of isolated clusters unchanged or causes an increase by one cluster only.

Let us turn now to the spectra of the network. Imagine an arbitrary edge permutation π and the process of turning the edges from *up* to *down*, moving along π from left to right. The first anchor designates the appearance of two clusters, the second anchor—of three clusters and the third—of four clusters. There are

Fig. 4.2 H_4 network (*Bold nodes are terminals*)

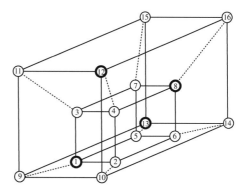

Fig. 4.3 Disintegration of H_4 into three clusters

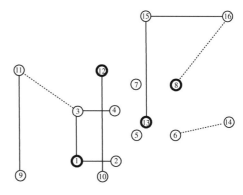

therefore three spectra. The corresponding cumulative marginal spectra are presented in Table 4.1. The results are obtained simulating 10^6 random permutations. A detailed description of Monte Carlo simulation algorithms is presented in (Gertsbakh and Shpungin 2009).

Let us recall the principal properties of these spectra. Suppose that $x = 20$ edges are chosen randomly from the set of 32 edges and are *destroyed*. Then, according to (4.5) (see also Remark 1), with probability $0.899883 \approx 0.9$ the network will not be in state 3, i.e. it will have two or more clusters. With probability ≈ 0.446 there will be three or four clusters, and with probability ≈ 0.0565 the network will be *DOWN*, i.e. there will be exactly four clusters.

Now suppose that edges in H_4 are, independent of each other, *up* with probability p and *down* with probability $q = 1 - p$. It is interesting to investigate how the network state probability depends on p. We will denote by R_1 the probability of having one cluster only and by $P(1) = 1 - R_1$. $P(2), P(3)$ and $P(4)$ are the probabilities of having *exactly* 2, 3 or 4 clusters respectively. The data are presented in Table 4.2.

From this Table we see, for example, that if $p = q = 0.5$, the network has one cluster with probability 0.5783, two clusters with probability 0.3092, three clusters

Table 4.1 The H_4 network cumulative marginal edge spectra

x	$F^{(1)}(x)$	$F^{(2)}(x)$	$F^{(3)}(x)$	x	$F^{(1)}(x)$	$F^{(2)}(x)$	$F^{(3)}(x)$
1	0	0	0	17	.506323	.059409	.000714
2	0	0	0	18	.664424	.135412	.003701
3	0	0	0	19	.802931	.270249	.016683
4	.000114	0	0	20	.899883	.445668	.056534
5	.000576	0	0	21	.956241	.623272	.135644
6	.001763	0	0	22	.983600	.771267	.252427
7	.004056	0	0	23	.994902	.875728	.392954
8	.008416	0	0	24	.998740	.940878	.538497
9	.015558	.000006	0	25	.999802	.975436	.672176
10	.026816	.000030	0	26	.999979	.991538	.785105
11	.044282	.000102	0	27	1	.997765	.871666
12	.070532	.000421	0	28	1	.999651	.932451
13	.109787	.001239	0	29	1	1	.971124
14	.166948	.003473	.000003	30	1	1	.991974
15	.249126	.009358	.000019	31	1	1	1
16	.362329	.024057	.000120	32	1	1	1

Table 4.2 R_1, $P(1)$, $P(2)$, $P(3)$, $P(4)$ as a function of p

p	R_1	$P(1)$	$P(2)$	$P(3)$	$P(4)$
0.9	0.9996	0.0004	0.0004	0.0000	0.0000
0.8	0.9926	0.0074	0.0074	0.0000	0.0000
0.7	0.9557	0.0443	0.0431	0.0012	0.000
0.6	0.8345	0.1655	0.1481	0.0162	0.0012
0.5	0.5783	0.4217	0.3092	0.0951	0.0174
0.4	0.2637	0.7363	0.3618	0.2668	0.1077
0.3	0.0623	0.9377	0.2117	0.3730	0.3530
0.2	0.0051	0.9949	0.0498	0.2493	0.6958

with probability 0.0951 and four clusters—with rather small probability 0.0174. If we want to preserve the network undamaged (with a single cluster), with probability 0.95, the edge failure probability should not exceed 0.3.

References

Barlow RE, Proschan F (1975) Statistical theory of reliability and life testing. Holt, Rinehart and Winston, NY

Elperin T, Gertsbakh IB, Lomonosov M (1991) Estimation of network reliability using graph evolution models. IEEE Trans Reliab R-40:572–581

Gertsbakh IB, Shpungin Y (2004) Combinatorial approaches to Monte Carlo estimation of network lifetime distribution. Appl Stoch Models Business Ind 20:49–57

Gertsbakh IB, Shpungin Y (2008) Network reliability importance measures: combinatorics and monte carlo based computations. WSEAS Trans Comp 4(7):216–227

Gertsbakh IB, Shpungin Y (2009) Models of Network Reliability: Analysis Combinatorics and Monte Carlo. CRC Press, Boca Raton

Levitin G, Gertsbakh IB, Shpungin Y (2011) Evaluating the damage associated with intentional network disintegration. Rel Eng Sys Saf 96(4):433–439

Navarro J, Rubio R (2010) Computation of signatures of coherent systems with five components. Com Stat Sim Comp 39(1):68–84

Samaniego FJ (1985) On closure of the IFR under formation of coherent systems. IEEE Trans Reliab 34:69–72

Samaniego FJ (2007) System signatures and their applications in engineering reliability. Springer, NY

Chapter 5
Applications of Samaniego Signatures to Bounds on Variances of Coherent and Mixed System Lifetimes

Tomasz Rychlik

Abstract Using the notion of Samaniego signature, we present bounds on lifetime variances for coherent and mixed systems composed of items with exchangeable and i.i.d. lifetimes. In the exchangeable case, we present characterizations of lifetime distributions of k-out-of-n and some more general systems, dependent on the component lifetime distribution, and apply them for calculating bounds on system lifetime variances, dependent on either the component lifetime distribution or its variance. In the i.i.d. case, for arbitrary mixed system we provide upper bounds on its lifetime variance, expressed in the single component variance units. Moreover, for fixed lifetime distributions, we describe mixed systems with minimal and maximal lifetime variances.

Keywords Coherent system · Mixed system · k-out-of-n system · Samaniego signature · Order statistic · Exchangeable random variables · i.i.d. random variables

5.1 Systems and Signatures

Coherent systems are represented by their structure functions $\varphi : \{0, 1\}^n \mapsto \{0, 1\}$, where n denotes the number of system components, and φ completely describes the working status of the system in dependence of which its particular components do work or not. By the standard convention, 1 means that the respective components

T. Rychlik (✉)
Institute of Mathematics, Polish Academy of Sciences,
Chopina 12, 87100 Toruń, Poland
e-mail: trychlik@impan.gov.pl

A. Lisnianski and I. Frenkel (eds.), *Recent Advances in System Reliability*,
Springer Series in Reliability Engineering, DOI: 10.1007/978-1-4471-2207-4_5,
© Springer-Verlag London Limited 2012

and the whole system are operating, and 0 represents the failures. Throughout the paper, we assume that the lifetimes $X_1, \ldots, X_n$ of components are non-negative exchangeable random variables. The latter assumption refers to the practical situation that all the components are identical and changing their roles in the system does not affects the distribution of its lifetime

$$T = \int\limits_0^\infty \varphi\big(\mathbf{1}_{[0,t]}(X_1), \ldots, \mathbf{1}_{[0,t]}(X_n)\big)\, dt,$$

where $\mathbf{1}_A$ stands for the indicator function of set A. Later on, we add some more restrictive assumptions on the common distribution of $X_1, \ldots, X_n$, e.g., independence and finiteness of the second moments.

Our further investigations are grounded on the following representation of the system lifetime distribution

$$P(T \leq t) = \sum_{k=1}^n s_k P(X_{k:n} \leq t), \qquad (5.1)$$

which holds true under the condition of exchangeability of component lifetimes. Here $X_{1:n}, \ldots, X_{n:n}$ denote the order statistics of the lifetime vector $(X_1, \ldots, X_n)$ which is clearly independent of the system structure. On the other hand, vector $\mathbf{s} = (s_1, \ldots, s_n)$ which belongs to the simplex

$$\mathbb{S}^n = \left\{ \mathbf{s} = (s_1, \ldots, s_n) \in \mathbb{R}^n : s_k \geq 0, \ k = 1, \ldots, n, \ \sum_{k=1}^n s_k = 1 \right\} \qquad (5.2)$$

is uniquely determined by the system structure function φ as follows

$$s_k = \frac{1}{n!} \sum_{\sigma \in \Sigma^n} \mathbf{1}_{\{k\}}\left(\max_{1 \leq j \leq n} \left\{ j\varphi\left(\sum_{i=j}^n \mathbf{e}_{\sigma(i)} \right) \right\} \right), \qquad k = 1, \ldots, n,$$

where Σ^n stand for the family of all permutations of the set $\{1, \ldots, n\}$, and $\mathbf{e}_i$ denotes the ith standard basis vector in $\mathbb{R}^n$ for $i = 1, \ldots, n$. Suppose that $\sigma \in \Sigma^n$ represents the sequences of indices of consecutively failing components. Then the argument of φ above describes the living statuses of components before the jth subsequent failure. Their maximum corresponds to the number of component failure which results in the system collapse. So each s_k, $k = 1, \ldots, n$, defines the proportion of the number of permutations of consecutive component failures for which the kth failure implies the hypothetical damage of the system in relation to the number $n!$ of all possible permutations.

Representation (5.1) was first established by Samaniego (1985) under the assumption that the component lifetimes are i.i.d. with a common continuous marginal distribution. Vector $\mathbf{s} = \mathbf{s}(\phi) = (s_1, \ldots, s_n)$ is called the Samaniego signature of the system. The Samaniego formula (5.1) was proved in Navarro and

Rychlik (2007) for $X_1, \ldots, X_n$ being exchangeable and jointly continuously distributed, but the claim appeared implicitly in some earlier papers (see, e.g., Kochar et al. 1999). The extension to the discontinuous case was due to Navarro et al. (2008). Note that under the continuity assumption, which implies that the order statistics are different almost surely, the Samaniego signature has a natural probabilistic interpretation $s_k = \mathrm{P}(T = X_{k:n})$, $k = 1, \ldots, n$, which immediately follows from the relations $\mathrm{P}\big(X_{\sigma(1)} < \cdots < X_{\sigma(n)}\big) = \frac{1}{n!}$, $\sigma \in \Sigma^n$.

Otherwise, under the convention $X_{0:n} = 0$, $X_{n+1:n} = +\infty$, we only have

$$\mathrm{P}(X_{k-1:n} < T = X_{k:n} < X_{k+1:n}) \leq s_k \leq \mathrm{P}(T = X_{k:n}), \quad k = 1, \ldots, n.$$

It is a challenging task to determine all the coherent systems of a fixed size n. There is one trivial system of size 1, and two systems of size 2, the series and parallel ones. All the essentially different (up to renumbering of components) systems of sizes 3 and 4 were presented in Kochar et al. (1999) and Shaked and Suarez-Llorens (2003), respectively. Recently Navarro and Rubio (2010) proposed a general algorithm of determining all the systems of a fixed size, and wrote down the complete list of 180 systems with 5 components. They claimed there are 16180 six-item systems.

Observe that due to (5.1), the lifetime distribution of a coherent system is identical with the distribution of a randomly chosen among the $(n + 1 - k)$-out-of-n systems with the lifetimes $X_{k:n}$, $k = 1, \ldots, n$, when the choice probability is defined by the Samaniego signature. Since the distributions of all possible Samaniego signatures of coherent systems of fixed sizes (especially large ones) over the simplices (5.2) are hard to describe, for the mathematical convenience Boland and Samaniego (2004) introduced the notion of mixed systems of sizes n, $n = 2, 3, \ldots$ which are defined as randomly chosen $(n + 1 - k)$-out-of-n systems with the choice distributions determined by the arbitrary points of (5.2). Navarro et al. (2008) noticed that all the coherent and mixed systems of sizes up to n can be uniformly treated as mixed systems of size n. In particular, for an arbitrary system of size $n - 1$ we can write

$$\mathrm{P}(T \leq t) = \sum_{k=1}^{n-1} s_k \mathrm{P}(X_{k:n-1} \leq t) = \sum_{k=1}^{n} \left(\frac{k-1}{n} s_{k-1} + \frac{n-k}{n} s_k \right) \mathrm{P}(X_{k:n} \leq t).$$

For example, the system with one component can be formally treated as the uniform mixture of all k-out-of-n systems, $k = 1, \ldots, n$, with $n \geq 2$, because

$$\mathrm{P}(T = X_1 \leq t) = \frac{1}{n} \sum_{k=1}^{n} \mathrm{P}(X_k \leq t) = \frac{1}{n} \sum_{k=1}^{n} \mathrm{P}(X_{k:n} \leq t). \tag{5.3}$$

It is worth pointing out that Navarro et al. (2007) provided representations of system lifetime distributions different from (5.1), based on linear integer combinations of lifetime distributions of series (and alternatively parallel) systems with smaller sizes from 1 up to n. They are especially useful for examining asymptotic properties of the system lifetimes. An extensive study of various notions of signatures and their role in the reliability analysis was presented by Samaniego (2007).

We concentrate here on the Samaniego formula (5.1), which is useful for evaluations of moments of the system lifetime. Note that the system lifetime expectations coincide with the expectations of L-statistics $\mathrm{E}T = \mathrm{E}\sum_{k=1}^{n} s_k X_{k:n}$ with the Samaniego coefficients $s_1, \ldots, s_n$. There is a vast literature concerning the optimal bounds on the expected L-statistics (especially single order statistics) from i.i.d. and exchangeable populations, under various assumptions on the marginal distributions. Comprehensive reviews of the results achieved up to the end of the twentieth century are presented in Rychlik (1998, 2001) (see also, David and Nagaraja 2003; Chap. 4). By the time, a significant number of new evaluations appeared in the literature. In the paper, we use (5.1) in order to obtain some estimates of the system lifetime variances. Note that it is not possible to get analogous variance bounds for the L-statistics $\sum_{k=1}^{n} s_k X_{k:n}$, except for the particular case of the single order statistics, because knowledge of one-dimensional marginal distributions of order statistics is not sufficient for determining their covariances. Even in the simplest case of i.i.d. variables there are no tractable formulae for the distributions of linear combinations of two and more order statistics.

In Sect. 5.2, we consider coherent and mixed systems with exchangeable components. We first mention results characterizing the marginal distributions of order statistics from exchangeable samples with a known marginal distribution. Then we show a method of calculating lower and upper bounds on variances of k-out-of-n and some other systems, dependent on the single component lifetime distribution. Finally, we present more general bounds, conditioned only on the component lifetime variance. Section 5.3 deals with the i.i.d. case. Under the assumption, we deliver upper bounds on the variances of arbitrarily fixed coherent and mixed systems, expressed in terms of the component lifetime variance, and discuss conditions of their attainability. In Sect. 5.4, we determine the mixed systems whose lifetimes have minimal and maximal variances under the assumptions that the component lifetimes are i.i.d. and have a known parent distribution.

5.2 Systems with Exchangeable Component Lifetimes

We first recall a characterization of the vector of all marginal distribution functions of order statistics.

Theorem 1 (Rychlik 1993) *A sequence of distribution functions $F_1, \ldots, F_n$ are the distribution functions of consecutive order statistics from an exchangeable sample $X_1, \ldots, X_n$ with a common marginal F iff*

$$F_1 \geq \cdots \geq F_n, \tag{5.4}$$

$$\sum_{k=1}^{n} F_k = nF. \tag{5.5}$$

The first condition is evident, and the latter can be justified by the latter relation in (5.3). A simplest way of constructing an exchangeable sequence with marginal F and distribution functions of order statistics $F_1, \ldots, F_n$ is to take a single standard uniform random variable U, define ordered random variables $F_1^{-1}(U) \leq \cdots \leq F_n^{-1}(U)$, and permute them randomly. Then the arising sequence $X_1, \ldots, X_n$ is clearly exchangeable,

$$P(X_i \leq t) = \sum_{k=1}^{n} P(X_i = F_k^{-1}(U)) P(F_k^{-1}(U) \leq t) = \frac{1}{n} \sum_{k=1}^{n} F_k(t) = F(t),$$

whereas $X_{k:n} = F_k^{-1}(U)$, $k = 1, \ldots, n$, have respective distribution functions F_k, $k = 1, \ldots, n$.

Theorem 1 allows us to deduce necessary conditions which should satisfy a linear combination of distribution functions of order statistics.

Theorem 2 *Let* $G = \sum_{k=1}^{n} s_k F_k$ *for some* $F_1, \ldots, F_n$ *satisfying* (5.4) *and* (5.5), *and* $\mathbf{s} = (s_1, \ldots, s_n) \in \mathbf{R}^n$ *being arbitrarily fixed. Let* $\underline{S}, \overline{S} : [0, 1] \mapsto \mathbf{R}$ *be the greatest convex and smallest concave functions, respectively, satisfying* $\underline{S}(0) = \overline{S}(0) = 0$ *and* $\underline{S}(\frac{i}{n}) \leq \sum_{k=1}^{i} s_k \leq \overline{S}(\frac{i}{n})$, $i = 1, \ldots, n$.

Then

$$\underline{S} \circ F \leq G \leq \overline{S} \circ F, \tag{5.6}$$

$$n \min_{1 \leq k \leq n} s_k \leq \frac{dG}{dF} \leq n \max_{1 \leq k \leq n} s_k \quad F - a.e. \tag{5.7}$$

Relations (5.6) were proved in Rychlik (1993) by pointwise minimization and maximization of the combinations $\sum_{i=1}^{n} s_i F_i$ with fixed coefficients under conditions (5.4) and (5.5). Inequalities (5.7) are implied by

$$n \min_{1 \leq k \leq n} s_k [F(t) - F(s)] = \min_{1 \leq k \leq n} s_k \sum_{i=1}^{n} [F_i(t) - F_i(s)]$$

$$\leq \sum_{i=1}^{n} s_i [F_i(t) - F_i(s)] = G(t) - G(s)$$

$$\leq s_k \sum_{i=1}^{n} [F_i(t) - F_i(s)] = n \max_{1 \leq k \leq n} s_k [F(t) - F(s)]$$

for all $t < s$. Note that functions $\underline{S} \circ F$ and $\overline{S} \circ F$ are distribution functions of generally signed measures that satisfy both (5.6) and (5.7). For arbitrary $\mathbf{s} \in \mathbf{R}^n$, Rychlik (1993) determined sequences $\underline{F}_1, \ldots, \underline{F}_n$ and $\overline{F}_1, \ldots, \overline{F}_n$ such that $\sum_{k=1}^{n} s_k \underline{F}_k = \underline{S} \circ F$ and $\sum_{k=1}^{n} s_k \overline{F}_k = \overline{S} \circ F$. This was especially useful for providing optimal bounds on the expectations of L-statistics from exchangeable samples because it implies that the inequalities

$$\int x(\overline{S} \circ F)(\mathrm{d}x) \leq \int x \left(\sum_{k=1}^{n} s_k F_k \right)(\mathrm{d}x) = \mathrm{E} \sum_{k=1}^{n} s_k X_{k:n} \leq \int x(\underline{S} \circ F)(\mathrm{d}x)$$

are sharp. If $\mathbf{s} = (s_1, \ldots, s_n) \in S^n$, then Theorem 2 describes necessary conditions for the lifetime distribution of a mixed system with signature $\mathbf{s}$ and exchangeable components having common lifetime marginal F. Moreover, $\underline{S} \circ F$ and $\overline{S} \circ F$ are proper distribution functions which are uniformly extreme for the system lifetime. So the sharp bounds follow $\int x(\overline{S} \circ F)(\mathrm{d}x) \leq \mathrm{E}T \leq \int x(\underline{S} \circ F)(\mathrm{d}x)$.

It was shown that conditions (5.6) and (5.7) specified for the single order statistics (k-out-of-n system lifetimes) are also sufficient.

Theorem 3 (Rychlik 1994) *A distribution function G is the distribution function of kth order statistic based on n exchangeable random variables with marginal F iff*

$$\max\left\{ \frac{nF + 1 - k}{n + 1 - k}, 0 \right\} \leq G \leq \min\left\{ \frac{nF}{k}, 1 \right\}, \tag{5.8}$$

$$0 \leq \frac{\mathrm{d}G}{\mathrm{d}F} \leq n \quad F - a.e. \tag{5.9}$$

The sufficiency proof is constructive. It is enough to define

$$F_1 = \cdots = F_{k-1} = \min\left\{ \frac{nF - G}{k}, 1 \right\}, \quad (\text{if } k > 1), \tag{5.10}$$

$$F_k = G, \tag{5.11}$$

$$F_{k+1} = \cdots = F_n = \max\left\{ \frac{nF - G + 1 - k}{n + 1 - k}, 0 \right\}, \quad (\text{if } k < n), \tag{5.12}$$

and check that they are distribution functions satisfying (5.4) and (5.5). Jaworski and Rychlik (2008) established slightly more stringent necessary and sufficient conditions for the marginal distributions of k-out-of-n system lifetimes under the assumptions that the component lifetimes have an absolutely continuous exchangeable joint distribution. This was a generalization of results by Durante and Jaworski (2009) and Jaworski (2009) who presented analogous results for thew parallel systems with two and more components, respectively.

Theorem 3 can be slightly extended to the mixed systems whose signatures have all coordinates equal except for one.

Theorem 4   *Let $H = \sum_{i=1}^{n} sF_i + \delta F_k$ for some $0 < \delta < 1$ and $0 < s = \frac{1-\delta}{n} < \frac{1}{n}$ for $F_1, \ldots, F_n$ satisfying (5.4) and (5.5). Then G is characterized by the relations*

$$\max\left\{nsF, n\left[s+\frac{\delta}{n+1-k}\right](F-1)+1\right\}\leq H\leq\min\left\{n\left[s+\frac{\delta}{k}\right]F, ns(F-1)+1\right\},$$

$$(5.13)$$

$$ns\leq\frac{dH}{dF}\leq n(s+\delta)\quad F-a.e.\qquad\qquad(5.14)$$

Also, $H=\sum_{i=1}^{n}sF_i-\delta F_k$ for some $0<\delta<\frac{1}{n-1}$ and $\frac{1}{n}<s=\frac{1+\delta}{n-1}<\frac{1}{n-1}$ for $F_1,\ldots,F_n$ satisfying (5.4) *and* (5.5) iff

$$\max\left\{n\left[s-\frac{\delta}{k}\right]F, ns(F-1)+1\right\}\leq H\leq\min\left\{nsF, n\left[s-\frac{\delta}{n+1-k}\right](F-1)+1\right\},$$

$$n(s-\delta)\leq\frac{dH}{dF}\leq ns\quad F-a.e.$$

Proof When δ is positive, relations (5.13) and (5.14) are specifications of necessary conditions (5.6) and (5.7), respectively. For proving sufficiency, we consider constructions (5.10–5.12) for arbitrary G satisfying the conditions of Theorem 3. By (5.5), we can write $H=nsF+\delta G$. Applying (5.8) and (5.9), we obtain

$$nsF+\delta\max\left\{0,\frac{nF+1-k}{n+1-k}\right\}\leq H\leq nsF+\delta\min\left\{\frac{nF}{k},1\right\},$$

which can be rewritten as (5.13), and $ns\leq\frac{dH}{dF}\leq ns+n\delta$, $F-a.e.$, as desired. For $\delta<0$, we apply similar arguments for $H=nsF-\delta G$. $\qquad\square$

It can be deduced from the proof that Theorem 4 can be stated in a more general form, if we drop the assumptions that the combination coefficients are non-negative and sum up to one. The theorem provides a complete characterization of all mixed systems with two components. It is still an open problem if (5.6) and (5.7) are the necessary and sufficient conditions describing arbitrary linear (or at least convex) combinations of distributions of order statistics. So we are still not able to characterize the lifetime distributions of arbitrary mixed systems with exchangeable components.

Applying Theorem 3, we can determine a parametric family G_α, $0<\alpha<1$ of the most dispersed distributions of $(n+1-k)$-out-of-n systems composed of exchangeable components in the sense that for any distribution G satisfying (5.8) and (5.9), and for any support point c of F, there exists an element of the parametric family which is more dispersed about c than the original G. The family has the representation

$$G_\alpha = \begin{cases} \frac{nF}{k}, & \text{if } F < \frac{k}{n}\alpha, \\ \alpha, & \text{if } \frac{k}{n}\alpha \le F < \alpha + (1-\alpha)\frac{k-1}{n}, \\ \frac{nF+1-k}{n+1-k}, & \text{if } F \ge \alpha + (1-\alpha)\frac{k-1}{n}. \end{cases} \quad 0 < \alpha < 1, \quad (5.15)$$

Also, there are two parametric families

$$G_\alpha^U = \begin{cases} 0, & \text{if } F < \frac{k-1}{n}\alpha, \\ nF - (k-1)\alpha, & \text{if } \frac{k-1}{n}\alpha < F \le \frac{k}{n}\alpha, \\ \frac{nF}{k}\alpha, & \text{if } \frac{k}{n}\alpha < F \le \frac{k}{n}, \\ 1, & \text{if } F \ge \frac{k}{n}, \end{cases} \quad (5.16)$$

and

$$G_\alpha^L = \begin{cases} 0, & \text{if } F < \frac{k-1}{n}\alpha, \\ \frac{nF+1-k}{n+1-k}, & \text{if } \frac{k-1}{n} \le F < \alpha + \frac{k-1}{n}(1-\alpha), \\ nF - k + 1 - (n-k)\alpha, & \text{if } \alpha + \frac{k-1}{n}(1-\alpha) \le F < \alpha + \frac{k}{n}(1-\alpha), \\ 1, & \text{if } F \ge \alpha + \frac{k}{n}(1-\alpha)n, \end{cases}$$

$$(5.17)$$

for $0 \le \alpha \le 1$, which contain the least dispersed distributions satisfying Theorem 3 in the sense described above. Accordingly, (5.15) and (5.16, 5.17) are the candidates for the maximal and minimal variance, respectively, of the $(n + 1 - k)$-out-of-n system lifetime under the assumption that its components have exchangeable lifetime distribution with marginal F. For specific F, establishing the parameter which provides extreme variances among (5.15–5.17) is often a challenging task. Similar most and least dispersive candidates for extreme variance distributions can be distinguished for the systems whose signatures have all the elements equal except for one. E.g., if the exception is greater by δ than the other ones (cf. the first part of Theorem 4), we simply mix the necessary part nF and extremely dispersed components (5.15–5.17) with weights $ns = 1 - \delta$ and δ, respectively. We omit here huge final formulae.

Now we discuss bounds on variances of $(n + 1 - k)$-out-of-n systems with exchangeable components and arbitrary marginal component lifetime distribution which has a positive and finite variance.

Theorem 5 (Rychlik 2008) *If $X_1, \ldots, X_n$ are exchangeable with variance $0 < \sigma^2 < \infty$, then the following lower and upper evaluations of the order statistics variances*

$$0 \le \frac{\operatorname{Var} X_{k:n}}{\sigma^2} \le \max\left\{\frac{n}{k}, \frac{n}{n+1-k}\right\}, \quad 1 \le k \le n,$$

are optimal.

The trivial lower bound is attained by the exhaustive drawing without replacement scheme from n arbitrary numbers taking on at least two different values. Then the consecutive drawings are exchangeable and have a positive variance, whereas all the order statistics are deterministic. The arguments for the upper inequalities are more elaborate and (5.15) are used there. The sharpness proof is constructive. The upper bounds are attained in limit for sequences of properly constructed joint distributions which have three-point marginals. The passage to the limit is not needed only if $k = \frac{n+1}{2}$.

5.3 Systems with i.i.d. Component Lifetimes

Suppose that a mixed system is composed of n elements with i.i.d. lifetimes, and its signature is $\mathbf{s} = (s_1, \ldots, s_n) \in \mathbb{S}^n$. By (5.1), the distribution function of the system lifetime has the representation

$$P(T \leq t) = G_{\mathbf{s}}(F(t)),$$

where

$$G_{\mathbf{s}}(x) = \sum_{k=1}^{n} s_k \sum_{i=k}^{n} \binom{n}{i} x^i (1-x)^{n-i}, \quad 0 \leq x \leq 1,$$

is a polynomial of degree n at most, dependent of the signature $\mathbf{s}$, and F is the common distribution function of the component lifetime. Note that the polynomial is the system lifetime distribution if $X_1, \ldots, X_n$ are standard uniform. It follows that it is positive and increasing on $(0,1)$ with $G_{\mathbf{s}}(0) = 0$ and $G_{\mathbf{s}}(1) = 1$ at the ends. Consequently, functions

$$G_{1,\mathbf{s}}(x) = \frac{G_{\mathbf{s}}(x)}{x}, \quad 0 < x \leq 1, \tag{5.18}$$

$$G_{2,\mathbf{s}}(x) = \frac{1 - G_{\mathbf{s}}(x)}{1 - x}, \quad 0 \leq x < 1, \tag{5.19}$$

are positive polynomials of degree $n - 1$ at most which can be continuously extended on the whole unit interval $[0, 1]$ by writing

$$G_{1,\mathbf{s}}(0) = G_{\mathbf{s}}'(0+) = ns_1,$$
$$G_{2,\mathbf{s}}(1) = G_{\mathbf{s}}'(1-) = ns_n.$$

Theorem 6 (Jasiński et al. 2009) *Let T denote the lifetime of a mixed system with signature $\mathbf{s}$, whose components have i.i.d. lifetimes with a positive finite variance σ^2. Then*

$$\frac{\operatorname{Var} T}{\sigma^2} \leq \max_{0 \leq u \leq v \leq 1} G_{1,\mathbf{s}}(u) G_{2,\mathbf{s}}(v). \tag{5.20}$$

The evaluation follows from the Hoeffding integral representation of the variance. We can note that the above maximization problem in the two-dimensional variable (u, v) over the triangle $T = \{(u, v) : 0 \leq u \leq v \leq 1\}$ can be simplified to seeking for maxima of single-valued polynomials. Clearly the product function restricted to the diagonal $D = \{(u, u) : 0 \leq u \leq 1\}$ of the unit square is a polynomial of degree $2n - 2$ or less that can be evaluated numerically. It can be also shown that any local maximum over $T \backslash D$ may be only attained at the pairs (u, v) such that both polynomials $G_{1,s}$ and $G_{2,s}$ attain their local maxima at u and v, respectively. So looking for the right-hand side of (5.20), we determine all the local maxima of (5.18) and (5.19), take all the products of the maxima for which the first argument is less than the other, and compare them with the maximum over the triangle hypotenuse D. It is clear that we need to take into account a restricted number of pairs.

Jasiński et al. (2009) showed that bound (5.20) is sharp iff the maximum is attained at a point of the line segment D. The distributions that attain the bounds have two-point supports, and the probability of the smaller one coincides with the argument maximizing $G_{1,s}(u)G_{2,s}(u)$, $0 \leq u \leq 1$. It was checked that for all the coherent systems of sizes up to 6, function $G_{1,s}(u)G_{2,s}(v)$, $(u, v) \in T$, attains its maximum at some point of D which implies that the respective bounds of Theorem 6 are optimal. It is an open problem if the claim is true for the coherent systems composed of more items. The results of Theorem 6 in the special case of k-out-of-n systems, with respective attainability conditions, were established by Papadatos (1995). As in the exchangeable case, the respective variance evaluations are symmetric about $\frac{n+1}{2}$, small for the central order statistics, and increasing as $\left| k - \frac{n+1}{2} \right|$ increases. For $k = \frac{n+1}{2}$, we have $\operatorname{Var} X_{\frac{n+1}{2}:n} / \sigma^2 \leq 1$ which was first established by Yang (1982), and attainability was proved by Lin and Huang (1989). Note that the bound are almost twice less than for the sample median in the exchangeable case. On the other hand, for the parallel and series systems, the right-hand sides of (5.20) amount to n, which coincides with the evaluations of Theorem 5 for the exchangeable components. Papadatos (1997) presented evaluations of variances of order statistics from independent identically symmetrically distributed samples, expressed in terms of variances of the single variables. They are attained by symmetric two- or three-point distributions.

Example   Let T denote the lifetime of a mixed system with signature $\mathbf{s} = \left(0, \frac{1}{2}, 0, 0, \frac{1}{2}, 0\right) \in S^6$. No coherent system has the signature. Theorem 6 implies the following non-sharp estimate $\frac{\operatorname{Var} T}{\sigma^2} \leq G_{1,s}(0.39879)G_{2,s}(0.60121) = 1.01297$. Under the restriction to the two-point parent distributions, we have the sharp evaluation $G_{1,s}(0.38168)G_{2,s}(0.38168) = 1.00342$. For the symmetric three-point distributions, the sharp bound is $G_{1,s}(0.39879)/2 = 1.00960$. It follows that the actual sharp bound for the system lifetime variance with no restrictions on the component lifetime distribution is a number from the interval $[1.00960, 1.01297]$.

5.4 Mixed Systems with Extremely Dispersed Lifetimes

We still consider mixed systems with n components whose lifetimes are i.i.d. with a positive finite variance. However, here we assume that we know the distribution function F of component lifetime and we try to determine the systems whose structure is represented by the signature vector $\mathbf{s} = (s_1, \ldots, s_n) \in \mathbb{S}^n$ with the minimal and maximal lifetime variance. The assumptions allow us to determine strictly ordered, positive and finite first two raw moments of order statistics

$$\mu_k = EX_{k:n}, \tag{5.21}$$

$$\tau_k = EX^2_{k:n}, \tag{5.22}$$

as well as the variances

$$\sigma^2_k = \operatorname{Var} X_{k:n} = \tau_k - \mu^2_k, \quad k = 1, \ldots, n. \tag{5.23}$$

Theorem 7 (Beśka et al. 2011) *Let $X_1, \ldots, X_n$ be the i.i.d. component lifetimes of mixed systems represented by signatures $\mathbf{s} \in \mathbb{S}^\kappa$, with moments of order statistics given in (5.21–5.23). Then for the system lifetime T we have*

$$\operatorname{Var} T \geq \min_{1 \leq i \leq n} \sigma^2_i, \tag{5.24}$$

and

$$\operatorname{Var} T \leq \begin{cases} \sigma^2_1, & \text{if} \quad \sigma^2_1 - \sigma^2_n \geq (\mu_n - \mu_1)^2, \\ \sigma^2_n, & \text{if} \quad \sigma^2_n - \sigma^2_1 \geq (\mu_n - \mu_1)^2, \\ \dfrac{(\mu_n - \mu_1)^2}{4} + \dfrac{\sigma^2_1 + \sigma^2_n}{2} + \dfrac{1}{4}\left(\dfrac{\sigma^2_n - \sigma^2_1}{\mu_n - \mu_1}\right)^2, & \text{if} \quad |\sigma^2_n - \sigma^2_1| < (\mu_n - \mu_1)^2. \end{cases} \tag{5.25}$$

When the right-hand side of (5.24) amounts to σ^2_k, then the equality in (5.24) holds for the $(n + 1 - k)$-out-of-n system. In the consecutive relations of (5.25), the respective equalities are attained by the series, parallel, and the mixture of series and parallel systems with the signature

$$\mathbf{s} = \frac{1}{2}\left[1 - \frac{\sigma^2_n - \sigma^2_1}{(\mu_n - \mu_1)^2}\right]\mathbf{e}_1 + \frac{1}{2}\left[1 + \frac{\sigma^2_n - \sigma^2_1}{(\mu_n - \mu_1)^2}\right]\mathbf{e}_n.$$

The lower bound is a simple consequence of the dependence of the system lifetime variance on the signature

$$\sigma^2(\mathbf{s}) = \sum_{i=1}^n s_i \tau_i - \left(\sum_{i=1}^n s_i \mu_i\right)^2, \quad \mathbf{s} \in \mathbb{S}^\kappa.$$

This is a concave function on the convex compact set $\mathbb{S}^{\mathbb{K}}$, and so its minimum is attained at some of the extreme point $\mathbf{e}_k$, $k = 1, \ldots, n$, of the simplex. Since $\sigma^2(\mathbf{e}_k) = \sigma_k^2$, $k = 1, \ldots, n$, (5.24) follows.

The crucial tool allowing us to verify (5.25) is the lemma stating that for the i.i.d. sample of any size $n \geq 3$ with arbitrary parent distribution function F with non-zero and finite variance, the sequence

$$d(k) = d(n, F; k) = \frac{\tau_{k+1} - \tau_k}{\mu_{k+1} - \mu_k}, \quad k = 1, \ldots, n-1,$$

is nondecreasing. This provides the following representation of planar points corresponding to means and variances of the mixed system lifetimes

$$\Sigma = \left\{ \left(\sum_{i=1}^{n} s_i \mu_i, \sum_{i=1}^{n} s_i \tau_i - \left(\sum_{i=1}^{n} s_i \mu_i \right)^2 \right) : \mathbf{s} \in \mathbb{S}^n \right\}$$

$$= \left\{ (\mu, \sigma^2) : \frac{\tau_{k+1} - \tau_k}{\mu_{k+1} - \mu_k} (\mu - \mu_k) + \tau_k - \mu^2 \leq \sigma^2 \leq \frac{\tau_n - \tau_1}{\mu_n - \mu_1} (\mu - \mu_1) + \tau_1 - \mu^2, \right.$$

$$\left. \mu_k \leq \mu \leq \mu_{k+1}, \ k = 1, \ldots, n-1 \right\}. \tag{5.26}$$

This shows that for every system with arbitrarily fixed lifetime mean $\mu_1 \leq \mu \leq \mu_n$, the maximal variance amounts to the upper restriction on σ^2 in (5.26), and this is attained by the mixture of the series and parallel systems with weights $\alpha = \frac{\mu_n - \mu}{\mu_n - \mu_1}$ and $1 - \alpha = \frac{\mu - \mu_1}{\mu_n - \mu_1}$, respectively. In order to get the global maximum, it suffices to maximize

$$\varsigma^2(\alpha) = \sigma^2(\alpha \mathbf{e}_1 + (1 - \alpha)\mathbf{e}_n)$$
$$= \alpha \tau_1 + (1 - \alpha)\tau_n - [\alpha \mu_1 + (1 - \alpha)\mu_n]^2, \quad 0 \leq \alpha \leq 1. \tag{5.27}$$

This is a concave quadratic function with the global maximum attained at

$$\alpha_* = \frac{1}{2} \left[1 - \frac{\sigma_n^2 - \sigma_1^2}{(\mu_n - \mu_1)^2} \right].$$

If $\alpha_* \leq 0$, i.e. $\sigma_n^2 - \sigma_1^2 \geq (\mu_n - \mu_1)^2$, then the constrained minimum of (5.27) amounts to $\varsigma^2(0) = \sigma_n^2$. If $\alpha_* \geq 1$, and so $\sigma_1^2 - \sigma_n^2 \geq (\mu_n - \mu_1)^2$, then $\varsigma^2(1) = \sigma_1^2$ is the maximal variance. Otherwise the restricted maximum is equal to the global one $\varsigma^2(\alpha_*)$ which is represented by the last formula in (5.25).

Note that relation $\sigma_1^2 = \sigma_n^2$ (which holds for the symmetric distributions in particular) implies that the maximal variance is

$$\sigma^2 \left(\frac{\mathbf{e}_1 + \mathbf{e}_n}{2} \right) = \frac{(\mu_n - \mu_1)^2}{4} + \sigma_1^2.$$

E.g., for the standard uniform components lifetime, we have

$$\text{Var } T \leq \sigma^2 \left(\frac{\mathbf{e}_1 + \mathbf{e}_n}{2} \right) = \frac{1}{4} - \frac{n}{(n+1)(n+2)}.$$

By (5.24), it is easy to see that

$$\text{Var } T \geq \sigma_1^2 = \sigma_n^2 = \frac{n}{(n+1)^2(n+2)}.$$

The lower and upper bounds tend to 0 and $1/4$ as n increases.
If $X_1, \ldots, X_n$ are standard exponential, we have

$$\mu_k = \sum_{i=n+1-k}^{n} \frac{1}{i}, \tag{5.28}$$

$$\sigma_k^2 = \sum_{i=n+1-k}^{n} \frac{1}{i^2}, \quad k = 1, \ldots, n. \tag{5.29}$$

So Theorem 7 implies

$$\text{Var } T \geq \text{Var } X_{1:n} = \frac{1}{n^2},$$

and

$$\text{Var } T \leq \frac{1}{4} \left(\sum_{i=1}^{n-1} \frac{1}{i} \right)^2 + \frac{1}{2} \left(\sum_{i=1}^{n-1} \frac{1}{i^2} + \frac{2}{n^2} \right) + \frac{1}{4} \left(\frac{\sum_{i=1}^{n-1} \frac{1}{i^2}}{\sum_{i=1}^{n-1} \frac{1}{i}} \right)^2.$$

The maximal variance is attained by the random mixture of series and parallel systems with probabilities

$$s_1 = \frac{1}{2} \left[1 - \frac{\sum_{i=1}^{n-1} \frac{1}{i^2}}{\left(\sum_{i=1}^{n-1} \frac{1}{i} \right)^2} \right] < \frac{1}{2}$$

and $s_n = 1 - s_1 > 1/2$, respectively. If n increases, then both s_1 and s_n approach $1/2$, and the upper variance bounds tend to infinity at the rate $O\left((\ln n)^2 \right)$.

It can be shown that in the case of Pareto component lifetime distribution functions $F_\theta(x) = 1 - x^{-\theta}$, $x > 1$, $\theta > 2$, for every n the maximal variance in (5.25) amounts to σ_n^2 if θ is sufficiently close to 2. This means that there are component lifetime distributions for which maximal system lifetime variance is attained by proper coherent system, either parallel or series one, and no formula is redundant in (5.25).

We finally observe that under specification of the parent distribution function F, representation (5.26) makes it possible to evaluate system lifetime variances under various moments conditions. Suppose, e.g., that F is standard exponential. Then all the functions describing the lower bounds are concave increasing on their domains, and attain their global maxima at the right-end points. Combining (5.26) with (5.28) and (5.29), we obtain

$$\min_{ET \ge \mu} \operatorname{Var} T = \frac{\tau_{k+1} - \tau_k}{\mu_{k+1} - \mu_k}(\mu - \mu_k) + \tau_k - \mu^2 = \sigma_{k+1}^2 - (\mu_{k+1} - \mu)^2$$

$$= \sum_{i=n-k}^{n} \frac{1}{i^2} - \left(\sum_{i=n-k}^{n} \frac{1}{i} - \mu\right)^2,$$

and the minimum is attained by the system with signature

$$\mathbf{s} = (n-k)\left[\left(\sum_{i=n-k}^{n} \frac{1}{i} - \mu\right)\mathbf{e}_k + \left(\mu - \sum_{i=n+1-k}^{n} \frac{1}{i}\right)\mathbf{e}_{k+1}\right],$$

when $\mu_k = \sum_{i=n+1-k}^{n} \frac{1}{i} \le \mu \le \mu_{k+1} = \sum_{i=n-k}^{n} \frac{1}{i}$, $k = 1, \ldots, n-1$. Similarly, if $\sigma_k^2 = \sum_{i=n+1-k}^{n} \frac{1}{i^2} \le \sigma^2 \le \sigma_{k+1}^2 = \sum_{i=n-k}^{n} \frac{1}{i^2}$ for some $k = 1, \ldots, n-1$, then

$$\max_{\operatorname{Var} T \le \sigma^2} \operatorname{E} T = \mu_{k+1} - (\sigma_{k+1}^2 - \sigma^2)^{1/2} = \sum_{i=n-k}^{n} \frac{1}{i} - \left(\sum_{i=n-k}^{n} \frac{1}{i^2} - \sigma^2\right)^{1/2},$$

and the maximum is attained when

$$\mathbf{s} = (n-k)\left\{\left(\sum_{i=n-k}^{n} \frac{1}{i^2} - \sigma^2\right)^{1/2}\mathbf{e}_k + \left[1 - \left(\sum_{i=n-k}^{n} \frac{1}{i^2} - \sigma^2\right)^{1/2}\right]\mathbf{e}_{k+1}\right\}.$$

In the standard uniform case we simply have

$$\min_{ET \ge \mu} \operatorname{Var} T = \operatorname{Var} X_{n:n} = \frac{n}{(n+1)^2(n+2)}, \quad \frac{1}{n+1} \le \mu \le \frac{n}{n+1},$$

$$\max_{\operatorname{Var} T \le \sigma^2} \operatorname{E} T = \operatorname{E} X_{n:n} = \frac{n}{n+1}, \quad \frac{n}{(n+1)^2(n+2)} \le \sigma^2 \le \frac{1}{4} - \frac{n}{(n+1)(n+2)}.$$

References

Beska M, Jasinski K, Rychlik T, Spryszynski M (2011) Mixed systems with minimal and maximal lifetime variances. Metrika

Boland PJ, Samaniego F (2004) The signature of a coherent system and its applications in reliability. In: Soyer R, Mazzuchi T, Singpurwalla N (eds) Mathematical reliability: an expository perspective. Kluwer, Boston, pp 1–29

David HA, Nagaraja HN (2003) Order statistics, 3rd edn. Wiley, Hoboken

Durante F, Jaworski P (2009) Absolutely continuous copulas with given diagonal sections. Com Stat Theory Meth 37:2924–2942

Jasiński K, Navarro J, Rychlik T (2009) Bounds on variances of lifetimes of coherent and mixed systems. J Appl Probab 46:894–908

Jaworski P (2009) On copulas and their diagonals. Inform Sci 179:2863–2871

Jaworski P, Rychlik T (2008) On distributions of order statistics for absolutely continuous copulas with applications to reliability problems. Kybernetika 44:757–776

Kochar S, Mukerjee H, Samaniego FJ (1999) The "signature" of a coherent system and its application to comparison among systems. Naval Res Logist 46:507–523

Lin G, Huang J (1989) Variances of sample medians. Stat Probab Lett 8:143–146

Navarro J, Rubio R (2010) Computations of coherent systems with five components. Commun Stat Sim Comput 39:68–84

Navarro J, Ruiz JM, Sandoval CJ (2007) Properties of coherent systems with dependent components. Commun Stat Theory Meth 36:175–191

Navarro J, Rychlik T (2007) Reliability and expectation bounds for coherent systems with exchangeable components. J Multivariate Anal 98:102–113

Navarro J, Samaniego FJ, Balakrishnan N, Bhattacharya D (2008) On the application and extension of system signatures to problems in engineering reliability. Naval Res Logist 55:313–327

Papadatos N (1995) Maximum variance of order statistics. Ann Inst Stat Math 47:185–193

Papadatos N (1997) A note on maximum variance of order statistics from symmetric populations. Ann Inst Stat Math 49:117–121

Rychlik T (1993) Bounds for expectation of L-estimates for dependent samples. Statistics 24:1–7

Rychlik T (1994) Distributions and expectations of order statistics for possibly dependent random variables. J Multivar Anal 48:31–42

Rychlik T (1998) Bounds for expectations of L-estimates. In: Balakrishnan N, Rao CR (eds) Order statistics: theory and methods. North-Holland, Amsterdam, pp 105–145

Rychlik T (2001) Projecting statistical functionals. Springer, New York

Rychlik T (2008) Extreme variances of order statistics in dependent samples. Stat Probab Lett 78:1577–1582

Samaniego F (1985) On closure of the IFR class under formation of coherent systems. IEEE Trans Reliab TR-34:69–72

Samaniego F (2007) System signatures and their applications in engineering reliability. Springer, New York

Shaked M, Suarez-Llorens A (2003) On the comparison of reliability experiments based on the convolution order. J Amer Stat Assoc 98:693–702

Yang H (1982) On the variances of median and some other order statistics. Bull Inst Math Acad Sinica 10:197–204

Chapter 6
L_z-Transform for a Discrete-State Continuous-Time Markov Process and its Applications to Multi-State System Reliability

Anatoly Lisnianski

Abstract During last years a specific approach called the universal generating function (UGF) technique has been widely applied to MSS reliability analysis. The UGF technique allows one to algebraically find the entire MSS performance distribution through the performance distributions of its elements. However, the main restriction of this powerful technique is that theoretically it may be only applied to random variables and, so, concerning MSS reliability, it operates with only steady-states performance distributions. In order to extend the UGF technique application to dynamic MSS reliability analysis the paper introduces a special transform for a discrete-states continuous-time Markov process that is called L_z-transform. The transform was mathematically defined, its main properties were studied, and numerical example illustrating its benefits for dynamic MSS reliability assessment is presented.

Keywords Discrete-state continuous-time Markov process · Universal generating function · Multi-state system · Reliability

6.1 Introduction

During last years a specific approach called the universal generating function (UGF) technique has been widely applied to MSS reliability analysis. The UGF technique allows one to find the entire MSS performance distribution based on the

A. Lisnianski (✉)
Planning, Development and Technology Division, The System Reliability Department,
The Israel Electric Corporation Ltd., New Office Building, St. Nativ haor 1, Haifa, Israel
e-mail: anatoly-l@iec.co.il

A. Lisnianski and I. Frenkel (eds.), *Recent Advances in System Reliability*,
Springer Series in Reliability Engineering, DOI: 10.1007/978-1-4471-2207-4_6,
© Springer-Verlag London Limited 2012

performance distributions of its elements using algebraic procedures. The basic ideas of the method were primarily introduced by Ushakov (1986, 1987) in the mid 1980s. Since then, the method has been considerably expanded in numerous research works (Lisnianski et al. 1996; Levitin et al. 1998; Yeh 2006, 2009; Tian et al. 2008; Li et al. 2011 etc.). In the books (Lisnianski and Levitin 2003; Levitin 2005) one can find a historical overview, detailed UGF-method description and the method application to many practically important cases.

Generally the UGF approach allows one to obtain a system's steady-state output performance distribution based on the given steady-state performance distribution of system's elements and the system structure function.

Each multi-state element i, $i = 1, \ldots, m$ in MSS is presented by its steady-state performance as a discrete random variable X_i with probability mass function (pmf)

$$\{(x_{i1}, p_{i1}), (x_{i2}, p_{i2}), \ldots, (x_{in}, p_{in_i})\}, \tag{6.1}$$

where x_{ij} is performance of element i at level j and p_{ij} represents probability that element i is staying at level j, $j = 1, \ldots, n_i$.

The system structure function f determines steady-state output performance for the entire MSS as a random variable

$$Y = f(X_1, X_2, \ldots, X_m). \tag{6.2}$$

According to the UGF method for each element i that is represented by random variable X_i with pmf (6.1) its individual moment generating function is written as z-transform in the following form

$$u_{Xi}(z) = \sum_{j=1}^{n_i} p_{ij} z^{x_{ij}}, \quad i = 1, \ldots, m. \tag{6.3}$$

Then by using Ushakov's universal generating operator Ω_f, which produces z-transform for the resulting random variable $Y = f(X_1, X_2, \ldots, X_m)$, one obtains the resulting z-transform for MSS steady-state output performance distribution based on individual z-transforms of the MSS's elements

$$U_Y(z) = \Omega_f\{u_{X_1}(z), \ldots, u_{X_m}(z)\}. \tag{6.4}$$

The technique for computation operator Ω_f is well established for various system structure functions (Lisnianski and Levitin 2003; Levitin 2005). Based on this technique after some algebra, one can obtain the resulting z-transform for random variable Y, which represents the entire MSS's steady-state output performance

$$U_Y(z) = \sum_{k=1}^{K} p_k z^{y_k}, \tag{6.5}$$

where y_k is MSS's steady-state output performance at level k and p_k represents probability that the MSS in steady-state is staying at level k, $k = 1, \ldots, K$.

Based on one-to-one correspondence between discrete random variable and its z-transform or, in other words, based on the property of z-transform (or moment generating function) *uniqueness* for each discrete random variable (see, for example, Ross 2000; Feller 1970) one obtains *pmf* $\{(y_1, p_1), (y_2, p_2), \ldots, (y_K, y_K)\}$ of output random variable Y, when its z-transform $U_Y(z)$ (expression (6.5)) is known.

Based on steady-state output performance distribution many MSS's reliability measures can be easily found (Lisnianski and Levitin 2003).

A main restriction of the UGF technique's application to real-world MSS reliability analysis is caused by the fact that z-transform was defined only for **random variables**. As one can see, this fact causes to consider performance of each MSS's element as a random variable with *pmf* (6.1), in spite of the fact that in reality it is a discrete-state continuous-time stochastic process (Lisnianski and Levitin 2003; Lisnianski et al. 2010; Natvig 2011). In practice this important restriction leads to consider only steady-state parameters of MSS. So, now by using basic UGF technique it is impossible to analyze transient modes in MSS, aging MSS, MSS under increasing or decreasing stochastic demand and so on. At last time some efforts were performed in order to remove this restriction (Lisnianski 2004, 2007; Lisnianski and Ding 2009). The method suggested in these works was named as a combined method of stochastic process and UGF technique. Then Lisnianski et al. (2010) introduced z-transform associated with stochastic process. In these works z-transform for discrete-state continuous-time stochastic process was intuitively defined and existence and uniqueness of z-transform were only suggested and were not mathematically proven. But after these works it became clear that such kind of transform may be very useful and essentially extends an area of solving problems. In this case it is possible to work with generic MSS model (Lisnianski et al. 2010), which includes the performance stochastic processes $X_i(t), i = 1, 2, \ldots, m$ for each system element i and the system structure function that produces the resulting stochastic process corresponding to the output performance of the entire MSS $Y(t) = f(X_1(t), X_2(t), \ldots, X_m(t))$.

In order to remove this essential restriction a special transform should be mathematically defined (introduced) for discrete-state continuous-time stochastic process. This transform should be similar to z-transform for discrete random variable in sense that Ushakov's operator Ω_f can be applied. It should be the transform that can be applied to discrete-state continuous-time stochastic process and its important properties such as existence and uniqueness should be proven. Below we introduce such transform (named as L_z-transform), prove its existence and uniqueness, study its main properties and demonstrate benefits of its application to MSS reliability analysis by using some numerical examples.

At this stage only Markov stochastic process will be considered.

6.2 L_z-Transform: Definition, Existence and Uniqueness

6.2.1 Definition

We consider a discrete-state continuous-time (DSCT) Markov process (Trivedi 2002) $X(t) \in \{x_1, \ldots, x_K\}$, $t \geq 0$, which has K possible states $i, (i = 1, \ldots, K)$ where performance level associated with any state i is x_i. This Markov process is completely defined by set of possible states $\mathbf{x} = \{x_1, \ldots, x_K\}$, transitions intensities matrix $\mathbf{A} = \|a_{ij}(t)\|$, $i, j = 1, \ldots K$ and by initial states probability distribution that can be presented by corresponding set

$$\mathbf{p}_0 = [p_{10} = \Pr\{X(0) = x_1\}, \ldots, p_{K0} = \Pr\{X(0) = x_K\}].$$

From now on we shall use for such Markov process the following notation by using triplet:

$$X(t) = \langle \mathbf{x}, \mathbf{A}, \mathbf{p}_0 \rangle. \tag{6.6}$$

Remark 1 If functions $a_{ij}(t) = a_{ij}$ are constants, then the DSCT Markov process is said to be *time-homogeneous*. When $a_{ij}(t)$ are time dependent, then the resulting Markov process is *non-homogeneous*.

Definition 1 L_z-transform of a discrete-state continuous-time Markov process $X(t) = \langle \mathbf{x}, \mathbf{A}, \mathbf{p}_0 \rangle$ is a function $u(z, t, \mathbf{p}_0)$ defined as

$$L_z\{X(t)\} = u(z, t, \mathbf{p}_0) = \sum_{i=1}^{K} p_i(t) z^{x_i}, \tag{6.7}$$

where $p_i(t)$ is a probability that the process is in state i at time instant $t \geq 0$ for any given initial states probability distribution $\mathbf{p}_0$ and z in general case is a complex variable.

In the future we sometime shall omit symbol $\mathbf{p}_0$ and write simply $u(z, t)$ keeping in mind that L_z-transform will depend on initial probability distribution $\mathbf{p}_0$.

Example 1 Consider a simplest element which has only two states 1 and 2 with corresponding performance levels $x_1 = 0$ and $x_2 = x_{nom}$, respectively. It means that state 1 is a complete failure state and state 2 is a state with nominal performance. The element's failure rate is λ and repair rate is μ.

Suppose that at time instant $t = 0$ the element is in the state 2, so that initial states probability distribution is the following $\mathbf{p}_0 = \{p_{10}, p_{20}\} = \{p_1(0); p_2(0)\} = \{0, 1\}$.

Let us define L_z-transform for Markov process $X(t)$ that describes the element's behavior.

Solution The Markov process $X(t)$ for our example is defined by the triplet $X(t) = \langle \mathbf{x}, \mathbf{A}, \mathbf{p}_0 \rangle$, where $\mathbf{x}$, $\mathbf{A}$, $\mathbf{p}_0$ are defined by the following:

- set of possible states $\mathbf{x} = \{x_1, x_2\} = \{0, x_{nom}\}$;
- transitions intensities matrix $\mathbf{A} = (a_{ij}) = \begin{pmatrix} -\mu & \mu \\ \lambda & -\lambda \end{pmatrix}$, $i, j = 1, 2$;
- initial states probability distribution $\mathbf{p}_0 = \{p_{10}, p_{20}\} = \{0, 1\}$.

Therefore, states probabilities of the process $X(t)$ at any time instant $t \geq 0$ will be defined as a solution of the following system of differential equations

$$\begin{cases} \dfrac{dp_1(t)}{dt} = -\mu p_1(t) + \lambda p_2(t), \\ \dfrac{dp_2(t)}{dt} = \mu p_1(t) - \lambda p_2(t), \end{cases}$$

under initial conditions $p_1(0) = p_{10} = 0$; $p_2(0) = p_{20} = 1$.

After solving the system one obtains

$$p_2(t) = \frac{\mu}{\mu + \lambda} + \frac{\lambda}{\mu + \lambda} e^{-(\lambda + \mu)t},$$

$$p_1(t) = \frac{\lambda}{\mu + \lambda} - \frac{\lambda}{\mu + \lambda} e^{-(\lambda + \mu)t}.$$

So, in accordance with Definition 1 (6.7) one can obtain L_z-transform of the given Markov process as follows:

$$L_z\{X(t)\} = u(z, t, \mathbf{p}_0) = \sum_{i=1}^{2} p_i(t) z^{x_i}$$

$$= \left[\frac{\lambda}{\mu + \lambda} - \frac{\lambda}{\mu + \lambda} e^{-(\lambda + \mu)t} \right] z^0 + \left[\frac{\mu}{\mu + \lambda} + \frac{\lambda}{\mu + \lambda} e^{-(\lambda + \mu)t} \right] z^{x_{nom}}.$$

6.2.2 Existence and Uniqueness

Each discrete-state continuous-time Markov process under certain initial conditions has only one (unique) L_z-transform $u(z, t)$ and each L_z-transform $u(z, t)$ will have only one corresponding DSCT Markov process $X(t)$ developing from these initial conditions.

We will formulate this as an **existence and uniqueness property** of L_z-transform.

Proposition 1 *Each Discrete-state Continuous-time Markov Process $X(t)$ under certain initial conditions $\mathbf{p}_0$ has one and only one L_z-transform: $L_z\{X(t)\} = u(z, t, \mathbf{p}_0)$.*

Proof The proof is based on Picard theorem (Coddington and Levinson 1955). In the theory of differential equations, the Picard theorem is an important theorem on existence and uniqueness of solutions to system of differential equations with a given initial value problems.

In our case for discrete-state continuous-time Markov process $X(t) = \langle \mathbf{x}, \mathbf{A}, \mathbf{p}_0 \rangle$ states probabilities $p_i(t) = \Pr\{X(t) = x_i\}, i = 1, 2, \ldots, K$ are defined by the solution of the following linear system of K ordinary differential equations

$$\begin{cases} \dfrac{dp_1(t)}{dt} = a_{11}(t)p_1(t) + a_{12}(t)p_2(t) + \cdots + a_{1K}(t)p_K(t), \\ \dfrac{dp_2(t)}{dt} = a_{21}(t)p_1(t) + a_{22}(t)p_2(t) + \cdots + a_{2K}(t)p_K(t), \\ \cdots \\ \dfrac{dp_K(t)}{dt} = a_{K1}(t)p_1(t) + a_{K2}(t)p_2(t) + \cdots + a_{KK}(t)p_K(t) \end{cases} \qquad (6.8)$$

under initial conditions $\mathbf{p}_0 = \{p_{10}, p_{20}, \ldots, p_{K0}\}$, where

$$p_1(t_0) = p_{10}, p_2(t_0) = p_{20}, \ldots, p_K(t_0) = p_{K0}. \qquad (6.9)$$

In accordance with Picard's theorem, **if coefficients** $a_{kl}(t), (k, l = 1, 2, \ldots, K)$ **are continuous functions of time t,** then the system (6.8) has a unique solution

$$p_1(t), p_2(t), \ldots, p_K(t) \qquad (6.10)$$

satisfied initial conditions (6.9).

So, in accordance with definition of L_z-transform (6.7) for discrete-state continuous-time Markov process $X(t) = \langle \mathbf{x}, \mathbf{A}, \mathbf{p}_0 \rangle$ we shall have

$$L_z\{X(t)\} = u(z, t, \mathbf{p}_0) = \sum_{i=1}^{K} p_i(t)z^{x_i},$$

where $p_1(t), p_2(t), \ldots, p_K(t)$ are determined as a unique solution of the system (6.8) under initial conditions (6.9).

Therefore, for discrete-state continuous-time Markov process $X(t) = \langle \mathbf{x}, \mathbf{A}, \mathbf{p}_0 \rangle$, where transition intensities $a_{ij}(t)$ are *continuous functions of time*, there exists only one (unique) L_z-transform.

Remark 2 The inverse statement is also true: if it is known $u(z, t, \mathbf{p}_0) = \sum_{i=1}^{K} p_i(t)z^{x_i}$, where $p_i(t)$ are defined as a solution of the system (6.8) (where coefficients $a_{kl}(t), k, l = 1, 2, \ldots K$ are continuous functions of time t) under initial conditions (6.9), then there exists only unique DSCT Markov process $X(t)$, for which $L_z\{X(t)\} = u(z, t, \mathbf{p}_0) = \sum_{i=1}^{K} p_i(t)z^{x_i}$.

Remark 3 In reliability interpretation L_z-transform may be applied to an aging system and to a system at burn-in period as well as to a system with constant failure and repair rates. The unique condition that should be fulfilled is a continuity of transitions intensities $a_{ij}(t)$.

6.3 L_z-Transform's Properties

Below we consider some important properties of L_z-transform:

Property 1 *Multiplying DSCT Markov process on constant value a is equal to multiplying corresponding performance level x_i at each state i on this value.*

$$L_z\{aX(t)\} = \sum_{i=1}^{K} p_i(t)z^{ax_i} \qquad (6.11)$$

Property 2 *L_z-transform from a single-valued function $f[G(t), W(t)]$ of two independent DSCT Markov processes $G(t)$ and $W(t)$ can be found by applying Ushakov's universal generating operator Ω_f to L_z-transform from $G(t)$ and $W(t)$ processes over all time points $t \geq 0$*

$$L_z\{f[G(t), W(t)]\} = \Omega_f\{L_z[G(t)], L_z[W(t)]\}.$$

Proof Consider two independent DSCT Markov processes: $G(t) = \langle \mathbf{g}, \mathbf{A}, \mathbf{p}_{g0}\rangle$ and $W(t) = \langle \mathbf{w}, \mathbf{B}, \mathbf{p}_{w0}\rangle$, where

$$\begin{aligned}
\mathbf{g} &= \{g_1, \ldots, g_k\}; & \mathbf{w} &= \{w_1, \ldots, w_m\}; \\
\mathbf{A} &= \|a_{ij}(t)\|, i,j = 1, \ldots, k; & \mathbf{B} &= \|b_{ij}(t)\|, i,j = 1, \ldots, m; \\
\mathbf{P}_{g0} &= \left\{p_{10}^{(g)}, p_{20}^{(g)}, \ldots, p_{k0}^{(g)}\right\}; & \mathbf{P}_{w0} &= \left\{p_{10}^{(w)}, p_{20}^{(w)}, \ldots, p_{m0}^{(w)}\right\}.
\end{aligned}$$

In accordance with the main definition (6.2) one can write the following L_z-transform:

$$\begin{aligned}
L_z\{G(t)\} &= \sum_{i=1}^{k} p_{gi}(t)z^{g_i}, \\
L_z\{W(t)\} &= \sum_{i=1}^{m} p_{wi}(t)z^{w_i}.
\end{aligned} \qquad (6.12)$$

The problem is to find L_z-transform for the resulting DSCT Markov process $Y(t) = f[G(t), W(t)]$ that is the single-valued function of these two independent processes $G(t)$ and $W(t)$

$$L_z\{Y(t)\} = L_z\{f[G(t), W(t)]\}. \qquad (6.13)$$

For any time instant $t_r \geq 0$ a DSCT Markov process $G(t)$ is a discrete random variable $G_r = G(t_r)$ with a corresponding *pmf*

$$\Pr\{G(t_r) = g_1\} = p_{g1}(t_r), \ldots, \Pr\{G(t_r) = g_k\} = p_{gk}(t_r).$$

For DSCT Markov process $G(t) = \langle \mathbf{g}, \mathbf{A}, \mathbf{p}_{g0} \rangle$, states probabilities $p_{gi}(t) = \Pr\{G(t) = g_i\}, i = 1, 2, \ldots, K$ are defined for any instant $t \geq 0$ by the solution of the following linear system of ordinary differential equations

$$
\begin{cases}
\dfrac{dp_{g1}(t)}{dt} = a_{11}(t)p_{g1}(t) + a_{12}(t)p_{g2}(t) + \cdots + a_{1k}(t)p_{gk}(t), \\
\dfrac{dp_{g2}(t)}{dt} = a_{21}(t)p_{g1}(t) + a_{22}(t)p_{g2}(t) + \cdots + a_{2k}(t)p_{gk}(t), \; \bullet \\
\cdots \\
\dfrac{dp_k(t)}{dt} = a_{k1}(t)p_{g1}(t) + a_{k2}(t)p_{g2}(t) + \cdots + a_{kk}(t)p_{gk}(t)
\end{cases}
\tag{6.14}
$$

under initial conditions $\{\mathbf{p}_{g0} = \left\{ p_{10}^{(g)}, p_{20}^{(g)}, \ldots, p_{k0}^{(g)} \right\},$

$$
p_{g1}(t_0) = p_{10}^{(g)}, p_{g2}(t_0) = p_{20}^{(g)}, \ldots, p_{gk}(t_0) = p_{k0}^{(g)}.
\tag{6.15}
$$

Analogously, for any time instant $t_r \geq 0$ a DSCT Markov process $W(t)$ is a discrete random variable $W_r = W(t_r)$ with a corresponding *pmf*

$$
\Pr\{W(t_r) = w_1\} = p_{w1}(t_r), \ldots, \Pr\{W(t_r) = w_k\} = p_{wk}(t_r).
$$

For DSCT Markov process $W(t) = \langle \mathbf{w}, \mathbf{A}, \mathbf{p}_{w0} \rangle$, states probabilities $p_{wi}(t) = \Pr\{W(t) = w_i\}, i = 1, 2, \ldots, m$ are defined for any instant $t \geq 0$ by the solution of the following linear system of ordinary differential equations

$$
\begin{cases}
\dfrac{dp_{w1}(t)}{dt} = b_{11}(t)p_{w1}(t) + b_{12}(t)p_{w2}(t) + \cdots + b_{1m}(t)p_{wm}(t), \\
\dfrac{dp_{w2}(t)}{dt} = b_{21}(t)p_{w1}(t) + b_{22}(t)p_{w2}(t) + \cdots + b_{2m}(t)p_{wm}(t), \\
\cdots \\
\dfrac{dp_{wm}(t)}{dt} = b_{m1}(t)p_{w1}(t) + b_{m2}(t)p_{w2}(t) + \cdots + b_{mm}(t)p_{wm}(t)
\end{cases}
\tag{6.16}
$$

under initial conditions $\mathbf{p}_{w0} = \left\{ p_{10}^{(w)}, p_{20}^{(w)}, \ldots, p_{m0}^{(w)} \right\},$

$$
p_{w1}(t_0) = p_{10}^{(w)}, p_{w2}(t_0) = p_{20}^{(w)}, \ldots, p_{wk}(t_0) = p_{m0}^{(w)}
\tag{6.17}
$$

In order to define random variable $Y(t_r) = f[G(t_r), W(t_r)]$ at any time instant $t_r \geq 0$, one has to evaluate its *pmf*. In other words, one has to evaluate the vector $\mathbf{y} = \{y_1, y_2, \ldots, y_K\}$ of all of the possible values of the resulting random variable $Y(t_r)$ and the vector $\mathbf{q} = \{q_1, q_2, \ldots, q_K\}$ of probabilities that variable $Y(t_r)$ takes these values.

Each possible value $y_1, y_2, \ldots, y_K$ of resulting random variable $Y_r = Y(t_r)$ corresponds to a combination of the values of variables $G_r = G(t_r)$, $W_r = W(t_r)$, and, since the variables are statistically independent, probability of each value $y_i, i = 1, \ldots, K$ is equal to a product of probabilities of the corresponding values $G(t_r), W(t_r)$. The *pmf* of resulting random variable $Y(t_r) = f[G(t_r), W(t_r)]$ is represented in Table 6.1.

Table 6.1 Probability mass function of random variable $Y(t_r) = f[G(t_r), W(t_r)]$

Values $\mathbf{y} = \{y_1, y_2, \ldots, y_K\}$	Probabilities $\mathbf{q} = \{q_1(t_r), q_2(t_r), \ldots, q_K(t_r)\}$
$y_1 = f\{g_1, w_1\}$	$q_1(t_r) = \Pr\{Y(t_r) = y_1\} = p_{g_1}(t_r)p_{w_1}(t_r)$
$\ldots$	$\ldots$
$y_m = f\{g_1, w_m\}$	$q_m(t_r) = \Pr\{Y(t_r) = y_m\} = p_{g_1}(t_r)p_{w_m}(t_r)$
$y_{m+1} = f(g_2, w_1)$	$q_{m+1}(t_r) = \Pr\{Y(t_r) = y_{m+1}\} = p_{g_2}(t_r)p_{w_1}(t_r)$
$\ldots$	$\ldots$
$y_{2m} = f(g_2, w_m)$	$q_{2m}(t_r) = \Pr\{Y(t_r) = y_{2m}\} = p_{g_2}(t_r)p_{w_m}(t_r)$
$y_{2m+1} = f(g_3, w_1)$	$q_{2m+1}(t_r) = \Pr\{Y(t_r) = y_{2m+1}\} = p_{g_3}(t_r)p_{w_1}(t_r)$
$\ldots$	$\ldots$
$y_{2m+m} = f(g_3, w_m)$	$q_{2m+m}(t_r) = \Pr\{Y(t_r) = y_{2m+m}\} = p_{g_3}(t_r)p_{w_m}(t_r)$
$\ldots$	$\ldots$
$y_{m(k-1)+1} = f(g_k, w_1)$	$q_{mk-m+1}(t_r) = \Pr\{Y(t_r) = y_{mk-m+1}\} = p_{g_k}(t_r)p_{w_1}(t_r)$
$\ldots$	$\ldots$
$y_{mk} = f(g_k, w_m)$	$q_{mk}(t_r) = \Pr\{Y(t_r) = y_{mk}\} = p_{g_k}(t_r)p_{w_m}(t_r)$

As one can see the total number K of possible values of random variable $Y(t_r) = f[G(t_r), W(t_r)]$ is $K = km$.

Therefore, vectors $\mathbf{y} = \{y_1, y_2, \ldots, y_K\}$ and $\mathbf{q} = \{q_1(t_r), q_2(t_r), \ldots, q_K(t_r)\}$ presented in the Table 6.1 completely determine resulting *pmf* of random variable $Y_r = Y(t_r) = f[G(t_r), W(t_r)] = f[G_r, W_r]$ at any time instant $t_r \geq 0$.

Now the resulting L_z-transform for the random variable $Y(t_r) = f[G(t_r), W(t_r)]$ can be written at any time instant $t_r \geq 0$

$$u_{Y_r}(z, t_r) = \sum_{j=1}^{K} q_j(t_r)z^{y_j}. \tag{6.18}$$

For any time instant $t_r \in [0, t]$ based on the Table 6.1 one can write the following

$$u_{Y_r}(z, t_r) = \sum_{j=1}^{K} q_j(t_r)z^{y_j} = \sum_{i=1}^{k}\sum_{j=1}^{m} p_{gi}(t_r)p_{wj}(t_r)z^{f(g_i, w_j)}. \tag{6.19}$$

By using Ushakov's universal generating operator (UGO) (Levitin 2005; Lisnianski et al. 2010) one can re-write expression (6.19) in the following form:

$$u_{Y_r}(z, t_r) = \sum_{i=1}^{k}\sum_{j=1}^{m} p_{gi}(t_r)p_{wj}(t_r)z^{f(g_i, w_j)} = \Omega_f\{u_{G_r}(z, t_r), u_{W_r}(z, t_r)\}, \tag{6.20}$$

where Ω_f—is Ushakov's universal generating operator, $u_{G_r}(z, t_i) = \sum_{i=1}^{k} p_{gi}(t_r)z^{g_i}$ and $u_{W_r}(z, t_r) = \sum_{i=1}^{m} p_{wi}(t_r)z^{w_i}$ are L_z-transform for random variables $G_r = G(t_r)$ and $W_r = W(t_r)$.

Time instant t_r is an arbitrary point in the interval $[0, \infty)$, therefore, for any time instant $t \geq 0$ expression (6.20) may be re-written as the following

$$L_z\{Y(t)\} = u_Y(z,t) = \sum_{i=1}^{k}\sum_{j=1}^{m} p_{gi}(t)p_{wj}(t)z^{f(g_i,w_j)}$$

$$= \Omega_f\{u_G(z,t), u_W(z,t)\} = \Omega_f\{L_z[G(t)], L_z[W(t)]\}. \qquad (6.21)$$

Therefore, in order to find L_z-transform of the resulting DSCT Markov process $Y(t)$, which is the single-valued function $Y(t) = f[G(t), W(t)]$ of two independent DSCT Markov processes $G(t), W(t)$, one can apply Ushakov's UGO to L_z-transform of $G(t)$ and $W(t)$ processes over all time points $t \geq 0$.

From computational point of view instead of summarization in expression (6.21) it is better to use the following matrix notation. We suppose that $k \leq m$.

Let us designate a column matrix of state probabilities for process $G(t)$ as $\mathbf{P_g}(t)$, and a row matrix of state probabilities for process $W(t)$ as $\mathbf{P_w}(t)$:

$$\mathbf{P_g}(t) = \begin{vmatrix} p_{g1}(t) \\ p_{g2}(t) \\ \dots \\ p_{gk}(t) \end{vmatrix}, \quad \mathbf{P_w}(t) = |p_{w1}(t), p_{w2}(t), \dots, p_{wm}(t)|.$$

Note that a matrix with minimal dimension should be chosen as a column matrix. In our case we have supposed that $k \leq m$ and, so, matrix $\mathbf{P_g}(t)$ was written as a column matrix. If $k > m$, then matrix $\mathbf{P_w}(t)$ should be a column matrix and matrix $\mathbf{P_g}(t)$ should be a row matrix.

Then the matrix of states probabilities for the resulting stochastic process $Y(t)$ or in other words, the matrix $\mathbf{Q}(t)$ that defines all coefficients $q_j(t), j = 1,\dots, mk$ in the resulting L_z-transform

$$u_Y(z,t) = \sum_{j=1}^{K} q_j(t)z^{y_j} \qquad (6.22)$$

in expression (6.21) can be obtained as a product of matrices $\mathbf{P_g}(t)$ and $\mathbf{P_w}(t)$

$$\mathbf{Q}(t) = \mathbf{P_g}(t) \cdot \mathbf{P_w}(t) = \begin{vmatrix} p_{g1}(t)p_{w1}(t)p_{g1}(t)p_{w2}(t), \dots, p_{g1}(t)p_{wm}(t) \\ p_{g2}(t)p_{w1}(t)p_{g2}(t)p_{w2}(t), \dots, p_{g2}(t)p_{wm}(t) \\ \dots \\ p_{gk}(t)p_{w1}(t)p_{gk}(t)p_{w2}(t), \dots, p_{gk}(t)p_{wm}(t) \end{vmatrix} \qquad (6.23)$$

and $K = mk$.

The corresponding matrix $\mathbf{Y}$ that defines all powers of z (values $y_j, j = 1,\dots,mk$) in the resulting L_z-transform $u_Y(z,t)$ will be the following

$$\mathbf{Y} = \begin{vmatrix} f(g_1, w_1) & f(g_1, w_2) & \dots & f(g_1, w_m) \\ f(g_2, w_1) & f(g_2, w_2) & \dots & f(g_2, w_m) \\ \dots & \dots & \dots & \dots \\ f(g_k, w_1) & f(g_k, w_2) & \dots & f(g_k, w_m) \end{vmatrix} \qquad (6.24)$$

Such matrix notation is especially useful when MATLAB is used for computation.

Remark 4 Expression (6.21) can be extended to general case where the resulting DSCT Markov process $Y(t)$ is the single-valued function $Y(t) = f[X_1(t), X_2(t), \ldots, X_n(t)]$ of n independent DSCT Markov processes $X_1(t), X_2(t), \ldots, X_n(t)$:

$$
\begin{aligned}
L_z\{Y(t)\} = u_Y(z,t) &= \Omega_f\{u_{X_1}(z,t), u_{X_2}(z,t), \ldots u_{X_n}(z,t)\} \\
&= \Omega_f\{L_z[X_1(t)], L_z[X_2(t)], \ldots, L_z[X_n(t)]\}.
\end{aligned}
\tag{6.25}
$$

For computation in accordance with expression (6.25) may be used as ordinary UGF technique (like-terms collection, recursive procedures, etc.) over any time instant $t \geq 0$.

Remark 5 As it follows from the properties 1 and 2 L_z-transform has no a property of linearity.

At first, $L_2\{aX(t)\} \neq aL_2\{X(t)\}$ because

$$
L_2\{aX(t)\} = \sum_{i=1}^{K} p_i(t)z^{ax_i} \neq a\sum_{i=1}^{K} p_i(t)z^{x_i} = aL_2\{X(t)\}.
$$

At second, $L_z\{X_1(t) + X_2(t)\} \neq L_z\{X_1(t)\} + L_z\{X_2(t)\}$ because

$$
L_z\{X_1(t) + X_2(t)\} = \sum_{i=1}^{k_1}\sum_{j=1}^{k_2} p_{1i}(t)p_{2j}(t)z^{x_{1i}+x_{2j}}
$$

and

$$
L_z\{X_1(t)\} + L_z\{X_2(t)\} = \sum_{i=1}^{K_1} p_{1i}(t)z^{x_{1i}} + \sum_{i=1}^{K_2} p_{2i}(t)z^{x_{2i}}.
$$

Thus, $L_z\{X_1(t) + X_2(t)\} \neq L_z\{X_1(t)\} + L_z\{X_2(t)\}$.

6.4 L_z-Transform Application to MSS Reliability Analysis

If L_z-transform

$$
U_Y(z,t) = \sum_{k=1}^{K} p_i(t)z^{y_k}
\tag{6.26}
$$

of entire MSS's output stochastic process $Y(t) \in \{y_1, \ldots, y_K\}$ is known, then important system's reliability measures can be easily found.

The system availability at instant $t \geq 0$

$$A(t) = \sum_{y_i \geq 0} p_i(t). \tag{6.27}$$

In other words, in order to find MSS's instantaneous availability one should summarize all probabilities from terms where powers of z are positive or equal to 0.

The system instantaneous mean expected performance at instant $t \geq 0$

$$E(t) = \sum_{k=1}^{K} p_i(t)y_i. \tag{6.28}$$

The system average expected performance for a fixed time interval $[0, T]$

$$E_T = \frac{1}{T} \int_0^T E(t)dt = \frac{1}{T} \sum_{i=1}^{K} y_i \int_0^T p_i(t)dt. \tag{6.29}$$

The system instantaneous performance deficiency

$$D(t) = \sum_{i=1}^{K} p_i(t) \min(y_i, 0). \tag{6.30}$$

The system accumulated performance deficiency for a fixed time interval $[0, T]$

$$D_f = \int_0^T D(t)dt = \sum_{i=1}^{K} \min(y_i, 0) \int_0^T p_i(t)dt. \tag{6.31}$$

In order to illustrate L_z-transform application to MSS reliability analysis, we consider the following example.

Example 2 Consider an aging production system that is characterized by 3 possible productivity levels: complete failure, when the system productivity $g_1 = 0$, reduced productivity $g_2 = 400$ units/year, and nominal productivity $g_3 = 800$ units/year. So, the corresponding set of the system's states performances is the following $g = \{g_1, g_2, g_3\} = \{0, 400, 800\}$.

The transition intensities for the system are represented by the following matrix

$$\mathbf{A}_g = \begin{pmatrix} -(a_{12}^{(g)} + a_{13}^{(g)}) & a_{12}^{(g)} & a_{13}^{(g)} \\ a_{21}^{(g)} & -(a_{21}^{(g)} + a_{23}^{(g)}) & a_{23}^{(g)} \\ a_{31}^{(g)} & a_{32}^{(g)} & -(a_{31}^{(g)} + a_{32}^{(g)}) \end{pmatrix}$$

where $a_{12}^{(g)} = 100 \, \text{year}^{-1}, a_{13}^{(g)} = 200 \, \text{year}^{-1}, a_{21}^{(g)} = 0.7 \, \text{year}^{-1}, a_{23}^{(g)} = 300 \, \text{year}^{-1}, a_{31}^{(g)} = 0.5 + 0.1t \, \text{year}^{-1}, a_{32}^{(g)} = 1 \, \text{year}^{-1}$.

Fig. 6.1 State-transition
diagrams for Markov
processes $G(t)$ (**a**) and
$W(t)$ (**b**)

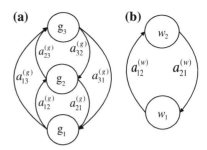

As one can see the system aging is expressed by increasing function
$a_{31}^{(g)} = 0.5 + 0.1t\,\text{year}^{-1}$.

The state with nominal productivity is the system's initial state $\mathbf{p}_{g0} = \left\{ p_{g1}(0), p_{g2}(0), p_{g3}(0) \right\} = \{0, 0, 1\}$.

So, the system's stochastic performance $G(t) \in \{g_1, g_2, g_3\}$ is defined by DSCT Markov process $G(t) = \langle \mathbf{g}, \mathbf{A}, \mathbf{p}_{g0} \rangle$.

State-transition diagram for Markov process $G(t)$ is presented in Fig. 6.1a.

The system has to satisfy a stochastic demand that is described by two-state DSCT Markov process with minimum performance level $w_1 = 0$ and maximum performance level $w_2 = 500$ units/year. Therefore, the corresponding set of demand's states performances is the following $\mathbf{w} = \{w_1, w_2\} = \{0, 500\}$.

The transition intensities for the demand are represented by the following matrix

$$\mathbf{A}_w = \begin{pmatrix} -a_{12}^{(w)} & a_{12}^{(w)} \\ a_{21}^{(w)} & -a_{21}^{(w)} \end{pmatrix}$$

where $a_{12}^{(w)} = 487 + 200\sin(2\pi t)\,\text{year}^{-1}, a_{21}^{(w)} = 1{,}095\,\text{year}^{-1}$.

As one can see the demand is seasonally changing – the maximum's duration is increasing at summer and decreasing at winter.

The state with maximum demand is the initial state $\mathbf{p}_{w0} = \left\{ p_{w1}(0), p_{w2}(0) \right\} = \{0, 1\}$.

The demand's stochastic performance is defined by DSCT Markov process $W(t) = \langle \mathbf{w}, \mathbf{A}, \mathbf{p}_{w0} \rangle$.

State-transition diagram for Markov process $W(t)$ is presented in Fig. 6.1b.

When the resulting stochastic process $Y(t) = f\{G(t), W(t)\} = G(t) - W(t)$ falls down to level zero such event is treated as a failure. Processes $G(t), W(t)$ are independent.

The problem is to find instantaneous availability and instantaneous mean performance deficiency for this aging multi-state production system under the seasonally changing stochastic demand.

Solution In Fig. 6.2 one can see a block-diagram for the MSS reliability computation. States probabilities $p_{gi}(t), i = 1, 2, 3$ for Markov processes $G(t)$ can be obtained by solving the following system of differential equations:

Fig. 6.2 Block-diagram for the MSS reliability computation

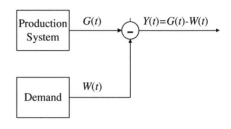

$$\begin{cases} \dfrac{\mathrm{d}p_{g1}(t)}{\mathrm{d}t} = -(a_{12}^{(g)} + a_{13}^{(g)})p_{1g}(t) + a_{21}^{(g)}p_{2g}(t) + a_{31}^{(g)}p_{3g}(t), \\[2mm] \dfrac{\mathrm{d}p_{g2}(t)}{\mathrm{d}t} = a_{12}^{(g)}p_{1g}(t) - (a_{21}^{(g)} + a_{23}^{(g)})p_{2g}(t) + a_{32}^{(g)}p_3(t), \\[2mm] \dfrac{\mathrm{d}p_{g3}(t)}{\mathrm{d}t} = a_{13}^{(g)}p_{1g}(t) - a_{23}^{(g)}p_{2g}(t) - (a_{31}^{(g)} + a_{32}^{(g)})p_3(t), \end{cases}$$

under initial conditions $\mathbf{p}_{g0} = \left\{ p_{g1}(0), p_{g2}(0), p_{g3}(0) \right\} = \{0, 0, 1\}$.

States probabilities $p_{wi}(t), i = 1, 2$ for Markov processes $W(t)$ can be obtained by solving the following system of differential equations:

$$\begin{cases} \dfrac{\mathrm{d}p_{w1}(t)}{\mathrm{d}t} = -a_{12}^{(w)}p_{1g}(t) - a_{21}^{(w)}p_{2w}(t), \\[2mm] \dfrac{\mathrm{d}p_{w2}(t)}{\mathrm{d}t} = a_{12}^{(w)}p_{1g}(t) - a_{21}^{(w)}p_{2w}(t), \end{cases}$$

under initial conditions $\mathbf{p}_{w0} = \left\{ p_{w1}(0), p_{w2}(0) \right\} = \{0, 1\}$.

In accordance with the L_Z–transform definition we obtain

$$L_z\{G(t)\} = p_{g1}(t)z^{g_1} + p_{g2}(t)z^{g_2} + p_{g3}(t)z^{g_3}$$
$$= p_{g1}(t)z^0 + p_{g2}(t)z^{400} + p_{g3}(t)z^{800};$$
$$L_z\{W(t)\} = p_{w1}(t)z^{w_1} + p_{w2}(t)z^{w_2} = p_{w1}(t)z^0 + p_{w2}(t)z^{500}.$$

Now, we have

$$L_z\{Y(t)\} = L_Z\{G(t) - W(t)\} = \Omega_f\{L_z\{G(t)\}, L_z\{W(t)\}\}$$
$$= \Omega_f\left\{ p_{g1}(t)z^0 + p_{g2}(t)z^{400} + p_{g3}(t)z^{800}; p_{w1}(t)z^0 + p_{w2}(t)z^{500} \right\}$$
$$= \sum_{i=1}^{3}\sum_{j=1}^{2} p_{gi}(t)p_{wj}(t)z^{f(g_i, w_j)} = \sum_{i=1}^{3}\sum_{j=1}^{2} p_{gi}(t)p_{wj}(t)z^{(g_i - w_j)}$$
$$= p_{g1}(t)p_{w2}(t)z^{-500} + p_{g2}(t)p_{w2}(t)z^{-100} + p_{g1}(t)p_{w1}(t)z^0$$
$$+ p_{g3}(t)p_{w2}(t)z^{300} + p_{g2}(t)p_{w1}(t)z^{400} + p_{g3}(t)p_{w1}(t)z^{800}.$$

Based on the last expression we obtain:

- MSS's instantaneous availability is calculated in accordance with expression (6.27) (see Fig. 6.3)

Fig. 6.3 MSS's
instantaneous availability

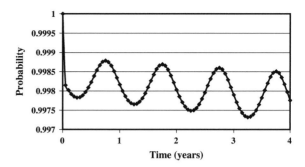

$$AV(t) = p_{g1}(t)p_{w1}(t) + p_{g3}(t)p_{w2}(t) + p_{g2}(t)p_{w1}(t) + p_{g3}(t)p_{w1}(t),$$

MSS's mean performance deficiency is calculated in accordance with expression (6.30) (see Fig. 6.4)

$$D(t) = 500p_{g1}(t)p_{w2}(t) + 100p_{g2}(t)p_{w2}(t).$$

Note that in order to solve this problem by using straightforward Markov methods one should solve the system of 6 differential equations with variable coefficients.

Remark 6 In general case, if one will use a straightforward Markov method for MSS consisting of n different multi-states elements where every element j has k_j different performance levels, he will obtain a model with $K = \prod_{j=1}^{n} k_j$ states.

Therefore, one should solve the system of K differential equations. This number K can be very large even for relatively small MSSs. So, by using straightforward Markov methods one will be faced with "dimension curse". State-space diagram building or model construction for complex MSSs is a difficult no formalized process that may cause numerous mistakes. In addition, solving models with hundreds of states can challenge the available computer resources. If one will use L_z-transform, he has to build and solve only n separate (relatively simple) Markov models for each element where maximum number of states is defined as $\max\{k_1, \ldots, k_j, \ldots, k_n\}$. All other computations are formalized algebraic procedures. So, by using L_z-transform method one should solve n systems of $k_1, \ldots, k_n$ differential equations respectively and the total number of differential equations will be $\sum_{i=1}^{n} k_i$. It means that computational complexity decreases drastically when L_z-transform is used.

6.5 Conclusions

In this paperL_z-transform for discrete-state continuous-time Markov process was introduced and mathematically defined, and its main properties were studied.

L_z-transform application to MSS reliability analysis essentially extends a circle of problems that can be solved by using universal generating function technique.

Fig. 6.4 MSS's mean
performance deficiency

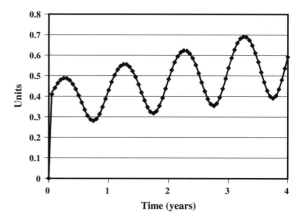

Based on L_Z-transform it is possible to use Ushakov's universal generating operator and corresponding universal generating functions technique in order to perform reliability analysis for MSS in transient modes, for aging MSS and MSS under stochastic increasing and decreasing demand.

It was shown that computational complexity is essentially less when L_Z-transform is used compared with traditional Markov method.

Numerical example illustrates the presented approach.

Acknowledgments The author is pleased to thank Professor I. Gertsbakh for his valuable and helpful comments.

Reference

Coddington E, Levinson N (1955) Theory of ordinary differential equations. McGraw-Hill, NY

Feller W (1970) An introduction to probability theory and its applications. Wiley, NY

Levitin G (2005) Universal generating function in reliability analysis and optimization. Springer, London

Levitin G, Lisnianski A, Ben-Haim H, Elmakis D (1998) Redundancy optimization for series–parallel multi-state systems. IEEE Trans Reliab 47:165–172

Li CY, Chen X, Yi XS, Tao JY (2011) Interval–valued reliability analysis of multi-state systems. IEEE Trans Reliab 60(1):321–330

Lisnianski A (2004) Universal generating function technique and random process methods for multi-state system reliability analysis. In: Proceedings of the 2nd International Workshop in Applied Probability (IWAP2004). Piraeus, Greece, pp 237–242

Lisnianski A (2007) Extended block diagram method for a multi-state system reliability assessment. Reliab Eng Syst Saf 92(12):1601–1607

Lisnianski A, Ding Y (2009) Redundancy analysis for repairable multi-state system by using combined stochastic process methods and universal generating function technique. Reliab Eng Syst Saf 94:1788–1795

Lisnianski A, Levitin G (2003) Multi-state system reliability: assessment. optimization and applications, World Scientific, Singapore

Lisnianski A, Levitin G, Ben-Haim H, Elmakis D (1996) Power system structure optimization subject to reliability constraints. Electr Power Syst Research 39:145–152

Lisnianski A, Frenkel I, Ding Y (2010) Multi-state system analysis and optimization for engineers and industrial managers. Springer, London

Natvig B (2011) Multistate systems reliability. Theory with Applications, Wiley, NY

Ross S (2000) Introduction to probability models. Academic, Boston

Tian Z, Zuo M, Huang H (2008) Reliability-redundancy allocation for multi-state series–parallel systems. IEEE Trans Reliab 57(2):303–310

Trivedi K (2002) Probability and statistics with reliability queuing and computer science applications. Wiley, NY

Ushakov I (1986) A universal generating function. Sov J Comput Syst Sci 24:37–49

Ushakov I (1987) Optimal standby problem and a universal generating function. Sov J Comput Syst Sci 25:61–73

Yeh W (2006) The k-out-of-n acyclic multistate-mode network reliability evaluation using the generating function method. Reliab Eng Sys Saf 91:800–808

Yeh W (2009) A convolution universal generating function method for evaluating the symbolic one-to-all target subset reliability function the acyclic multi-state information networks. IEEE Trans Reliab 58(3):476–484

Chapter 7
Reliability Decisions for Supermarket Refrigeration System by using Combined Stochastic Process and Universal Generating Function Method: Case Study

Ilia Frenkel and Lev Khvatskin

Abstract This chapter presents a case study of combined stochastic process and universal generating function method application to calculation of reliability measures for complex supermarket refrigeration system. The system and its components can have different performance levels ranging from perfect functioning to complete failure and, so it is treated as a multi-state system. Decision making about the system structure is based on calculated reliability measures. Combined universal generating functions and stochastic processes method is used for computation of availability, output performance and performance deficiency for such refrigeration system.

Keywords Reliability measures · Multi-state system · Combined universal generating functions and stochastic process method · Availability · Output performance · Performance deficiency

I. Frenkel (✉)
Shamoon College of Engineering,
Industrial Engineering and Management Department,
Center for Reliability and Risk Managemet,
Bialik/Basel Sts., 84100 Beer Sheva, Israel
e-mail: iliaf@sce.ac.il

L. Khvatskin
Industrial Engineering and Management Department,   ·
Sami Shamoon College of Engineering,
Center for Reliability and Risk Management,
Beer Sheva, Israel
e-mail: khvat@sce.ac.il

A. Lisnianski and I. Frenkel (eds.), *Recent Advances in System Reliability*,   97
Springer Series in Reliability Engineering, DOI: 10.1007/978-1-4471-2207-4_7,
© Springer-Verlag London Limited 2012

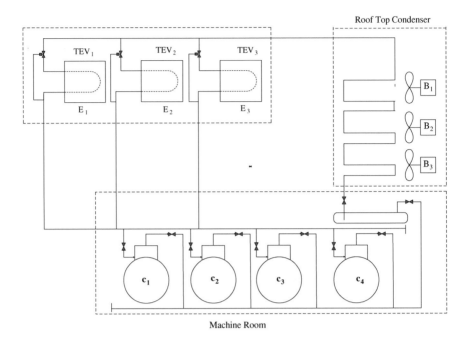

Fig. 7.1 Refrigeration system scheme

7.1 Supermarket Refrigeration System

Supermarkets suffer serious financial losses because of problems with their refrigeration systems. A typical supermarket may contain more than one hundred individual refrigerated cabinets, cold store rooms and items of plant machinery which interact as part of a complex integrated refrigeration system within the store. Things very often go wrong with individual units (icing up of components, electrical or mechanical failure and so forth…) or with components which serve a network of units (coolant tanks, pumps, compressors and so on).

The most commonly used refrigeration system for supermarkets today is the multiplex direct expansion system (Fig. 7.1). Heat rejection is usually done with compressors and air-cooled condensers with simultaneously working axial blowers. All display cases and cold store rooms use direct expansion air-refrigerant coils that are connected to the system compressors in a remote machine room located in the back or on the roof of the store.

Principal refrigeration system includes 4 basic elements: compressors (C), evaporators (E), condensers and thermo-expansion valves (TEV).

In the case of big supermarket commercial refrigeration system it is used installed in parallel 4 compressors that could support 4 levels of refrigeration capacity (performance) depended on demanded refrigeration load level. Using multiple compressors in parallel provides a capacity control, since the compressors can be selected and cycled as needed to meet the refrigeration load.

Condenser unit includes refrigeration coil and two axial blowers (B), which could support two performance levels depended on demanded refrigeration load level. It is possible to extend the condenser unit by additional axial blower. In this case the condenser unit will support three performance levels.

All compressors and condenser unit are situated outside of commerce supermarket area and are piped by refrigeration lines with a large number of display cases and cold store rooms, situated inside of commerce supermarket area and include two other basic parts: evaporators and termo-expansion valves.

Due to the system's nature, a fault in a single unit or item of machinery cannot have detrimental effects on the entire store, only decrease of system cool capacity. Failure of compressor or axial condenser blower leads to partial system failure (degradation of output cooling capacity) as well as to complete failures of the system. We treat refrigeration system as multi-state system (MSS), where components and systems have an arbitrary finite number of states. According to the generic MSS model (Lisnianski and Levitin 2003, Lisnianski et al. 2010), the system can have different states corresponding to the system's performance rates. The performance rate of the system at any instant is a discrete-state continuous-time stochastic process.

7.2 Brief Description of the Method

In this paper, a generalized approach was applied for decision-making for multi-state supermarket refrigeration system structure. The approach is based on the combined Universal Generating Functions (UGF) and stochastic processes method for computation of availability, output performance and performance deficiency for multi-state system. Primarily the method was described in (Lisnianski 2004) and one can find more detailed description in (Lisnianski 2007) and in (Lisnianski et al. 2010). Shortly, the method description is as follows.

In general case any element j in MSS can have k_j different states corresponding to different performance, represented by the set $\mathbf{g}_j = \{g_{j1}, \ldots, g_{jk_j}\}$, where g_{ji} is the performance rate of element j in the state i, $i \in \{1, 2, \ldots, k_j\}$.

At first stage a model of stochastic process should be built for each multi-state element in MSS. Based on this model state probabilities

$$p_{ji}(t) = \Pr\{G_j(t) = g_{ji}\}, \ i \in \{1, \ldots, k_j\},$$

for every MSS's element can be obtained. These probabilities define output stochastic process $G_j(t)$ for each element j in the MSS.

At the next stage the output performance distribution for the entire MSS at each time instant t should be defined based on previously determined states probabilities for all elements and system structure function that produces the stochastic process corresponding to the output performance of the entire MSS: $G(t) = f(G_1(t), \ldots, G_n(t))$.

Then individual universal generating function (UGF) for each element should be written. For each element j it will be UGF $u_j(z, t)$ associated with corresponding stochastic processes $G_j(t)$. Then by using composition operators over UGF of individual elements and their combinations in the entire MSS structure, one can obtain the resulting UGF $U(z, t)$ associated with output performance stochastic process $G(t)$ of the entire MSS by using simple algebraic operations. This UGF $U(z,t)$ defines the output performance distribution for the entire MSS at each time instant t. MSS reliability measures can be easily derived from this output performance distribution.

7.3 Reliability Block-Diagram for the Refrigeration System

We consider here a typical refrigeration system that is used in one of Israeli supermarkets (Frenkel et al. 2009, 2010). The system consists of 2 identical subsystems: main and reserved. Each subsystem consists from 2 elements: the first element is block of 4 compressors, situated in the machine room and the second one is block of 2 axial condenser blowers. Structure scheme of the system is presented in the Fig. 7.2.

The way to growth availability of the system is to replace block of 2 axial condenser blowers on the block of 3 axial condenser blowers.

One should compute reliability indices for these two possible structures and make the decision—what structure is more appropriate.

7.3.1 System with 2 Condenser Blowers

Series–parallel refrigerating multi-state system with two blowers is presented in the Fig. 7.2. State-space diagram of the elements of this system is presented in the Fig. 7.3.

The performance of the elements is measured by their produced cold capacity (BTU per year). Times to failures and times to repairs are distributed exponentially for all elements. Elements are repairable. Minimal repair is only possible. All elements are multi-state elements with minor failures and minor repairs. The first and the second elements can be in one of five states: a state of total failure corresponding to a capacity of 0, states of partial failures corresponding to capacities of $2.6 \cdot 10^9$, $5.2 \cdot 10^9$, $7.9 \cdot 10^9$ BTU per year and a fully operational state with a capacity of $10.5 \cdot 10^9$ BTU per year. For simplification we will present system capacity in 10^9 BTU per year units. Therefore,

$$
\begin{aligned}
G_1(t) \in \{g_{11}, g_{12}, g_{13}, g_{14}, g_{15}\} &= \{0, 2.6, 5.2, 7.9, 10.5\}, \\
G_2(t) \in \{g_{21}, g_{22}, g_{23}, g_{24}, g_{25}\} &= \{0, 2.6, 5.2, 7.9, 10.5\}.
\end{aligned}
\tag{7.1}
$$

Fig. 7.2 Series–parallel refrigerating multi-state system with two blowers

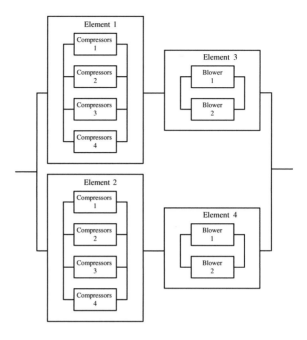

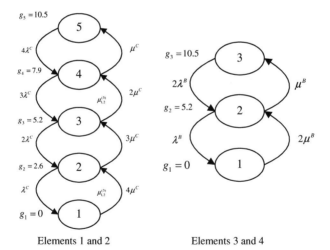

Elements 1 and 2   Elements 3 and 4

Fig. 7.3 State-space diagram of the multi-state system with two blowers

The failure rates and repair rates corresponding to the first element are $\lambda^C = 1\,\text{year}^{-1}$, $\mu^C = 12\,\text{year}^{-1}$.

The third and fourth elements can be in one of three states: a state of total failure corresponding to a capacity of 0, state of partial failure corresponding to capacity of $5.2 \cdot 10^9$ BTU per year and a fully operational state with a capacity of $10.5 \cdot 10^9$ BTU per year. Therefore,

$$
\begin{aligned}
G_3(t) &\in \{g_{31}, g_{32}, g_{33}\} = \{0, 5.2, 10.5\}, \\
G_4(t) &\in \{g_{41}, g_{42}, g_{43}\} = \{0, 5.2, 10.5\}.
\end{aligned}
\tag{7.2}
$$

The failure rate and repair rate corresponding to the second element are $\lambda^B = 10\,\text{year}^{-1}$, $\mu^B = 365\,\text{year}^{-1}$.

The MSS structure function is:

$$
\begin{aligned}
G_s(t) &= f(G_1(t), G_2(t), G_3(t), G_4(t)) \\
&= \max\{\min(G_1(t), G_3(t)), \min(G_2(t), G_4(t))\}.
\end{aligned}
\tag{7.3}
$$

The demand is constant: $w = 7.0 \cdot 10^9$ BTU per year.

Using combined UGF and stochastic process method (Lisnianski et al. 2010), we will find MSS availability $A(t, w)$, expected output performance $E(t)$ and expected performance deficiency $D(t, w)$.

Applying the procedure, described in Lisnianski et al. (2010), we proceed as follows.

According to the Markov method we build the following systems of differential equations for each element separately (using the state-space diagrams presented in Fig. 7.3):

- For elements 1 and 2 ($i = 1, 2$):

$$
\begin{cases}
\dfrac{dp_{i1}(t)}{dt} = -4\mu^C p_{i1}(t) + \lambda^C p_{i2}(t) \\[2mm]
\dfrac{dp_{i2}(t)}{dt} = 4\mu^C p_{i1}(t) - (\lambda^C + 3\mu^C) p_{i2}(t) + 2\lambda^C p_{i3}(t) \\[2mm]
\dfrac{dp_{i3}(t)}{dt} = 3\mu^C p_{i2}(t) - (2\lambda^C + 2\mu^C) p_{i3}(t) + 3\lambda^C p_{i4}(t) \\[2mm]
\dfrac{dp_{i4}(t)}{dt} = 2\mu^C p_{i3}(t) - (3\lambda^C + \mu^C) p_{i4}(t) + 4\lambda^C p_{i5}(t) \\[2mm]
\dfrac{dp_{i5}(t)}{dt} = \mu^C p_{i4}(t) - 4\lambda^C p_{i5}(t).
\end{cases}
\tag{7.4}
$$

Initial conditions are: $p_{i1}(0) = p_{i2}(0) = p_{i3}(0) = p_{i4}(0) = 0$; $p_{i5}(0) = 1$.

- For elements 3 and 4 ($i = 3, 4$):

$$
\begin{cases}
\dfrac{dp_{i1}(t)}{dt} = -2\mu^B p_{i1}(t) + \lambda^B p_{i2}(t), \\[2mm]
\dfrac{dp_{i2}(t)}{dt} = 2\mu^B p_{i1}(t) - (\lambda^B + \mu^B) p_{i2}(t) + 2\lambda^B p_{i3}(t), \\[2mm]
\dfrac{dp_{i3}(t)}{dt} = \mu^B p_{i2}(t) - 2\lambda^B p_{i3}(t).
\end{cases}
\tag{7.5}
$$

Initial conditions are: $p_{i1}(0) = p_{i2}(0) = 0$; $p_{i3}(0) = 1$.

A closed form solution can be obtained for each of these 4 systems of differential equations. All calculations were made using MATLAB®. Corresponding expressions for states probabilities are the following:

- For elements 1 and 2 ($i = 1, 2$):

$$p_{i1}(t) = 3.5 \cdot 10^{-5} - 1.4 \cdot 10^{-4}e^{-13t} + 2.1 \cdot 10^{-4}e^{-26t} - 1.4 \cdot 10^{-4}e^{-39t}$$
$$+ 3.5 \cdot 10^{-5}e^{-52t},$$
$$p_{i2}(t) = 1.68 \cdot 10^{-3} - 4.9 \cdot 10^{-3}e^{-13t} + 4.62 \cdot 10^{-3}e^{-26t} - 1.26 \cdot 10^{-3}e^{-39t}$$
$$+ 1.4 \cdot 10^{-4}e^{-52t},$$
$$p_{i3}(t) = 3.025 \cdot 10^{-2} - 5.546 \cdot 10^{-2}e^{-13t} + 2.038 \cdot 10^{-2}e^{-26t} - 4.62 \cdot 10^{-3}e^{-39t}$$
$$+ 2.1 \cdot 10^{-4}e^{-52t},$$
$$p_{i4}(t) = 0.242 - 0.182e^{-13t} + 0.055e^{-26t} - 4.9 \cdot 10^{-3}e^{-39t} + 1.4 \cdot 10^{-4}e^{-52t},$$
$$p_{i5}(t) = 0.762 - 0.242e^{-13t} + 0.030e^{-26t} - 1.68 \cdot 10^{-3}e^{-39t} + 3.5 \cdot 10^{-5}e^{-52t}.$$

$$(7.6)$$

- For elements 3 and 4 ($i = 3, 4$):

$$p_{i1}(t) = 7.111 \cdot 10^{-4} + 7.111 \cdot 10^{-4}e^{-750t} - 1.422 \cdot 10^{-3}e^{-375t},$$
$$p_{i2}(t) = 0.0519 - 1.422 \cdot 10^{-3}e^{-750t} - 0.0505e^{-375t}, \qquad (7.7)$$
$$p_{i3}(t) = 0.9474 + 7.111 \cdot 10^{-4}e^{-750t} + 0.0519e^{-375t}.$$

Therefore, one obtains the following output performance stochastic processes:

- element 1 : $\begin{cases} \mathbf{g}_1 = \{g_{11}, g_{12}, g_{13}, g_{14}, g_{15}\} = \{0, 2.6, 5.2, 7.9, 10.5\}, \\ \mathbf{p}_1(t) = \{p_{11}(t), p_{12}(t), p_{13}(t), p_{14}(t), p_{15}(t)\}; \end{cases}$

- element 2 : $\begin{cases} \mathbf{g}_2 = \{g_{21}, g_{22}, g_{23}, g_{24}, g_{25}\} = \{0, 2.6, 5.2, 7.9, 10.5\}, \\ \mathbf{p}_2(t) = \{p_{21}(t), p_{22}(t), p_{23}(t), p_{24}(t), p_{25}(t)\}; \end{cases}$

- element 3 : $\begin{cases} \mathbf{g}_3 = \{g_{31}, g_{32}, g_{33}\} = \{0, 5.2, 10.5\}, \\ \mathbf{p}_3(t) = \{p_{31}(t), p_{32}(t), p_{33}(t)\}; \end{cases}$

- element 4 : $\begin{cases} \mathbf{g}_4 = \{g_{41}, g_{42}, g_{43}\} = \{0, 5.2, 10.5\}, \\ \mathbf{p}_4(t) = \{p_{41}(t), p_{42}(t), p_{43}(t)\}. \end{cases}$

Having the sets $\mathbf{g}_j$, $\mathbf{p}_j(t)$ for $j = 1, 2, 3, 4$ one can define for each individual element j the u-function associated with the element's output performance stochastic process:

$$u_1(z,t) = p_{11}(t)z^{g_{11}} + p_{12}(t)z^{g_{12}} + p_{13}(t)z^{g_{13}} + p_{14}(t)z^{g_{14}} + p_{15}(t)z^{g_{15}}$$
$$= p_{11}(t)z^0 + p_{12}(t)z^{2.6} + p_{13}(t)z^{5.2} + p_{14}(t)z^{7.9} + p_{15}(t)z^{10.5},$$
$$u_2(z,t) = p_{21}(t)z^{g_{21}} + p_{22}(t)z^{g_{22}} + p_{23}(t)z^{g_{23}} + p_{24}(t)z^{g_{24}} + p_{25}(t)z^{g_{25}}$$
$$= p_{21}(t)z^0 + p_{22}(t)z^{2.6} + p_{23}(t)z^{5.2} + p_{24}(t)z^{7.9} + p_{25}(t)z^{10.5},$$
$$u_3(z,t) = p_{31}(t)z^{g_{31}} + p_{32}(t)z^{g_{32}} + p_{33}(t)z^{g_{33}}$$
$$= p_{31}(t)z^0 + p_{32}(t)z^{5.2} + p_{33}(t)z^{10.5},$$
$$u_4(z,t) = p_{41}(t)z^{g_{41}} + p_{42}(t)z^{g_{42}} + p_{43}(t)z^{g_{43}}$$
$$= p_{41}(t)z^0 + p_{42}(t)z^{5.2} + p_{43}(t)z^{10.5}.$$

$$(7.8)$$

Using the composition operator $\Omega_{f_{ser}}$ for pairs of MSS elements 1 and 3, and 2 and 4, connected in series, one obtains the UGFs for the series subsystems:

$$U_1(z,t) = \Omega_{f_{ser}}(u_1(z,t), u_3(z,t))$$
$$= \Omega_{f_{ser}}\big(p_{11}(t)z^0 + p_{12}(t)z^{2.6} + p_{13}(t)z^{5.2} + p_{14}(t)z^{7.9} + p_{15}(t)z^{10.5},$$
$$p_{31}(t)z^0 + p_{32}(t)z^{5.2} + p_{33}(t)z^{10.5}\big)$$
$$= p_{11}(t)p_{31}(t)z^0 + p_{11}(t)p_{32}(t)z^0 + p_{11}(t)p_{33}(t)z^0$$
$$+ p_{12}(t)p_{31}(t)z^0 + p_{12}(t)p_{32}(t)z^{2.6} + p_{12}(t)p_{33}(t)z^{2.6}$$
$$+ p_{13}(t)p_{31}(t)z^0 + p_{13}(t)p_{32}(t)z^{5.2} + p_{13}(t)p_{33}(t)z^{5.2}$$
$$+ p_{14}(t)p_{31}(t)z^0 + p_{14}(t)p_{32}(t)z^{5.2} + p_{14}(t)p_{33}(t)z^{7.9}$$
$$+ p_{15}(t)p_{31}(t)z^0 + p_{15}(t)p_{32}(t)z^{5.2} + p_{15}(t)p_{33}(t)z^{10.5}.$$

$$(7.9)$$

$$U_2(z,t) = \Omega_{f_{ser}}(u_2(z,t), u_4(z,t))$$
$$= \Omega_{f_{ser}}\big(p_{21}(t)z^0 + p_{22}(t)z^{2.6} + p_{23}(t)z^{5.2} + p_{24}(t)z^{7.9} + p_{25}(t)z^{10.5},$$
$$p_{41}(t)z^0 + p_{42}(t)z^{5.2} + p_{43}(t)z^{10.5}\big)$$
$$= p_{21}(t)p_{41}(t)z^0 + p_{21}(t)p_{42}(t)z^0 + p_{21}(t)p_{43}(t)z^0$$
$$+ p_{22}(t)p_{41}(t)z^0 + p_{22}(t)p_{42}(t)z^{2.6} + p_{22}(t)p_{43}(t)z^{2.6}$$
$$+ p_{23}(t)p_{41}(t)z^0 + p_{23}(t)p_{42}(t)z^{5.2} + p_{23}(t)p_{43}(t)z^{5.2}$$
$$+ p_{24}(t)p_{41}(t)z^0 + p_{24}(t)p_{42}(t)z^{5.2} + p_{24}(t)p_{43}(t)z^{7.9}$$
$$+ p_{25}(t)p_{41}(t)z^0 + p_{25}(t)p_{42}(t)z^{5.2} + p_{25}(t)p_{43}(t)z^{10.5}.$$

$$(7.10)$$

In the resulting UGFs $U_1(z, t)$ and $U_2(z, t)$ the powers of z are found as minimum of powers of corresponding terms.

Taking into account that

$$p_{11}(t) + p_{12}(t) + p_{13}(t) + p_{14}(t) + p_{15}(t) = 1,$$
$$p_{21}(t) + p_{22}(t) + p_{23}(t) + p_{24}(t) + p_{25}(t) = 1,$$
$$p_{31}(t) + p_{32}(t) + p_{33}(t) = 1 \text{ and } p_{41}(t) + p_{42}(t) + p_{43}(t) = 1,$$

one can simplify the expressions for $U_1(z,t)$ and $U_2(z,t)$, and obtain the resulting UGFs in the following form

$$
\begin{aligned}
U_1(z,t) =& [p_{11}(t) + (1 - p_{11}(t))p_{31}(t)]z^0 + p_{12}(t)(p_{32}(t) + p_{33}(t))z^{2.6} \\
& + [(p_{13}(t) + p_{14}(t) + p_{15}(t))p_{32}(t) + p_{13}(t)p_{33}(t)]z^{5.2} \\
& + p_{14}(t)p_{33}(t)z^{7.9} + p_{15}(t)p_{33}(t)z^{10.5}.
\end{aligned}
\tag{7.11}
$$

$$
\begin{aligned}
U_2(z,t) =& [p_{21}(t) + (1 - p_{21}(t))p_{41}(t)]z^0 + p_{22}(t)(p_{42}(t) + p_{43}(t))z^{2.6} \\
& + [(p_{23}(t) + p_{24}(t) + p_{25}(t))p_{42}(t) + p_{23}(t)p_{43}(t)]z^{5.2} \\
& + p_{24}(t)p_{43}(t)z^{7.9} + p_{25}(t)p_{43}(t)z^{10.5}.
\end{aligned}
\tag{7.12}
$$

Using the composition operator $\Omega_{f_{par}}$ for MSS subsystems 1 and 2, one obtains the UGF for the entire MSS associated with the output performance stochastic process $\mathbf{g}$, $\mathbf{p}(t)$ in the following manner:

$$
\begin{aligned}
U(z,t) =& \Omega_{f_{par}}(U_1(z,t), U_2(z,t)) \\
=& \Omega_{f_{par}}\big(\pi_{11}(t)z^0 + \pi_{12}(t)z^{2.6} + \pi_{13}(t)z^{5.2} + \pi_{14}(t)z^{7.9} + \pi_{15}(t)z^{10.5}, \\
& \pi_{21}(t)z^0 + \pi_{22}(t)z^{2.6} + \pi_{23}(t)z^{5.2} + \pi_{24}(t)z^{7.9} + \pi_{25}(t)z^{10.5}\big) \\
=& \pi_{11}(t)\pi_{21}(t)z^0 + (2\pi_{11}(t)\pi_{22}(t) + \pi_{12}(t)\pi_{22}(t))z^{2.6} \\
& + (2\pi_{11}(t)\pi_{23}(t) + 2\pi_{12}(t)\pi_{23}(t) + \pi_{13}(t)\pi_{23}(t))z^{5.2} \\
& + (2\pi_{11}(t)\pi_{24}(t) + 2\pi_{12}(t)\pi_{24}(t) + 2\pi_{13}(t)w_{24}(t) + \pi_{14}(t)\pi_{24}(t))z^{7.9} \\
& + (2\pi_{11}(t)\pi_{25}(t) + 2\pi_{12}(t)\pi_{25}(t) + 2\pi_{13}(t)\pi_{25}(t) \\
& + 2\pi_{14}(t)\pi_{25}(t) + \pi_{15}(t)\pi_{25}(t)z^{10.5}.
\end{aligned}
\tag{7.13}
$$

where

$$
\begin{aligned}
\pi_{11}(t) =& [p_{11}(t) + (1 - p_{11}(t))p_{31}(t)], \quad \pi_{12}(t) = p_{12}(t)(p_{32}(t) + p_{33}(t)), \\
\pi_{13}(t) =& [(p_{13}(t) + p_{14}(t) + p_{15}(t))p_{32}(t) + p_{13}(t)p_{33}(t)], \\
\pi_{14}(t) =& p_{14}(t)p_{33}(t), \quad \pi_{15}(t) = p_{15}(t)p_{33}(t), \\
\pi_{21}(t) =& [p_{21}(t) + (1 - p_{21}(t))p_{41}(t)], \quad \pi_{22}(t) = p_{22}(t)(p_{42}(t) + p_{43}(t)), \\
\pi_{23}(t) =& [(p_{23}(t) + p_{24}(t) + p_{25}(t))p_{42}(t) + p_{23}(t)p_{43}(t)], \\
\pi_{24}(t) =& p_{24}(t)p_{43}(t), \quad \pi_{25}(t) = p_{25}(t)p_{43}(t).
\end{aligned}
$$

And finally

$$
U(z,t) = \sum_{i=1}^{5} p_i(t)z^{g_i}
\tag{7.14}
$$

where

$g_1 = 0,$ $p_1(t) = \pi_{11}(t)\pi_{21}(t),$

$g_2 = 2.6 \cdot 10^9$ BTU/year, $p_2(t) = (2\pi_{11}(t)\pi_{22}(t) + \pi_{12}(t)\pi_{22}(t)),$

$g_3 = 5.2 \cdot 10^9$ BTU/year, $p_3(t) = (2\pi_{11}(t)\pi_{23}(t) + 2\pi_{12}(t)\pi_{23}(t) + \pi_{13}(t)\pi_{23}(t)),$

$g_4 = 7.9 \cdot 10^9$ BTU/year, $p_4(t) = (2\pi_{11}(t)\pi_{24}(t) + 2\pi_{12}(t)\pi_{24}(t) + 2\pi_{13}(t)\pi_{24}(t)$
$$+\pi_{14}(t)\pi_{24}(t)),$$

$g_5 = 10.5 \cdot 10^9$ BTU/year, $p_5(t) = (2\pi_{11}(t)\pi_{25}(t) + 2\pi_{12}(t)\pi_{25}(t) + 2\pi_{13}(t)\pi_{25}(t)$
$$+2\pi_{14}(t)\pi_{25}(t) + \pi_{15}(t)\pi_{25}(t)).$$

These two sets

$\mathbf{g} = \{g_1, g_2, g_3, g_4, g_5\}$ and $\mathbf{p}(t) = \{p_1(t), p_2(t), p_3(t), p_4(t), p_5(t)\}$
completely define output performance stochastic process for the entire MSS.

Based on resulting UGF $U(z,t)$ of the entire MSS, one can obtain the MSS reliability indices. The instantaneous MSS availability for the constant demand level $w=7.0 \cdot 10^9$ BTU per year

$$A(t) = \delta_A(U(z,t), w) = \delta_A(\sum_{i=1}^{5} p_i(t)z^{g_i}, 7)$$

$$= \sum_{i=1}^{5} p_i(t)1(F(g_i, 7) \geq 0) = p_4(t) + p_5(t). \tag{7.15}$$

The instantaneous mean output performance at any instant $t>0$

$$E(t) = \delta_E(U(z,t)) = \sum_{i=1}^{5} p_i(t)g_i = 2.6p_2(t) + 5.2p_3(t) + 7.9p_4(t) + 10.5p_5(t).$$
$$\tag{7.16}$$

The instantaneous performance deficiency $D(t)$ at any time t for the constant demand $w = 7.0 \cdot 10^9$ BTU per year:

$$D(t) = \delta_D(U(z), w) = \sum_{i=1}^{5} p_i(t) \cdot \max(7 - g_i, 0)$$
$$= p_1(t)(7 - 0) + p_2(t)(7 - 2.6) + p_3(t)(7 - 5.2) \tag{7.17}$$
$$= 7p_1(t) + 4.4p_2(t) + 1.8p_3(t).$$

Calculated reliability indices $A(t)$, $E(t)$ and $D(t)$ are presented on the Figs. 7.5–7.7.

Note that instead of solving the system of $K=5*3=15$ differential equations (as it should be done in the straightforward Markov method) here we solve just two systems. The derivation of the entire system states probabilities and reliability indices is based on using simple algebraic equations.

7.3.2 System with 3 Condenser Blowers

To increase the reliability level of the refrigeration system the supermarket's administration decided to replace the block with 2 axial condenser blowers by a block with 3 axial condenser blowers.

Our goal is to compare reliability indices for both these structures.

As in the previous case the analyzed system consists of two subsystems: the first subsystem consists of elements 1 and 2 (blocks with 4 compressors) and the second—of elements 3 and 4 (blocks with 3 blowers).

Note that in this case the third and the fourth elements can be in one of 4 states: a state of total failure corresponding to a capacity of 0, state of partial failure corresponding to capacity of $5.2 \cdot 10^9$ BTU per year and two fully operational states with a capacity of $10.5 \cdot 10^9$ BTU per year. State-space diagram of the block of 3 blowers (elements 3 and 4) of this system is presented in Fig. 7.4.

Hence,

$$G_3(t) \in \{g_{31}, g_{32}, g_{33}, g_{34}\} = \{0, 5.2, 10.5, 10.5\},$$
$$G_4(t) \in \{g_{41}, g_{42}, g_{43}, g_{44}\} = \{0, 5.2, 10.5, 10.5\}. \tag{7.18}$$

The MSS structure function as in previous case is:

$$G_s(t) = f(G_1(t), G_2(t), G_3(t), G_4(t)) = $$
$$\max\{\min(G_1(t), G_3(t)), \min(G_2(t), G_4(t))\}. \tag{7.19}$$

with $G_1(t)$ and $G_2(t)$ given by (7.1). Applying again the Lisnianski et al. (2010) procedure, we obtain for elements 3 and 4 ($i = 3, 4$) the system of differential equations as follows:

$$\begin{cases} \dfrac{dp_{i1}(t)}{dt} = -3\mu^B p_{i1}(t) + \lambda^B p_{i2}(t), \\[2mm] \dfrac{dp_{i2}(t)}{dt} = 3\mu^B p_{i1}(t) - (\lambda^B + 2\mu^B)p_{i2}(t) + 2\lambda^B p_{i3}(t), \\[2mm] \dfrac{dp_{i3}(t)}{dt} = 2\mu^B p_{i3}(t) - (2\lambda^B + \mu^B)p_{i3}(t) + 3\lambda^B p_{i3}(t), \\[2mm] \dfrac{dp_{i4}(t)}{dt} = \mu^B p_{i3}(t) - 3\lambda^B p_{i4}(t). \end{cases} \tag{7.20}$$

with initial conditions $p_{i1}(0) = p_{i2}(0) = p_{i3}(0) = 0$; $p_{i4}(0) = 1$ and states probabilities ($i = 3, 4$) given by

$$p_{i1}(t) = 1.896 \cdot 10^{-5} + 5.689 \cdot 10^{-5} e^{-750t} - 5.689 \cdot 10^{-5} e^{-375t} - 1.896 \cdot 10^{-5} e^{-1125t},$$

$$p_{i2}(t) = 2.076 \cdot 10^{-4} + 1.963 \cdot 10^{-3} e^{-750t} - 4.096 \cdot 10^{-3} e^{-375t} - 5.689 \cdot 10^{-5} e^{-1125t},$$

$$p_{i3}(t) = 0.076 - 4.096 \cdot 10^{-3} e^{-750t} - 0.0716 e^{-375t} - 5.689 \cdot 10^{-5} e^{-1125t},$$

$$p_{i4}(t) = 0.922 + 2.076 \cdot 10^{-4} e^{-750t} + 0.0758 e^{-375t} + 1.896 \cdot 10^{-5} e^{-1125t}.$$

$$\tag{7.21}$$

Fig. 7.4 State-space diagram
of the block of 3 blowers

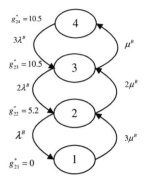

The output performance stochastic processes for elements 1 and 2 remain the same as in the previous case while the ones for elements 3 and 4 are:

- element3 : $\begin{cases} \mathbf{g}_3 = \{g_{31}, g_{32}, g_{33}, g_{34}\} = \{0, 5.2, 10.5, 10.5\}, \\ \mathbf{p}_3(t) = \{p_{31}(t), p_{32}(t), p_{33}(t), p_{34}(t)\}; \end{cases}$

- element4 : $\begin{cases} \mathbf{g}_4 = \{g_{41}, g_{42}, g_{43}, g_{44}\} = \{0, 5.2, 10.5, 10.5\}, \\ \mathbf{p}_4(t) = \{p_{41}(t), p_{42}(t), p_{43}(t), p_{44}(t)\}. \end{cases}$

Then the u-functions associated with the element's output performance stochastic process are:

$$
\begin{aligned}
u_1(z,t) &= p_{11}(t)z^{g_{11}} + p_{12}(t)z^{g_{12}} + p_{13}(t)z^{g_{13}} + p_{14}(t)z^{g_{14}} + p_{15}(t)z^{g_{15}} \\
&= p_{11}(t)z^0 + p_{12}(t)z^{2.6} + p_{13}(t)z^{5.2} + p_{14}(t)z^{7.9} + p_{15}(t)z^{10.5}, \\
u_2(z,t) &= p_{21}(t)z^{g_{21}} + p_{22}(t)z^{g_{22}} + p_{23}(t)z^{g_{23}} + p_{24}(t)z^{g_{24}} + p_{25}(t)z^{g_{25}} \\
&= p_{21}(t)z^0 + p_{22}(t)z^{2.6} + p_{23}(t)z^{5.2} + p_{24}(t)z^{7.9} + p_{25}(t)z^{10.5}, \\
u_3(z,t) &= p_{31}(t)z^{g_{31}} + p_{32}(t)z^{g_{32}} + p_{33}(t)z^{g_{33}} + p_{34}(t)z^{g_{34}} \\
&= p_{31}(t)z^0 + p_{32}(t)z^{5.2} + p_{33}(t)z^{10.5} + p_{34}(t)z^{10.5}, \\
u_4(z,t) &= p_{41}(t)z^{g_{41}} + p_{42}(t)z^{g_{42}} + p_{43}(t)z^{g_{43}} + p_{44}(t)z^{g_{43}} \\
&= p_{41}(t)z^0 + p_{42}(t)z^{5.2} + p_{43}(t)z^{10.5} + p_{44}(t)z^{10.5}.
\end{aligned}
\tag{7.22}
$$

Using again the composition operator $\Omega_{f_{ser}}$ for MSS elements 1 and 3, and 2 and 4, connected in series, one obtains the UGFs for the series subsystems:

$$U_1(z,t) = \Omega_{f_{ser}}(u_1(z,t), u_3(z,t))$$
$$= \Omega_{f_{ser}}\big(p_{11}(t)z^0 + p_{12}(t)z^{2.6} + p_{13}(t)z^{5.2} + p_{14}(t)z^{7.9} + p_{15}(t)z^{10.5},$$
$$p_{31}(t)z^0 + p_{32}(t)z^{5.2} + (p_{33}(t) + p_{34}(t))z^{10.5}\big)$$
$$= [p_{11}(t) + (1 - p_{11}(t))p_{31}(t)]z^0 + p_{12}(t)[1 - p_{32}(t)]z^{2.6} +$$
$$+ [p_{13}(t)(1 - p_{31}(t)) + (p_{14}(t) + p_{15}(t))p_{32}(t)]z^{5.2} +$$
$$+ p_{14}(t)[p_{33}(t) + p_{34}(t)]z^{7.9} + p_{15}(t)[p_{33}(t) + p_{34}(t)]z^{10.5}. \tag{7.23}$$

$$U_2(z,t) = \Omega_{f_{ser}}(u_2(z,t), u_4(z,t))$$
$$= \Omega_{f_{ser}}\big(p_{21}(t)z^0 + p_{22}(t)z^{2.6} + p_{23}(t)z^{5.2} + p_{24}(t)z^{7.9} + p_{25}(t)z^{10.5},$$
$$p_{41}(t)z^0 + p_{42}(t)z^{5.2} + (p_{43}(t) + p_{44}(t))z^{10.5}\big)$$
$$= [p_{21}(t) + (1 - p_{21}(t))p_{41}(t)]z^0 + p_{22}(t)[1 - p_{42}(t)]z^{2.6} +$$
$$+ [p_{23}(t)(1 - p_{41}(t)) + (p_{24}(t) + p_{25}(t))p_{42}(t)]z^{5.2} +$$
$$+ p_{24}(t)[p_{43}(t) + p_{44}(t)]z^{7.9} + p_{25}(t)[p_{43}(t) + p_{44}(t)]z^{10.5}. \tag{7.24}$$

Using the composition operator $\Omega_{f_{par}}$ for MSS subsystems 1 and 2, one obtains the UGF for the entire MSS associated with the output performance stochastic process $\mathbf{g}, \mathbf{p}(t)$ in the following manner:

$$U(z,t) = \Omega_{f_{par}}(U_1(z,t), U_2(z,t))$$
$$= \Omega_{f_{par}}\big(\pi_{11}(t)z^0 + \pi_{12}(t)z^{2.6} + \pi_{13}(t)z^{5.2} + \pi_{14}(t)z^{7.9} + \pi_{15}(t)z^{10.5},$$
$$\pi_{21}(t)z^0 + \pi_{22}(t)z^{2.6} + \pi_{23}(t)z^{5.2} + \pi_{24}(t)z^{7.9} + \pi_{25}(t)z^{10.5}\big)$$
$$= \pi_{11}(t)\pi_{21}(t)z^0 + (2\pi_{11}(t)\pi_{22}(t) + \pi_{12}(t)\pi_{22}(t))z^{2.6}$$
$$+ (2\pi_{11}(t)\pi_{23}(t) + 2\pi_{12}(t)\pi_{23}(t) + \pi_{13}(t)\pi_{23}(t))z^{5.2}$$
$$+ (2\pi_{11}(t)\pi_{24}(t) + 2\pi_{12}(t)\pi_{24}(t) + 2\pi_{13}(t)\pi_{24}(t) + \pi_{14}(t)\pi_{24}(t))z^{7.9}$$
$$+ (2\pi_{11}(t)\pi_{25}(t) + 2\pi_{12}(t)\pi_{25}(t) + 2\pi_{13}(t)\pi_{25}(t)$$
$$+ 2\pi_{14}(t)\pi_{25}(t) + \pi_{15}(t)\pi_{25}(t))z^{10.5}. \tag{7.25}$$

where

$$\pi_{11}(t) = [p_{11}(t) + (1 - p_{11}(t))p_{31}(t)], \quad \pi_{12}(t) = p_{12}(t)(1 - p_{32}(t)),$$
$$\pi_{13}(t) = [p_{13}(t)(1 - p_{31}(t))] + [(p_{14}(t) + p_{15}(t))p_{32}(t)],$$
$$\pi_{14}(t) = p_{14}(t)(p_{33}(t) + p_{34}(t)), \quad \pi_{15}(t) = p_{15}(t)(p_{33}(t) + p_{34}(t)),$$
$$\pi_{21}(t) = [p_{21}(t) + (1 - p_{21}(t))p_{41}(t)], \quad \pi_{22}(t) = +p_{22}(t)(1 - p_{42}(t)),$$
$$\pi_{23}(t) = [p_{23}(t)(1 - p_{41}(t))] + [(p_{24}(t) + p_{25}(t))p_{42}(t)],$$
$$\pi_{24}(t) = p_{24}(t)(p_{43}(t) + p_{44}(t)), \quad \pi_{25}(t) = p_{25}(t)(p_{43}(t) + p_{44}(t)).$$

Fig. 7.5 MSS instantaneous availability for different types of systems

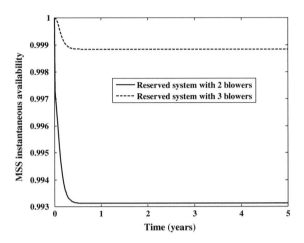

Finally the system UGF is as follows:

$$U(z,t) = \sum_{i=1}^{5} p_i(t)z^{g_i}, \qquad (7.26)$$

where

$$g_1 = 0, \qquad p_1(t) = \pi_{11}(t)\pi_{21}(t),$$
$$g_2 = 2.6 \cdot 10^9 \text{ BTU/year}, \qquad p_2(t) = (2\pi_{11}(t)\pi_{22}(t) + \pi_{12}(t)\pi_{22}(t)),$$
$$g_3 = 5.2 \cdot 10^9 \text{ BTU/year}, \qquad p_3(t) = (2\pi_{11}(t)\pi_{23}(t) + 2\pi_{12}(t)\pi_{23}(t) + \pi_{13}(t)\pi_{23}(t)),$$
$$g_4 = 7.9 \cdot 10^9 \text{ BTU/year}, \qquad p_4(t) = (2\pi_{11}(t)\pi_{24}(t) + 2\pi_{12}(t)\pi_{24}(t) + 2\pi_{13}(t)\pi_{24}(t)$$
$$\qquad\qquad\qquad + \pi_{14}(t)\pi_{24}(t)),$$
$$g_5 = 10.5 \cdot 10^9 \text{ BTU/year}, \qquad p_5(t) = (2\pi_{11}(t)\pi_{25}(t) + 2\pi_{12}(t)\pi_{25}(t) + 2\pi_{13}(t)\pi_{25}(t)$$
$$\qquad\qquad\qquad + 2\pi_{14}(t)\pi_{25}(t) + \pi_{15}(t)\pi_{25}(t)).$$

These two sets $\mathbf{g} = \{g_1, g_2, \ldots, g_5\}$ and $\mathbf{p}(t) = \{p_1(t), p_2(t), \ldots, p_5(t)\}$ completely define output performance stochastic process for the entire MSS. Using the above formulas the MSS reliability indices $A(t)$, $E(t)$ and $D(t)$ are obtained again by Equations (7.15–7.17).

Calculated reliability indices $A(t)$, $E(t)$ and $D(t)$ for both systems are presented on Figs. 7.5–7.7. Observe that the curves in these figures support the engineering decision-making and determine the areas where required performance deficiency level of the refrigeration system can be provided by configuration "Reserved system with 2 blowers" or "Reserved system with 3 blowers". For example, from Fig. 7.5 one can conclude that the configuration "Reserved system with 2 blowers" cannot provide the required average availability, if it must be greater than 0.995. Figure 7.6 shows that after half a year capacity of the system with 3 blowers block little greater than capacity of the system with 2 blowers block and the difference is

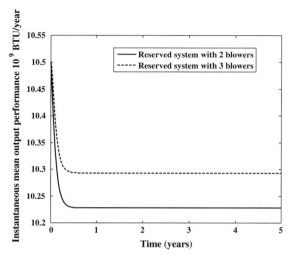

Fig. 7.6 MSS instantaneous mean output performance for different types of systems

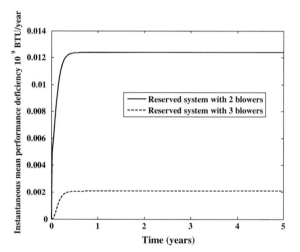

Fig. 7.7 MSS instantaneous mean performance deficiency for different types of systems

0.68%. From the Fig. 7.7 one can conclude that the system with 2 blowers block cannot provide the required $0.01 \cdot 10^9$ BTU per year level of performance deficiency.

7.4 Conclusion

The combined stochastic process and UGF method was applied to compute reliability measures for supermarket refrigerating system: system availability, output performance and performance deficiency.

It was demonstrated that the combined method is well formalized and suitable for practical application in reliability engineering for real-world MSSs analysis. It supports the engineering decision-making and determines different system structures providing a required reliability/availability level for the multi-state system.

In the considered case study it was shown that configuration "Reserved system with 2 blowers" cannot provide the required availability, if it must be greater than 0.995. In addition the system with 2 blowers block does not provide the required $0.01 \cdot 10^9$ BTU per year level of performance deficiency.

Acknowledgments This work was partially supported by the Internal Funding Program of the Shamoon College of Engineering (SCE).

References

Frenkel I, Khvatskin L, Lisnianski A (2009) Markov reward model for performance deficiency calculation of refrigeration system. In: Bris R, Soares CG, Martorell S (eds) Reliability, risk and safety: theory and applications. CRC Press, Taylor & Francis Group, London, pp 1591–1596

Frenkel I, Khvatskin L, Lisnianski A (2010) Management decision making based on Markov reward models for refrigeration system. J Pol Saf Reliab Assoc 1:89–98

Lisnianski A (2004) Universal generating function technique and random process methods for multi-state system reliability analysis. In: Proceedings of the 2nd international workshop in applied probability (IWAP2004). Piraeus, Greece, pp 237–242

Lisnianski A (2007) Extended block diagram method for a multi-state system reliability assessment. Reliab Eng Syst Saf 92(12):1601–1607

Lisnianski A, Levitin G (2003) Multi-state system reliability. Assessment, Optimization, Applications. World Scientific, Singapore

Lisnianski A, Frenkel I, Ding Y (2010) Multi-state system reliability analysis and optimization for engineers and industrial managers. Springer, London

Lisniansky A, Frenkel I, Khvatskin L, Ding Y (2007) Markov reward model for multi-state system reliability assessment. In: Vonta F, Nikulin M, Limnios N, Huber-Carol C (eds) Statistical models and methods for biomedical and technical systems. Birkhauser, Boston, pp 153–168

Chapter 8
Importance Analysis of a Multi-State System Based on Multiple-Valued Logic Methods

Elena Zaitseva

Abstract Importance analysis allows to identify vulnerabilities within a system and to quantify criticality (importance) of system components. Importance measures estimate peculiarities of the particular system component influence on a system. New methodology based on logical differential calculus for importance analysis of a multi-state system is discussed. Algorithms for calculation of multi-state system importance measures (IM) are proposed. These algorithms allow to compute traditional IMs as Birnbaum and Fussell-Vesely importance, reliability achievement worth, reliability reduction worth and a new type of IM as dynamic reliability indices.

8.1 Introduction

A detailed analysis of the current situation in Reliability Engineering was presented in paper (Zio 2009). Zio shows that a complex system with a multifarious component and an involved connection is the principal investigated object in reliability engineering. Reliability analysis of such system implies investigation of causes of system failure, estimation of reliable system development and quantification of system reliability in design, operation and management. Traditional problems, such as system representation and modelling, or quantification of this model require new solutions. According to Zio (2009), application of the *multi-state system* (MSS) is one of the possible ways to solve these problems.

E. Zaitseva (✉)
Department of Informatics, University of Žilina, Žilina, Slovakia
e-mail: elena.zaitseva@fri.uniza.sk

A. Lisnianski and I. Frenkel (eds.), *Recent Advances in System Reliability*,
Springer Series in Reliability Engineering, DOI: 10.1007/978-1-4471-2207-4_8,
© Springer-Verlag London Limited 2012

MSS is a mathematical model in reliability analysis that represents a system with some (more than two) levels of performance (availability, reliability) (Zio 2009; Lisnianski and Levitin 2003). MSS makes it possible to present an analysable system in more detail than in a traditional Binary-State System with two possible states defined as working and failing. MSSs have been used in manufacturing, production, water distribution, power generation and gas and oil transportation (Natvig et al. 2009; Pham 2003; Zio et al. 2007). An MSS reliability analysis is a complex problem that includes different directions how to estimate MSS behaviour. Importance analysis is one of them (Meng 1993; Levitin et al. 2003; Ramirez-Marquez and Coit 2005).

These methods allow to examine different aspects of changes in MSS performance level and MSS failure caused by changes in component states. Importance analysis is used in particular for MSS estimation depending on the system structure and its component states. Various evaluations of MSS component importance are called *importance measure* (IM). IM quantifies criticality of a particular component within an MSS. It has been widely used as a tool to identify system weaknesses and to prioritise activities improving reliability.

Theoretical aspects of MSS importance analysis have been extensively investigated. Different methods and algorithms are discussed in many papers. Authors such as Levitin and Lisnianski 1999; Levitin et al. 2003 discuss a basic IM for a system with two performance levels and multi-state components including their definitions by an output performance measure. Universal generating function methods are the principal approach for calculation in (Levitin et al. 2003). zRamirez-Marquez and Coit generalized this result for an MSS and proposed a new type of IM, referred to as composite importance measures (Ramirez-Marquez and Coit 2005). Markov process was used in (Van et al. 2008) for importance analysis both for a binary-state system and MSS. New methods based on logical differential calculus for MSS importance analysis are discussed in Zaitseva and Levashenko (2006), Zaitseva (2010) and a new type of IM has been proposed. These measures are called *dynamic reliability indices* (DRIs). Logical differential calculus is one of the mathematical approaches in *multiple-valued logic* (MVL). Another mathematical approach using MVL for MSS importance analysis was described in Zaitseva and Puuronen (2009). It is an MSS analysis based on a *multi-valued decision diagram* (MDD). Use of MDDs in reliability analysis is a development of *binary decision diagram* (BDD) application in analysing both MSS and BSS (Andrews and Dunnett 2000). BDD is an efficient method to manipulate the Boolean expression and, in most cases, BDDs use less memory to represent large Boolean expressions than representing them explicitly (Andrews and Dunnett 2000; Chang et al. 2004). The BDD method is widely used in reliability analysis because a system is either in the operational state or in the fail state and for fault tree in the first place BDD. There are some papers on BDDs adaptations for MSS reliability analysis (Chang et al. 2004; Zang et al. 2003). MDD is a natural extension of BDD to the multi-state case (Zaitseva and Puuronen 2009; Xing and Dai 2009). However, methods of reliability analysis in Zaitseva and Puuronen (2009) have been defined for a MSS with equal performance levels for all system components, whose structure function can be interpreted as a Multiple-Valued Logic function (see Appendix A). A detailed example of system reliability estimation by MDD in paper (Xing and Dai 2009) is discussed for a system with two performance levels and multi-state elements. A new methodology for MSS importance analysis based on Logical Differential Calculus and MDD is proposed below. In this paper, Logical Differential Calculus is used in importance analysis that includes calculation of IMs, such as Structural importance, Criticality importance, Birnbaum importance, Fussell-Vesely importance, reliability achievement worth, reliability reduction worth and DRIs.

8.2 Basic Concepts

8.2.1 MSS Structure Function

MSS is a mathematical model under consideration assuming that a system consists of n components and has M levels of performance rate from complete failure (this level corresponds to 0) to perfect functioning (this level is interpreted as $M - 1$. Each of the n components in the MSS is characterized by a different performance level. The ith component has m_i possible states: from complete failure (it is 0) to the perfect functioning (it is $m_i - 1$). The vector $\mathbf{x} = (x_1, \ldots, x_n)$ is a vector of MSS component states and the vector $\mathbf{m} = (m_1, \ldots, m_n)$ defines a range of change in components. The ith component state x_i in a given time t is characterized by the probability of performance rate (component state probability):

$$p_{i,s_i} = \Pr\{x_i = s_i\}, \quad s_i = 0, \ldots, m_i - 1. \tag{8.1}$$

Correlation between the MSS performance level and the system component states is defined by the *structure function* (Zaitseva and Levashenko 2006):

$$\begin{aligned}
\phi(x_1, \ldots, x_n) = \phi(\mathbf{x}) : & \{0, \ldots, m_1 - 1\} \times \cdots \times \{0, \ldots, m_n - 1\} \\
& \rightarrow \{0, \ldots, M - 1\}.
\end{aligned} \tag{8.2}$$

In this paper, a coherent MSS is considered (Lisnianski and Levitin 2003; Meng 1993) and the following assumptions are used for structure functions (8.2):

a) the structure function is monotone;
b) all the components are s-independent and relevant to the system;
c) performance levels of the system and its components show a gradual change: from level a to level $a - 1$ or $(a + 1)$ only.

For example, consider a MSS (Fig. 8.1) with three components ($n = 3$ and $M = 3$, $(m_1, m_2, m_3) = (2, 2, 4)$. The structure function of this MSS is presented in Table 8.1.

Zaitseva and Levashenko (2006) and Zaitseva and Puuronen (2009) have shown that the MSS structure function (8.2) with $m_i = m_z = M$ can be interpreted as an

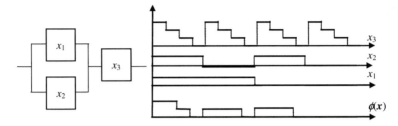

Fig. 8.1 MSS ($n = 3$, $M = 3$ and $\mathbf{m} = (2, 2, 4)$)

Table 8.1 Structure function $\phi(x)$ of the MSS in Fig. 8.1

x_3		0	1	2	3
x_1	x_2				
0	0	0	0	0	0
0	1	0	1	1	1
1	0	0	1	1	1
1	1	0	1	2	2

MVL function (A1) and MVL mathematical tools can be used to analyse it. These tools make it possible to analyse different aspects of MSS reliability. However, MSSs with equal performance levels for every system component and the system are not often used in engineering practice. An important problem is how to generalise reliability analysis methods of such MSSs to get an MSS with varied performance range for the components and the system.

From the point of view of MVL, structure function can be represented: (a) algebraically, as an expression in terms of algebra; (b) in tabular form, as a decision table; (c) in graphic form, as a graph. One of the graphic forms is a decision diagram of MVL function. This form of MVL function representation is often used in MVL (Miller and Drechsler 2002). There are different types of MVL function decision diagram that differ in construction, complexity, problem application, etc. A basic type of decision diagram referred to as an MDD is used in this paper.

An MDD is a directed acyclic graph of MVL function representation (Miller and Drechsler 2002). For the structure function (8.2) this graph has M sink nodes, labelled from 0 to $(M - 1)$, representing M corresponding constants from 0 to $(M - 1)$. Each non-sink node is labelled with a structure function variable x_i and has m_i outgoing edges. In an MSS reliability analysis, the sink node is interpreted as the system reliability state from 0 to $(M - 1)$ and the non-sink node presents either system component. Each non-sink node has m_i edges (Fig. 8.2) and the first (left) one is labelled as the "0" edge corresponding to component fail, while the m_ith last outgoing edge is labelled as the "$m_i - 1$" edge and presents the perfect operation state of the system component.

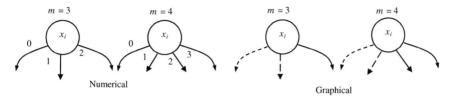

Fig. 8.2 Examples of non-sink nodes

Fig. 8.3 MDD for structure function of the MSS in Fig. 8.1

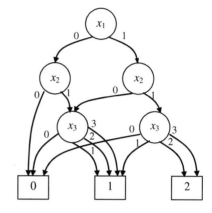

The paths from the top non-sink node to the zero-sink node are analysed for MSS failure. The paths from the top non-sink node to another sink node are considered for system repair by means of MDD. There is a special kind of software for MDD that allows to calculate necessary measures (Miller and Drechsler 2002). Paths for system failure and system repair in particular can be determined using this software. For example, the MSS representation in Fig. 8.1 by MDD is shown in Fig. 8.3.

8.2.2 Logical Differential Calculus in MSS Reliability Analysis

The logical differential calculus of an MVL function (see Appendix) is one of the MVL mathematical tools that can be used in MSS reliability analysis. This mathematical tool allows analysing changes in the MVL function depending on changes in its variables. Therefore logical derivatives can be used to evaluate the influence of change in every system component's state on the performance level of the MSS. This methodology was discussed for the first time in (Zaitseva and Levashenko 2006). Application of logical derivatives for binary-state system has been proposed by Moret and Thomason (1984) and Schneeweiss (2009). There are different types of logical derivates in MVL (Zaitseva and Levashenko 2006; Zaitseva and Puuronen 2009).

Table 8.2 DPLDs of the MSS structure function with respect to the first variable

x_2	x_3	$\frac{\partial\phi(1\to0)}{\partial x_1(1\to0)}$	$\frac{\partial\phi(2\to1)}{\partial x_1(1\to0)}$	$\frac{\partial\phi(0\to1)}{\partial x_1(1\to0)}$	$\frac{\partial\phi(1\to2)}{\partial x_1(1\to0)}$
0	0	0	0	0	0
0	1	1	0	1	0
0	2	1	0	1	0
0	3	1	0	1	0
1	0	0	0	0	0
1	1	0	0	0	0
1	2	0	1	0	1
1	3	0	1	0	1

One of these types is a *direct partial logic derivative* (DPLD) of MVL function (Tapia et al. 1991; Levashenko et al. 1999). These derivatives reflect the change in the value of the underlying function when the values of variables change (A2). Consider the definition of DPLD for an MSS structure function (8.2) based on Equation (A2). A DPLD with respect to ith variable for an MSS structure function is defined as:

$$\frac{\partial\phi(j\to\tilde{j})}{\partial x_i(s_i\to\tilde{s}_i)} = \begin{cases} 1, & \text{if } \phi(s_i,\mathbf{x})=j \text{ and } \phi(\tilde{s}_i,\mathbf{x})=\tilde{j}, \\ 0, & \text{other,} \end{cases} \tag{8.3}$$

where $\phi(s_i,\mathbf{x}) = \phi(x_1,\ldots,x_{i-1},s_i,x_{i+1},\ldots,x_n)$ is the value of the structure function; $\phi(\tilde{s}_i,\mathbf{x}) = \phi(s+1,\mathbf{x})$ or $\phi(\tilde{a}_i,\mathbf{x}) = \phi(a-1,\mathbf{x})$; $\tilde{j}=j+1$ or $\tilde{j}=j-1$; $j,\tilde{j}\in\{0,\ldots,M-1\}$ and $s_i,\tilde{s}_i\in\{0,\ldots,m_i-1\}$.

Because the DPLD (A2) has only two values, zero and $m-1$, the derivatives for the structure function (8.3) are defined as a binary function with values zero and one.

The value of the variable in the MSS structure function in (8.3) changes from s to $(s-1)$ or from s to $(s+1)$ only because no fast changes are permitted in the performance level of the working component. The MSS state in the DPLD (8.3) changes similarly, therefore the change in value of the structure function is analysed from j to $(j-1)$ or from j to $(j+1)$ (according to the assumed structure function).

For example, consider the MSS in Fig. 8.1. DPLDs for this system with respect to x_1 are shown in Table 8.2 and permit to describe all the changes in system performance levels depending on the changes in the state of the first component, taking into account the assumption (c). So, DPLD $\partial\phi(1\to0)/\partial x_i(1\to0)$ allows to determine a system state for which skip down of the first component causes a system failure. These states are defined by the variables vector values $\mathbf{x}=(x_1,x_2,x_3)=(\underline{1},0,1)$, $\mathbf{x}=(x_1,x_2,x_3)=(\underline{1},0,2)$, $\mathbf{x}=(x_1,x_2,x_3)=(\underline{1},0,3)$. Decrease in the system performance level from 2 to 1 caused by the failure of the first system component is analysed based on DPLD $\partial\phi(2\to1)/\partial x_1(1\to0)$. This derivative has two non-zero values that correspond to the MSS state $\mathbf{x}=(x_1,x_2,x_3)=(\underline{1},1,2)$ and $\mathbf{x}=(x_1,x_2,x_3)=(\underline{1},1,3)$. Therefore, the failure of the first system component causes the decrease of the MSS performance level if the second component is working ($x_2=1$) and the performance level of the third component is 2 or 3. Analysis of a system

Table 8.3 DPLDs of the MSS structure function with respect to third variable

x_2	x_3	$\dfrac{\partial\phi(1\to0)}{\partial x_3(1\to0)}$	$\dfrac{\partial\phi(1\to0)}{\partial x_3(2\to1)}$	$\dfrac{\partial\phi(1\to0)}{\partial x_3(3\to2)}$	$\dfrac{\partial\phi(2\to1)}{\partial x_3(1\to0)}$	$\dfrac{\partial\phi(2\to1)}{\partial x_3(2\to1)}$	$\dfrac{\partial\phi(2\to1)}{\partial x_3(3\to2)}$
0	0	0	0	0	0	0	0
0	1	1	0	0	0	0	0
1	0	1	0	0	0	0	0
1	1	1	0	0	0	1	0

restore is similar and derivatives $\partial\phi(0\to1)/\partial x_1(0\to1)$ and $\partial\phi(1\to2)/\partial x_1(0\to1)$ are used in this case (Table 8.2).

DPLDs with respect to the third variable allow analysing the influence of the change in the third system component's state on the change of the system performance level (Table 8.3). Only two derivatives have non-zero values in Table 8.3. The first of them is $\partial\phi(1\to0)/\partial x_3(1\to0)$, therefore a breakdown of the third component will cause an MSS failure if one or both the first and second components are working ($\mathbf{x}=(x_1, x_2, x_3)=(0, 1, \underline{1})$, $\mathbf{x}=(x_1, x_2, x_3)=(1, 0, \underline{1})$, $\mathbf{x}=(x_1, \quad x_2, \quad x_3)=(1, \quad 1, \quad \underline{1})$). The second non-zero derivative is $\partial\phi(2\to1)/\partial x_3(2\to1)$, indicating that the system state of the MSS performance level decreases from "2" to "1" depending on the decrease of the third component state from "2" to "1": $\mathbf{x}=(x_1, x_2, x_3)=(1, 1, \underline{2})$.

Note that in this example, the first and the second variables of the MSS structure function are symmetric, therefore the DPLDs of this function with respect to the first and the second variables are similar (Zaitseva and Levashenko 2006).

A DPLD with respect to the x_i variable for the MSS structure function (8.2) permits to analyse the change in system performance level from j to $\tilde{j}$ when the ith component state changes from s to $\tilde{s}_i$. However, a reliability analysis in MSS investigation of system failure is more important. This change is defined by the DPLD

$$\frac{\partial\varphi(1\to0)}{\partial x_i(s_i\to s_i-1)}=\begin{cases}1, & \text{if } \phi(s_i,\mathbf{x})=1 \text{ and } \phi((s_i-1),\mathbf{x})=0 \\ 0, & \text{other}\end{cases} \tag{8.4}$$

Therefore, an MSS failure will be analysed below, but analyses of other changes are similar and differ only by DPLDs.

8.2.3 MSS Reliability Function

Reliability function for a Binary-State System is defined as the probability of system function without failure during a given period of time. However, the reliability function has some interpretation for MSS.

In Lisnianski and Levitin 2003, an MSS reliability function $R(t)$ is the probability of the system being operational throughout the interval $[0, t)$:

$$R(t)=\Pr\{T\ \ge t,\ \phi(\mathbf{x})>0\} \text{ or } R=\Pr\{\phi(\mathbf{x})>0\}. \tag{8.5}$$

Table 8.4 Component state probabilities

State Component x_i	0	1	2	3
1	0.2	0.8	–	–
2	0.3	0.7	–	–
3	0.1	0.3	0.4	0.2

However, for MSS there are some levels of system performance and reliability analysis concerning this system that need to include estimation of the probability that the system will be in each of these performance states. Therefore, some definitions of reliability function for MSS have been proposed. One of them allows to present the probability of the MSS being in the state that is not less than the performance level j ($0 \le j \le M - 1$) (Meng 1993):

$$R(j) = \Pr\{\phi(\mathbf{x}) \ge j\}.$$

There is one more interpretation of the MSS reliability function in (Zaitseva 2010). In Lisnianski and Levitin 2003, this interpretation is referred to as the probability of MSS state. In this case, the MSS reliability function is defined as the probability of system reliability that is equal to the performance level j:

$$R(j) = \Pr\{\phi(\mathbf{x}) = j\}, \quad j = 1, \ldots, M - 1. \tag{8.6}$$

MSS unreliability according to this methodology is defined as:

$$F = R(0) = 1 - \sum_{j=1}^{M-1} R(j). \tag{8.7}$$

The reliability function $R(j)$ of the MSS in Fig. 8.1 according to the definition (8.6) is:

$$R(j) = \sum_{\varphi(x)=j} p_{1,s_1} p_{2,s_2} p_{3,s_3}.$$

For example, the reliability function $R(1)$ of this MSS is calculated as:

$$R(1) = p_{1,0}p_{2,1}p_{3,1} + p_{1,0}p_{2,1}p_{3,1} + p_{1,0}p_{2,1}p_{3,3} + p_{1,1}p_{2,0}p_{3,1}$$
$$+ p_{1,1}p_{2,1}p_{3,1} + p_{1,1}p_{2,0}p_{3,1} + p_{1,1}p_{2,1}p_{3,1} = 0.510,$$

if the component state probabilities in Table 8.4 are used for quantification of this MSS.

Calculation of the Reliability Function $R(j)$ based on the MDD is simpler. Paths from the top non-sink node to the "j" sink node are analysed to calculate values of the Reliability Function $R(j)$ by MDD. The MDD is therefore divided into M sub-diagrams that have one sink node and include paths to this node from the top non-sink node of the MDD. Every edge from the x_i variable with the s_i label is marked by the ith component state probability p_{i,s_i} (Fig. 8.4), because the MDD is an

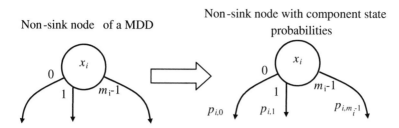

Fig. 8.4 Interpretation of component state probability in MDD

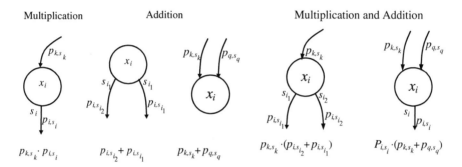

Fig. 8.5 Basic computational rules in MDD

orthogonal form to represent the structure function (Miller and Drechsler 2002). Probability of the MSS performance level j is calculated by trivial rules for MDD (Fig. 8.5).

For example, the $R(1)$, $R(2)$ reliability functions and unreliability function of the MSS concerned (Fig. 8.1) are calculated on the basis of the sub-diagram in Fig. 8.6 and these functions are:

$$R(1) = p_{1,1}p_{2,1}p_{3,1} + (p_{3,1} + p_{3,2} + p_{3,3})(p_{1,0}p_{2,1} + p_{1,1}p_{2,0}) = 0.510$$
$$R(2) = p_{1,1}p_{2,1}(p_{3,2} + p_{3,3}) = 0.336$$
$$F = p_{1,0}p_{2,0} + p_{3,0}(p_{1,0}p_{2,1} + p_{1,1}) = 0.154$$

8.3 Importance Measures

DPLDs allow to present changes in MSS performance level depending on system component state change. Thus this mathematical approach can be used for MSS importance analysis because an importance analysis quantifies the influence of every system component's state change on some MSS performance level or

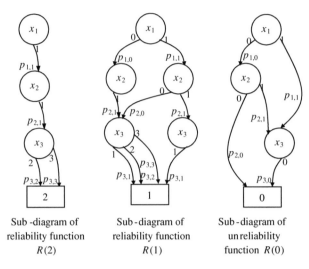

Sub-diagram of reliability function R(2)

Sub-diagram of reliability function R(1)

Sub-diagram of unreliability function R(0)

Fig. 8.6 Calculation of the $R(j)$ Reliability Function of the MSS in Fig. 8.1 by MDD

failure. This paper provides new algorithms for calculation commonly used in IM, such as *structural importance* (SI), *criticality importance* (CI), *Birnbaum importance* (BI), *Fussell-Vesely importance* (FVI), *reliability achievement worth* (RAW), *reliability reduction worth* (RRW) and *dynamic reliability indices* (DRIs). Algorithms used to calculate these measures are implemented based on the Logical Differential Calculus and allow quantifying probabilities of MSS failure depending on changes of the ith system component states.

8.3.1 Structural Importance

SI is one of the simplest measures of component importance and this measure focuses on the topological aspects of the system. According to the definition in (Meng 2009; Wu 2005), this measure determines the proportion of working states of the system in which the working state of the ith component makes the difference between a system failure and its functioning. SI of an MSS failure for the ith component state s is the probability of a failure of this system if the component state changes from s to $s_i - 1$ depending on the topological properties of the system:

$$I_S(s_i) = \frac{\rho_{i,s_i}}{m_1 \ldots m_{i-1} m_{i+i} \ldots m_n} \qquad (8.8)$$

where ρ_{i,s_i} is the number of system states when the component's state changes from s to $s_i - 1$, resulting in the system failure, and this number is calculated as numbers of non-zero values of DPLDs (8.4) $\partial \phi(1 \to 0)/\partial x_i(s_i \to s_i - 1)$.

Every system component of the MSS is characterised by $(m_i - 1)$ measures (8.7). Average SI is used to estimate all the changes of the ith system component with influence on a MSS failure:

$$I_S(x_i) = \frac{\sum_{s_i=1}^{m_i-1} \rho_{i,s_i}}{m_1 \ldots m_{i-1} m_{i+i} \ldots m_n} = \sum_{s_i=1}^{m_i-1} I_S(s_i) \tag{8.9}$$

There is one more definition of SI (Zaitseva 2010). It is a modified SI that represents the ith system component state change in an MSS failure for the boundary system state. In terms of DPLD (8.4), the modified SI is determined as:

$$I_{MS}(s_i) = \frac{\rho_{i,s_i}}{\rho_i^{(s_i,1)}} \tag{8.10}$$

where ρ_{i,s_i} is defined in (8.7); $\rho_i^{(s_i,1)}$ is the number of boundary system states when $\phi(s_i, \mathbf{x}) = 1$ (it is computed using the structure function of the MSS (8.2)).

The average modified SI has a meaning similar to the average SI (8.9) and is computed as:

$$I_{MS}(x_i) = \sum_{s_i=1}^{m_i-1} \frac{\rho_{i,s_i}}{\rho_i^{(s_i,1)}} = \sum_{s_i=1}^{m_i-1} I_{MS}(s_i). \tag{8.11}$$

Therefore, SI I_S conforms to the probability of an MSS failure among all possible system states. The modified SI I_{MS} is the probability of an MSS failure depending on the ith component state change and the boundary system states. A system component with the maximal value of the SI measure (I_S and I_{MS}) has the biggest influence on an MSS failure, or a failure of this component causes a high possibility of an MSS failure (Meng 2009, Zaitseva 2010).

SI and modified SI measures do not depend on component state probability (8.1) and characterize only topological aspects of MSS performance. These measures are used in prevention system analysis or reliability analysis in the stage of initial system design.

8.3.2 Birnbaum Importance

BI of a given component is defined as the probability that the component is critical to MSS functioning (Ramirez-Marquez and Coit 2005, Meng 2009, Fricks and Trivedi 2003). This measure has been defined for a Binary-State System first as:

$$I_B(x_i) = |\Pr\{\phi(\mathbf{x}) = 1, \ x_i = 1\} - \Pr\{\phi(\mathbf{x}) = 1, \ x_i = 0\}|.$$

Mathematical and logical generalization of this measure for MSS has some interpretations. Levitin et al. (2003), propose a definition of BI for a system with two performance levels that consists of multi-state components. Ramirez-Marquez and Coit (2005), authors propose a definition of BI for MSS failure analysis as:

$$I_B(x_i) = \frac{\sum_{k=1}^{m_i-1} |\Pr\{\phi(\mathbf{x}) \geq j|\, x_i = k\} - \Pr\{\phi(\mathbf{x}) \geq j\}|}{m_i - 1}.$$

Furthermore, new modifications of BI and algorithms for calculations based on different methodological approaches were proposed in (Wang et al. 2004, Wu 2005, Zio et al. 2007). One more interpretation of BI for MSS in terms of Logical Differential Calculus was presented in Zaitseva (2010). According to this definition, BI is a probabilistic measure that can be interpreted as the rate at which the MSS fails as the ith system component state decreases:

$$I_B(s_i) = |\Pr\{\phi(\mathbf{x}) = 1|x_i = s_i\} - \Pr\{\phi(\mathbf{x}) = 1|x_i = s_i - 1\}|, \qquad (8.12)$$

where $\Pr\{\phi(\mathbf{x}) = j|x_i = s_i\} = \sum p_{1,a_1} \cdots p_{i-1,a_{i-1}} p_{i+1,a_{i+1}} \cdots p_{n,a_n}$ if $\phi(\mathbf{x}) = 1$ and $x_i = s_i$ for $a_w = \{0, \ldots, m_w-1\}$, $w = 1,\ldots, n$ and $w \neq i$; $s_i = \{1, \ldots, m_i-1\}$.

The assumption (c) in Eq. (8.12) is taken into account, allowing to consider only the changes in component state from s_i to $(s_i - 1)$. The BI (8.12) analyses an MSS failure for all the aspects of the ith system component changes and the averaged value of BI can be calculated as:

$$I_B(x_i) = \frac{\sum_{s_i=1}^{m_i-1} I_B(s_i)}{m_i - 1}. \qquad (8.13)$$

Note that the BI based on (8.11) can be calculated by MDD. Two parts of Eq. (8.11) $\Pr\{\phi(\mathbf{x}) = j|x_i = s_i\}$ and $\Pr\{\phi(\mathbf{x}) = j|x_i = s_i - 1\}$ are computed based on an MDD in the form of a sub-diagram with paths from the top non-sink node to the sink node 1 that includes edges labelled s_i and $s_i - 1$ accordingly (however, probabilities that correspond to the edges with these labels are not considered). Rules for determining component states probabilities and their multiplication and addition are similar to the computation of the reliability function (Figs. 8.4 and 8.5).

8.3.3 Criticality Importance

BI measures (8.11)–(8.13) depend on the structure of the system and states of other components, but are independent on the actual state of the ith component. Consider the definition of CI as the probability that the ith system component is relevant to an MSS failure if it failed (Fricks and Trivedi 2003). For MSS, this measure can be defined as the probability of the MSS failure if the state of the ith system component has changed from s_i to $s_i - 1$, $s_i = \{1, \ldots, m_i - 1\}$:

$$I_C(s_i) = I_B(s_i) \cdot \frac{p_{i,s_i-1}}{F}, \qquad (8.14)$$

where $I_B(s_i)$ is the ith system component BI measure (8.12); P_{i,s_i} is the probability of the ith system component state s_i (8.1) and F is the probability of a system failure (system unreliability) that is defined in accordance with (8.7).

The CI measure (8.14) corrects BI for unreliability of the ith component relative to the unreliability of the whole system. This measure is useful if the component has a high BI and low unreliability with respect to MSS unreliability. In this case the ith component CI is low. The average CI measure is calculated as:

$$I_C(x_i) = \frac{\sum_{s_i=1}^{m_i-1} I_C(s_i)}{m_i - 1}. \tag{8.15}$$

8.3.4 Fussell-Vesely Importance

FVI measure quantifies the maximum decrement in system reliability caused by the ith system component state deterioration (Levitin et al. 2003; Ramirez-Marquez and Coit 2005). In other words, this measure represents the contribution of each component to the probability of a system failure and it is calculated for a Binary-State System (if $M = 2$ and $m_i = m_k = 2$) by the following equation (Chang et al. 2004):

$$I_{FV}(x_i) = \frac{\Pr\{\phi(\mathbf{x}) = 0\} - \Pr\{\phi(\mathbf{x}) = 0 | x_i = 1\}}{\Pr\{\phi(\mathbf{x}) = 0\}}.$$

FVI for an MSS represents a probabilistic measure of the ith component state deterioration influence on failure of the system:

$$I_{FV}(s_i) = 1 - \frac{\Pr\{\phi(s_i, \mathbf{x}) = 0\}}{F}, \tag{8.16}$$

where $\Pr\{s_i, \phi(\mathbf{x}) = 0\} = \sum p_{1,a_1} \cdots p_{i,s_i} \cdots p_{n,a_n}$ if $\phi(\mathbf{x}) = 0$ and $x_i = s_i$ for $a_w = \{0, \ldots, m_w - 1\}$, $w = 1, \ldots, n$ and $w \neq i$; $s_i = \{1, \ldots, m_i - 1\}$.

Consider generalizations of FVI (8.16) as the average value of this measure:

$$I_{FV}(x_i) = \frac{\sum_{s_i=1}^{m_i-1} I_{FV}(s_i)}{m_i - 1}, \tag{8.17}$$

that quantifies the contribution of each component state change to the probability of a system failure.

The calculation of the FVI measure can be implemented by MDD and this algorithm is similar to the algorithm used to compute the BI measure.

8.3.5 Reliability Achievement Worth

RAW and RRW are two importance measures and both represent adjustments of the improvement potential of MSS unreliability. RAW for a binary-state system indicates the increase in the system unreliability when the ith component is a failure and this measure is defined as (Chang et al. 2004):

$$I_{\text{RAW}}(x_i) = \frac{\Pr\{\phi(0_i, \mathbf{x}) = 0\}}{F}.$$

According to Levitin et al. 2003; Ramirez-Marquez and Coit 2005, RAW for MSS is defined as the ratio of MSS unreliability if the ith component state decreases:

$$I_{\text{RAW}}(s_i) = \frac{\Pr\{\phi(s_i - 1, \mathbf{x}) = 0\}}{F}, \tag{8.18}$$

where $s_i = \{1, \ldots, m_i - 1\}$.

The generalization of RAW (8.18) for all states of the ith MSS component is calculated as:

$$I_{\text{RAW}}(x_i) = \frac{\sum_{s_i=1}^{m_i-1} I_{\text{RAW}}(s_i)}{m_i - 1}. \tag{8.19}$$

8.3.6 Reliability Reduction Worth

RRW can be interpreted as the opposite importance measure to RAW and it is defined for a Binary-State System as (Chang et al. 2004):

$$I_{\text{RRW}}(x_i) = \frac{F}{\Pr\{\phi(1_i, \mathbf{x}) = 0\}}.$$

Generalization of this equation and representation of RRW in (Levitin et al. 2003; Ramirez-Marquez and Coit 2005) allows to define RRW for MSS as importance measure quantifies potential damage caused to the MSS by the ith system component:

$$I_{\text{RRW}}(s_i) = \frac{F}{\Pr\{\phi(s_i, \mathbf{x}) = 0\}}, \tag{8.20}$$

where $s_i = \{1, \ldots, m_i - 1\}$.

Average RRW for MSS is below:

$$I_{\text{RRW}}(x_i) = \frac{\sum_{s_i=1}^{m_i-1} I_{\text{RRW}}(s_i)}{m_i - 1}. \tag{8.21}$$

8.3.7 Dynamic Reliability Indices

There is one more type of IMs for MSS that are DRIs. These measures have been defined in paper (Zaitseva and Levashenko 2006; Zaitseva 2010). DRIs allow estimating a component relevant to the MSS and to quantify the influence of this component state change on the MSS performance. There are two groups of DRIs: Component Dynamic Reliability Indices (CDRIs) and Dynamic Integrated Reliability Indices (DIRIs).

CDRI indicates the influence of the ith component state change on change in MSS performance level (Zaitseva and Levashenko 2006). In case of MSS failure, it is the probability of MSS unreliability caused by a decrement of the ith component state. This definition of CDRI is similar to the definition of modified SI, but CDRIs for MSS failure take into consideration two probabilities: (a) the probability of MSS failure provided that the ith component state is reduced and (b) the probability of inoperative component state:

$$I_{\text{CDRI}}(s_i) = I_{\text{MS}}(s_i) \cdot p_{i,s_i-1}, \tag{8.22}$$

where $I_S(x_i)$ is the modified SI (8.10); p_{i,s_i} is the probability of component state (8.1).

The average CDRI measure indicates the probability of MSS failure caused by diminution or skip down of the ith system component:

$$I_{\text{CDRI}}(x_i) = I_{\text{MS}}(x_i) \cdot \sum_{s_i=1}^{m_i-1} p_{i,s_i-1} = \sum_{s_i=1}^{m_i-1} I_{\text{CDRI}}(s_i). \tag{8.23}$$

DIRI is the probability of an MSS failure caused by a decrement in one of the system component states. DIRIs allow estimating probability of an MSS failure caused by some system components (one of n):

$$I_{\text{DIRI}} = \sum_{i=1}^{n} I_{\text{CDRI}}(x_i) \prod_{\substack{q=1 \\ q \neq i}}^{n} \left(1 - I_{\text{CDRI}}(x_q)\right). \tag{8.24}$$

8.3.8 Hand Calculation Example

For example, consider the MSS in Fig. 8.1 ($n = 3$ and $M = 3, (m_1, m_2, m_3) = (2, 2, 4)$) and compute IM for this MSS. The structure function and values of component state probabilities of this MSS are shown in Tables 8.1 and 8.4 accordingly.

SI measures are calculated by (8.8)–(8.11) and are presented in Table 8.5 for all system components. These measures are computed by DPLD $\partial\phi(1 \rightarrow 0)/ \partial x_i(s_i \rightarrow s_i - 1)$. The second part of Table 8.5 includes values of $p_{i,s}$ and $\rho_i^{(1,1)}$

Table 8.5 Structural Importance and Modified Structural Importance measures of the MSS presented in Fig. 8.1

Component	$I_S(1_i)$	$I_S(2_i)$	$I_S(3_i)$	$I_S(x_i)$	$I_{MS}(1_i)$	$I_{MS}(2_i)$	$I_{MS}(3_i)$	$I_{MS}(x_i)$	$\rho_{i,1}$	$\rho_{i,2}$	$\rho_{i,3}$	$\rho_i^{(1,1)}$
1	0.375	–	–	0.375	0.750	–	–	0.750	3	–	–	4
2	0.375	–	–	0.375	0.750	–	–	0.750	3	–	–	4
3	0.750	0	0	0.750	1	0	0	1	3	0	0	3

Table 8.6 Birnbaum Importance measures of the MSS presented in Fig. 8.1

Component	$I_B(1_i)$	$I_B(2_i)$	$I_B(3_i)$	$I_B(x_i)$
1	0.090	–	–	0.090
2	0.300	–	–	0.300
3	0.940	0.460	0	0.750

numbers that are defined on the basis of DPLD in Table 8.2 and on the basis of the MSS structure function in Table 8.1. These numbers are used to calculate the SI (8.8) and modified SI (8.10). Average values of these measures are defined by (8.9) and (8.11).

For this example, the third component has the maximal value of the SI measure. Therefore, it is the component with the maximal influence on system failure and the system will not work in the event of this component's failure ($I_{MS} = 1$).

Continue with this example to analyse the MSS by BI based on MDD. The procedure for calculation of the BI for this MSS is shown in Fig. 8.7. Consider this procedure for the first component in more detail. First of all, the edges from the "x_1" non-sink nodes labelled 0 are removed in the sub-diagram $R(1)$ and then the "x_1" non-sink nodes are removed, too, and a new sub-diagram $\Pr\{\phi(\mathbf{x}) = 1|x_1 = 1\}$ is formed. The $\Pr\{\phi(\mathbf{x}) = 1|x_1 = 0\}$ sub-diagram is made similar to the previous sub-diagram, but the edges with label 1 are removed in this case. Probabilities $\Pr\{\phi(\mathbf{x}) = 1|x_1 = 1\}$ and $\Pr\{\phi(\mathbf{x}) = 1|x_1 = 0\}$ are calculated by the two sub-diagrams and BI $I_B(1_1)$ is determined based on (8.12). The procedure for computation of the BI $I_B(1_2)$, BI $I_B(1_3)$, BI $I_B(2_3)$ and BI $I_B(3_3)$ is similar to the procedure used to calculate the $I_B(1_1)$ and is presented in Fig. 8.7. All BI values for this MSS are shown in Table 8.6.

BI measures indicate the probability of MSS functioning if the ith component has unreliability. In this example, the third component has the maximal influence on functioning of the MSS, because the values of the BI measures for this component are maximal (for every component state and average).

The CI measures of components of the MSS in Fig. 8.1 are calculated by (8.14) and are presented in Table 8.7. The third component value of FVI is maximal but decreases in relation BI $I_B(1_3)$, because the probability of this component unreliability is minimal. The value of the average CI measure of the second component is of the highest value, because this component unreliability is maximal (see Table 8.4).

The FVI measures (8.16) and (8.17) for this MSS are shown in Table 8.8. One of the possible ways to compute these measures is to apply an MDD. This

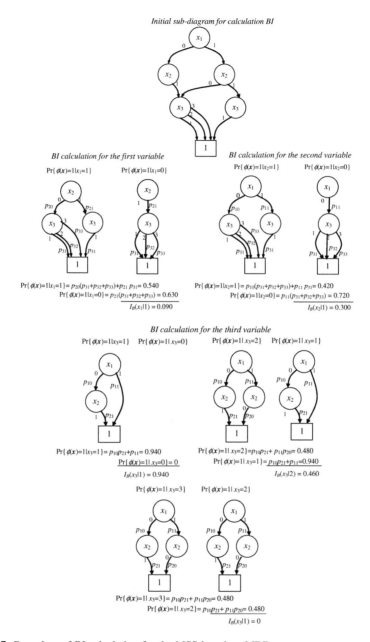

Fig. 8.7 Procedure of BI calculation for the MSS based on MDD

procedure is similar to the calculation of BI. Sub-diagrams for FVI measure calculation are shown in Fig. 8.8 and are used to compute probabilities $\Pr\{\phi(s_i, \mathbf{x}) = 0\}$. FVI measures present the probability of MSS unreliability

Table 8.7 Criticality Importance measures of the MSS in Fig. 8.1

Component	$I_C(1_i)$	$I_C(2_i)$	$I_C(3_i)$	$I_C(x_i)$
1	0.117	–	–	0.117
2	0.584	–	–	0.584
3	0.610	0.874	0	0.495

Table 8.8 Fussell-Vesely Importance measures of the MSS in Fig. 8.1

Component	$I_{FV}(1_i)$	$I_{FV}(2_i)$	$I_{FV}(3_i)$	$I_{FV}(x_i)$
1	0.481	–	–	0.481
2	0.545	–	–	0.545
3	0.883	0.844	0.922	0.883

Table 8.9 Reliability Achievement Worth and Reliability Reduction Worth measures of the MSS in Fig. 8.1

Component	$I_{RAW}(1_i)$	$I_{RAW}(2_i)$	$I_{RAW}(3_i)$	$I_{RAW}(x_i)$	$I_{RRW}(1_i)$	$I_{RRW}(2_i)$	$I_{RRW}(3_i)$	$I_{RRW}(x_i)$
1	0.481	–	–	0.481	1.925	–	–	1.925
2	0.545	–	–	0.545	2.200	–	–	2.200
3	0.649	0.117	0.156	0.307	8.556	6.417	12.833	9.286

Table 8.10 Component Dynamic Reliability and Dynamic Reliability Indices of the MSS in Fig. 8.1

Component	$I_{CDRI}(1_i)$	$I_{CDRI}(2_i)$	$I_{CDRI}(3_i)$	$I_{CDRI}(x_i)$	I_{DIRI}
1	0.150	–	–	0.150	
2	0.225	–	–	0.225	0.343
3	0.100	0	0	0.100	

caused by a decrease of the ith component state. According to the data in Table 8.7, the third component has the maximal influence on the MSS unreliability (the average FVI measure of this component is $I_{FV}(x_i) = 0.883$).

Continue with the analysis of the MSS in Fig. 8.1 and consider the RAW (8.18), (8.19) and RRW (8.20), (8.21) measures for this system (Table 8.9). According to RAW and RRW, the third system component has the maximal influence on the MSS. Unreliability of this MSS depends on diminution of the third component functional state, but primarily on its failure.

CDRIs (8.22) of the MSS in Fig. 8.1 for the ith component indicate the probabilities of a system failure caused by different changes in the component state (Table 8.10). DIRI is a similar measure but represents the probability of system failure depending on a decrease in one of the system component states and is calculated according to Eq. (8.24).

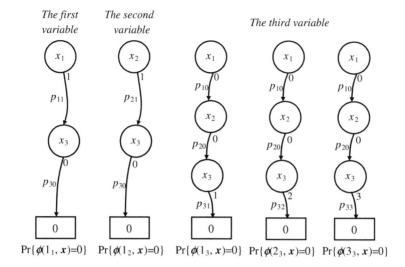

$$\text{Pr}\{\phi(1_1, \boldsymbol{x})=0\} \quad \text{Pr}\{\phi(1_2, \boldsymbol{x})=0\} \quad \text{Pr}\{\phi(1_3, \boldsymbol{x})=0\} \quad \text{Pr}\{\phi(2_3, \boldsymbol{x})=0\} \quad \text{Pr}\{\phi(3_3, \boldsymbol{x})=0\}$$

Fig. 8.8 Sub-diagram for FVI calculation

Fig. 8.9 Diagram of IM for the series–parallel MSS in Fig. 8.1

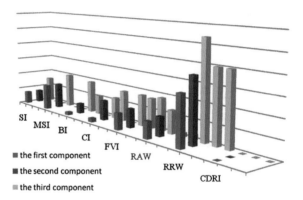

From the point of view of the topological aspects of this MSS and probabilities of component states, the second system component has the maximal influence on system failure. The probability of the MSS failure in the situation when one system component is in an inoperative state is defined by DIRI and equals to 0.343.

The CDRI is the probability of an MSS failure caused by the ith component breakdown. DIRI is a similar measure, but allows to analyse an MSS failure depending on unavailability of some system components. DRIs consider both the system topology and component performance (component state probability).

Graphical interpretation of all the IMs for the MSS concerned is shown in Fig. 8.9. The third system component has influence on the MSS function in all situations arising from this diagram. But the first and second components cause system functioning by high probability $p_{1,1}$ and $p_{2,1}$ of these component states.

8.4 Conclusion

This paper discusses the principal importance measures for MSS failure analysis and a new method of their calculation based on Logical Differential Calculation mathematical approach is proposed. These measures supplement importance analysis of MSS and allow to obtain alternative information about the component being investigated. System components with the maximal and minimal influence on changes in MSS performance level are revealed based on this analysis. This information is principal for reliability analysis of real-world system designs and their behaviour in time of exploitation. There are a number of applications for IMs in reliability analysis in nuclear power engineering (Marseguerra and Zio 2004; Eisenberg and Sagar 2000) or transport systems (Zio et al. 2007). IM is useful in designing complex systems (Zio and Podofillini 2007). Therefore, development of a mathematical approach to calculate these measures is crucial for reliability engineering.

8.5 Appendix

Multiple-Valued Logic and Logical Differential Calculus

A multiple-valued logic function is defined as:

$$f(x_1,\ldots,x_n) = f(\mathbf{x}) : \{0,\ldots,m-1\}^n \to \{0,\ldots,m-1\}. \tag{8.25}$$

Logical Differential Calculus is used for sensitivity analysis and investigation of dynamic properties of a MVL-function. There are different types of logical derivatives in MVL (Levashenko et al. 1999). One of the types of logical derivatives of the MVL function is *Direct Partial Logic Derivatives* (DPLDs). These derivatives reflect the change in the value of the underlying function when the values of variables change.

DPLD $\partial f(j \to h)/\partial x_i(a \to b)$ of a MVL function $f(\mathbf{x})$ of n variables with respect to the x_i variable reflects the fact of changing the function from j to h when the value of the x_i variable changes from a to b (Tapia et al. 1991; Shmerko et al. 1997):

$$\frac{\partial f(j \to h)}{\partial x_i(a \to b)} = \begin{cases} m-1, & \text{if } f(a,\mathbf{x}) = j \text{ and } f(b_i,\mathbf{x}) = h \\ 0, & \text{in theother case} \end{cases} \tag{8.26}$$

where $f(s_i,\mathbf{x}) = f(x_1,\ldots,x_{i-1},s,x_{i+1},\ldots,x_n)$, $s \in \{0,\ldots,m-1\}$.

For example, consider the MVL function ($m = 3$ and $n = 3$) in Table 8.11. Based on the truth table of this function compute the two DPLDs with respect to the second variable $\partial f(2 \to 1)/\partial x_2(1 \to 0)$ and with respect to the third variable $\partial f(2 \to 1)/\partial x_3(1 \to 0)$. The DPLD $\partial f(2 \to 1)/\partial x_2(1 \to 0)$ has four non-zero

Table 8.11 Example of the MVL function and its DPLDs $\partial f(2 \to 1)/\partial x_2(1 \to 0)$ and $\partial f(2 \to 1)/\partial x_3(1 \to 0)$

$x_1x_2x_3$	$f(x)$	$\frac{\partial f(2\to1)}{\partial x_2(1\to0)}$	$\frac{\partial f(2\to1)}{\partial x_3(1\to0)}$	$x_1x_2x_3$	$f(x)$	$\frac{\partial f(2\to1)}{\partial x_2(1\to0)}$	$\frac{\partial f(2\to1)}{\partial x_3(1\to0)}$	$x_1x_2x_3$	$f(x)$	$\frac{\partial f(2\to1)}{\partial x_2(1\to0)}$	$\frac{\partial f(2\to1)}{\partial x_3(1\to0)}$
0 0 0	0	–	–	1 0 0	0	–	–	2 0 0	0	–	–
0 0 1	0	–	0	1 0 1	1	–	0	2 0 1	1	–	0
0 0 2	0	–	–	1 0 2	1	–	–	2 0 2	1	–	–
0 1 0	0	0	–	1 1 0	0	0	–	2 1 0	0	0	–
0 1 1	1	0	0	1 1 1	2	2	0	2 1 1	2	2	0
0 1 2	1	0	–	1 1 2	2	2	–	2 1 2	2	2	–
0 2 0	0	–	–	1 2 0	0	–	–	2 2 0	0	–	–
0 2 1	1	–	0	1 2 1	2	–	0	2 2 1	2	–	0
0 2 2	1	–	–	1 2 2	2	–	–	2 2 2	2	–	–

values. This values correspond to the values of the variables vector $\mathbf{x} = (x_1, x_2, x_3) = (1, 1, 1)$, $\mathbf{x} = (x_1, x_2, x_3) = (1, 1, 2)$, $\mathbf{x} = (x_1, x_2, x_3) = (2, 1, 2)$ and $\mathbf{x} = (x_1, x_2, x_3) = (2, 1, 2)$. Therefore the change in value of the second variable x_2 from 1 to 0 causes the change in the function value from 2 to 0 for these values of the variables vector $\mathbf{x}$. The DPLD with respect to the third variable $\partial f(2 \to 1)/\partial x_3(1 \to 0)$ has no non-zero values. So this derivative is zero and the investigation of the MVL function has no values of variables vector for which the change in the third variable from 1 to 0 causes the change of the function value from 2 to 0.

The non-zero values of a DPLD are interpreted as critical or boundary values of a MVL function and are used in testing methods (Levashenko et al. 1999, Shmerko et al. 1997), logic design (Levashenko et al. 1999) and reliability analysis (Zaitseva and Levashenko 2006).

References

Andrews JD, Dunnett SJ (2000) Event-tree analysis using binary decision diagrams. IEEE Trans Reliab 49(2):230–238

Chang YR, Amari S, Kuo SY (2004) Computing system failure frequencies and reliability importance measures using OBDD. IEEE Trans Comp 53(1):54–68

Eisenberg NA, Sagar B (2000) Importance measures for nuclear waste repositories. Reliab Eng Sys Saf 70(3):217–239

Fricks RM, Trivedi KS (2003) Importance analysis with Markov chains. In: Proceedings of IEEE 49th annual reliability and maintainability symposium (RAMS), January 27–30, Tampa, USA, pp 89–95

Levashenko V, Moraga C, Shmerko V, Kholovinski G, Yanushkevich S (1999) Test algorithm for multiple-valued combinatorial circuits. Autom Remote Control 61(3):844–857

Levitin G, Lisnianski A (1999) Importance and sensitivity analysis of multi-state systems using the universal generating function method. Reliab Eng Sys Saf 65(3):271–282

Levitin G, Podofilini L, Zio E (2003) Generalised importance measures for multi-state elements based on performance level restrictions. Reliab Eng Sys Saf 82(3):287–298

Lisnianski A, Levitin G (2003) Multi-state system reliability. Assessment optimization and applications. World scientific, Singapore

Marseguerra M, Zio E (2004) Monte Carlo estimation of the differential importance measure: application to the protection system of a nuclear reactor. Reliab Eng Sys Saf 86(1):11–24

Meng FC (1993) Component-relevancy and characterization results in multistate systems. IEEE Trans Reliab 42(3):478–483

Meng FC (2009) On some structural importance of system components. J Data Sci 7:277–283

Miller M, Drechsler R (2002) On the construction of multiple-valued decision diagrams. In: Proceedings of IEEE 32nd international symposium on multiple-valued logic, Boston, USA, pp 264–269

Moret BME, Thomason MG (1984) Boolean difference techniques for time-sequence and common-cause analysis of fault-trees. IEEE Trans Reliab R-33(5):399–405

Natvig B, Eide KA, Gasemyr J, Huseby AB, Isaksen SL (2009) Simulation based analysis and an application to an offshore oil and gas production system of the Natvig measures of component importance in repairable systems. Reliab Eng Sys Saf 94(10):1629–1638

Pham H (2003) Handbook of reliability engineering. Springer, London

Ramirez-Marquez JE, Coit DW (2005) Composite importance measures for multi-state systems with multi-state components. IEEE Trans Reliab 54(3):517–529

Schneeweiss WG (2009) A short Boolean derivation of mean failure frequency for any (also non-coherent) system. Reliab Eng Sys Saf 94(8):1363–1367

Shmerko V, Yanushkevich S, Levashenko V (1997) Test pattern generation for combinational multi-valued networks based on generalized D-algorithm. In: Proceedings of 27th IEEE international symposium on multiple-valued logic, Nova Scotia, Canada, pp 139–144

Tapia MA, Guima TA, Katbab A (1991) Calculus for a multivalued-logic algebraic system. Appl Math Comput 42(3):255–285

Van PD, Barros A, Bérenguer C (2008) Reliability importance analysis of Markovian systems at steady state using perturbation analysis. Reliab Eng Sys Saf 93(11):1605–1615

Wang W, Loman J, Vassiliou P (2004) Reliability importance of components in complex system. In: Proceedings of IEEE 50th annual reliability and maintainability symposium (RAMS), Los Angeles, USA, pp 6–11

Wu S (2005) Joint importance of multistate systems. Com Ind Eng 49:63–75

Xing L, Dai Y (2009) A new decision-diagram-based method for efficient analysis on multistate systems. IEEE Trans Dependable Sec Comput 6(3):161–174

Zaitseva E (2010) Importance analysis of multi-state system by tools of differential logical calculus. In: Bris R, Guedes C, Martorell S (eds) Reliability, risk and safety. Theory and applications. CRC Press, London, pp 1579–1584

Zaitseva E, Levashenko V (2006) Dynamic reliability indices for parallel, series and k-out-of-n multi-state system. In: Proceedings of IEEE 52nd annual reliability and maintainability symposium (RAMS), January 23–26, 2006, Newport Beach, USA, pp 253–259

Zaitseva E, Puuronen S (2009) Representation and estimation of multi-state system reliability by decision diagrams. In: Martorells S, Guedes C, Barnett J (eds) Safety, reliability and risk analysis: Theory, methods and applications. CRC Press, London, pp 1995–2002

Zang X, Wang D, Sun H, Trivedi KS (2003) A BDD-based algorithm for analysis of multistate systems with multistate components. IEEE Trans Comp 52(12):1608–1618

Zio E (2009) Reliability engineering: Old problems and new challenges. Reliab Eng Sys Saf. 94(2):125–141

Zio E, Podofillini L (2007) Importance measures and genetic algorithms for designing a risk-informed optimally balanced system. Reliab Eng Sys Saf 92(10):1435–1447

Zio E, Marella M, Podofillini L (2007) Importance measures-based prioritization for improving the performance of multi-state systems: application to the railway industry. Reliab Eng Sys Saf 92(10):1303–1314

Chapter 9
Optimal Replacement and Protection Strategy for Parallel Systems

Rui Peng, Gregory Levitin, Min Xie and Szu Hui Ng

Abstract We consider a parallel system consisting of components with different characteristics. The components can be unavailable due to internal failures or external impacts. Each component has an increasing failure rate and is subjected to external impacts that can occur with fixed frequency. The system reliability is defined as the ability to maintain a specified performance level. It is assumed that the component destruction probability depends on the investment made in protection actions. In order to increase the system reliability, two measures can be taken: (1) increase of the components replacement frequency to improve their availability and (2) improvement of the components protection to reduce the probability of their destruction by the external impacts. The optimal maintenance and protection strategy which minimizes the total cost of maintenance, protection and damage caused by unsupplied demand is studied. The proposed approach is based on a universal generating function technique and a genetic algorithm.

Keywords Maintenance · Optimization · Parallel systems · Protection · Universal generating function · Genetic algorithm

R. Peng (✉) · M. Xie · S. H. Ng
Department of Industrial and Systems Engineering,
National University of Singapore, Kent Ridge Crescent,
Singapore 119260, Singapore
e-mail: g0700981@nus.edu.sg

M. Xie
e-mail: mxie@nus.edu.sg

S. H. Ng
e-mail: isensh@nus.edu.sg

G. Levitin
The Israel Electric Corporation Ltd., Haifa, Israel
e-mail: levitin@iec.co.il

A. Lisnianski and I. Frenkel (eds.), *Recent Advances in System Reliability*,
Springer Series in Reliability Engineering, DOI: 10.1007/978-1-4471-2207-4_9,
© Springer-Verlag London Limited 2012

9.1 Introduction

When a system is subjected to both internal failures and external impacts, maintenance and protection are two measures intended to enhance system availability. The
internal failures can be from component degradation or wear out, which accumulates
and accelerates over time. The external impacts can be from natural disasters, such
as avalanches, earthquakes, tsunamis, fire and disease. For multistate systems, the
system availability is a measure of the system's ability to meet the demand (required
performance level). In order to provide the required availability with minimum cost,
the optimal maintenance and protection strategy is studied.

For systems containing elements with increasing failure rates, preventive
replacement of the elements is an efficient measure to increase the system reliability (Levitin and Lisnianski 1999). Replacing elements that have a high risk of
failure, while reducing the chance of failure, can incur significant expenses,
especially in systems with high replacement rates. Minimal repair, the less
expensive option, enables the system element to resume its work after failure, but
does not affect its hazard rate (Beichelt and Fischer 1980; Beichelt and Franken
1983). Since the component replacement reduces its failure rate, the more frequently the component is replaced the higher the availability of the component is.

Besides internal failures, a component may also fail due to external impacts
(Zhuang and Bier 2007). In order to increase the survivability of a component
under external impacts, defensive investments can be made to protect the component. For example, installing anti-seismic devices can increase the survivability
of system components in the case of earthquakes. It is reasonable to assume that
the external impact frequency is constant over time and that the probability of the
component destruction by the external impact decreases with the increase of the
protection effort allocated on the component.

We consider a parallel system consisting of components with different characteristics (nominal performances, hazard functions, protection costs, etc.). The
objective is to minimize the total cost of the damage associated with unsupplied
demand and the costs of the system maintenance and protection. A universal
generating function technique is used to evaluate the system availability for any
maintenance and protection policy. A genetic algorithm is used for the optimization.

Section 9.2 formulates the problem. Section 9.3 describes the method of
calculating the system availability. Section 9.4 provides a description of the genetic
algorithm. Numerical examples are shown in Sect. 9.5. Section 9.6 concludes.

9.2 Problem Formulation and Description of System Model

Assumptions:

1. All N system components are independent.
2. The failures caused by the internal failures and external impacts are independent.

3. The time spent on replacement is negligible.
4. The time spent on a minimal repair is much less than the time between failures.

A system that consists of N components connected in parallel is considered. The lifetime for the system is denoted as T_c. For each component i, its nominal performance is denoted as G_i and the expected number of internal failures during time interval $(0, t)$ is denoted as $\lambda_i(t)$, which is an increasing function of t. Each component is subjected to internal failures and external impacts. The failures caused by internal failures and external impacts are fixed by minimal repairs. It is assumed that the following two kinds of maintenance actions can be taken (Sheu and Chang 2009):

- Preventive replacement: the ith component is replaced when it reaches an age T_i. The cost c_i of each replacement is constant. As the preventive replacement is planned action the average time for this replacement is negligible.
- Minimal repair: this action is used after internal failures or destructive external impacts and does not affect the hazard function of the component. The average cost for a minimal repair of component i is σ_i in the case of internal failure and θ_i in the case of external impact. The average time for a minimal repair of component i is t_i in the case of internal failure and τ_i in the case of external impact.

The repair costs and times differ for the two cases because in the case of external impact both the element and its protection are damaged.

The average number of internal failures during the period between replacements $\lambda_i(T_i)$ can be obtained by using the replacement interval T_i for each element. Having the replacement interval one can obtain the number of preventive replacements n_i during the system life cycle as

$$n_i = \frac{T_c}{T_i} - 1. \tag{9.1}$$

The total cost of the preventive replacements of component i during the system life cycle is

$$n_i C_i = \left(\frac{T_c}{T_i} - 1 \right) C_i. \tag{9.2}$$

Therefore, the total expected number of internal failures of the component i during the system life cycle is

$$(n_i + 1)\lambda_i(T_i) = \frac{\lambda_i(T_i)T_c}{T_i}. \tag{9.3}$$

We use x_i to denote the protection effort allocated on component i and a_i to denote the unit protection effort cost for component i. The total cost of component i protection is $a_i x_i$. It is assumed that the external impact frequency q is a constant. In this case the expected number of the external impacts during the system lifetime is qT_c.

The expected impact intensity is d. The component vulnerability (conditional probability of a component failure caused by an external impact) is evaluated using the contest function model (Hausken 2005; Tullock 1980; Skaperdas 1996) as

$$v(x_i, d) = \frac{d^m}{x_i^m + d^m},$$ (9.4)

where m is the contest intensity parameter. The expected number of the failures caused by the external impacts is therefore

$$q \cdot T_c \cdot v(x_i, d) = \frac{q \cdot T_c \cdot d^m}{x_i^m + d^m}.$$ (9.5)

Hausken and Levitin (2008) discussed the meaning of the contest intensity parameter m. A benchmark intermediate value is $m = 1$, which means that the investments into protection have proportional impact on the vulnerability reduction. $0 < m < 1$ corresponds to the low effective types of protections with component vulnerability less sensitive to variation of the protection effort. $m > 1$ corresponds to the highly effective types of protections with component vulnerability very sensitive to variation of the protection effort.

The total expected repair time of component i is

$$r_i = \frac{t_i \lambda_i(T_i) T_c}{T_i} + q \cdot T_c \cdot v(x_i, d) \cdot \tau_i.$$ (9.6)

The expected minimal repair cost of component i is

$$l_i = \frac{\sigma_i \lambda_i(T_i) T_c}{T_i} + q \cdot T_c \cdot v(x_i, d) \cdot \theta_i.$$ (9.7)

The availability of each element is

$$A_i = \frac{T_c - r_i}{T_c} = \frac{T_c - t_i \lambda_i(T_i) T_c / T_i - q T_c v(x_i, d) \tau_i}{T_c}.$$ (9.8)

The system capacity distribution must be obtained to estimate the entire system availability. It may be expressed by vectors $\boldsymbol{G}$ and $\boldsymbol{P}$. $\boldsymbol{G} = \{G_v\}$ is the vector of all the possible total system capacities, which corresponds to its V different possible states; $\boldsymbol{P} = \{p_v\}$ is the vector of probabilities, which corresponds to these states.

After obtaining the availability estimate for each system element, the entire system capacity distribution can be defined by using the algorithm presented in Sect. 9.3. If we denote the system demand as W, the unsupplied demand probability should be calculated as

$$P_{ud} = \sum_{v=1}^{V} p_v \cdot 1(W - G_v > 0).$$ (9.9)

The reliability of the entire system requires an availability index $A = 1 - P_{ud}$ that is not less than some preliminary specified level A^*.

The total unsupplied demand cost can be estimated with the following expression

$$C_{ud} = \alpha \sum_{v=1}^{V} p_v \cdot (W - G_v) \cdot 1(W - G_v > 0), \qquad (9.10)$$

where α is the cost of the unsupplied demand unit.

The general formulation of the system replacement versus protection optimization problem can be presented as follows:

Find the replacement intervals and protection efforts for system element $T = (T_1, T_2, \ldots, T_N)$ and $x = \{x_1, x_2, \ldots, x_N\}$ that minimize the sum of costs of the replacement, protection and unsupplied demand:

$$T, x = \arg\min\left\{ C_{ud}(T, x) + \sum_{i=1}^{N} a_i x_i + \sum_{i=1}^{N} (T_c/T_i - 1)C_i + \sum_{i=1}^{N} l_i \right\}; \qquad (9.11)$$

$$\text{subject to}: A \geq A^*.$$

9.3 System Availability Estimation Method

The entire system capacity distribution must be obtained in order to evaluate the availability index A and the total unsupplied demand cost C_{ud}. The procedure used in this paper for system capacity distribution evaluation is based on the universal z-transform, which was introduced in Ushakov (1986). The following is a brief introduction to this technique. The universal moment generating function (u-transform) of a discrete variable G is defined as a polynomial

$$u(z) = \sum_{j=1}^{J} p_j z^{g_j}, \qquad (9.12)$$

where the discrete random variable G has J possible values and p_j is the probability that G is equal to g_j. In our case, the polynomial $u(z)$ can define capacity distributions, meaning it represents all possible states of the system (or element) by relating the probabilities of each state p_j with capacity g_j of the system in this state.

Since each component i has a nominal performance g_i and its availability is A_i, the u-function of component i has only two terms and can be defined as

$$u_i(z) = (1 - A_i)z^0 + A_i z^{g_i}, \qquad (9.13)$$

The u-function of the whole system $u(z)$ can be obtained by using the π operator

$$u(z) = \pi(u_1(z), u_2(z), \ldots, u_N(z)) = \prod_{i=1}^{N} u_i(z). \qquad (9.14)$$

The π operator is a product of polynomials representing the individual u-functions. Each term of the resulting polynomial is obtained by multiplying the probabilities that correspond to different states of elements and by reaching a summation of the elements' capacities that correspond to these states.

The simple operator should be used to evaluate the probability that the random variable G represented by polynomial $u(z)$ defined in (9.12) does not exceed the value W:

$$P_{ud} = p(G \le W) = \delta(u(z), W) = \sum_{g_j \le W} p_j. \qquad (9.15)$$

Furthermore the availability index A of the entire system can be obtained as

$$A = 1 - P_{ud} = 1 - \sum_{g_j \le W} p_j. \qquad (9.16)$$

The total unsupplied demand cost can be estimated as

$$C_{ud} = \alpha \sum_{g_j \le W} p_j(W - g_j). \qquad (9.17)$$

9.4 Optimization Technique

Equation (9.11) formulates a complicated combinatorial optimization problem. An exhaustive examination of all possible solutions is not realistic, considering reasonable time limitations. In this work, the genetic algorithm (GA) is used to solve the optimal (T, x). Detailed information on GA and its basic operators can be found in Goldberg (1989), Gen and Chen (1997) and Lisnianski and Levitin (2003). The basic structure of the version of GA referred to as GENITOR is as follows (Whitley 1989).

First, an initial population of N_s randomly constructed solutions (strings) is generated. Within this population, new solutions are obtained during the genetic cycle by using crossover, and mutation operators. The crossover produces a new solution (offspring) from a randomly selected pair of parent solutions, facilitating the inheritance of some basic properties from the parents by the offspring. Mutation results in slight changes to the offspring's structure, and maintains a diversity of solutions.

Each new solution is decoded, and its objective function (fitness) values are estimated. These values, which are a measure of quality, are used to compare different solutions. The comparison is accomplished by a selection procedure that determines which solution is better: the newly obtained solution, or the worst solution in the population. The better solution joins the population, while the other is discarded. If the population contains equivalent solutions following selection, redundancies are eliminated, and the population size decreases as a result.

After new solutions are produced N_{rep} times, new randomly constructed solutions are generated to replenish the shrunken population, and a new genetic cycle begins.

The GA is terminated after N_c genetic cycles. The final population contains the best solution achieved. It also contains different near-optimal solutions which may be of interest in the decision-making process.

9.4.1 Solution Representation and Decoding Procedures

To apply the GA to a specific problem the solution representation and the decoding procedure must be defined.

Each solution is represented by string $S = \{s_1, s_2, ..., s_N\}$, where s_i corresponds to component i for each $i = 1, 2, ..., N$.

Each number s_i determines both the replacement interval of component i (T_i) and the protection effort allocated on component i (x_i). To provide this property all the numbers s_i are generated in the range

$$0 \le s_i < (M + 1) \cdot \Lambda, \tag{9.18}$$

where M is the maximum protection effort allowed to be allocated on a component and Λ is the total number of considered replacement frequency alternatives.

The solutions are decoded in the following manner:

$$x_i = [s_i/\Lambda], \tag{9.19}$$

$$v_i = 1 + \mathrm{mod}_\Lambda s_i, \tag{9.20}$$

where v_i is the number of replacement frequency alternative for component i, $[x]$ is the maximal integer not greater than x, and $\mathrm{mod}_x y = y - [y/x]x$. For given v_i and x_i the corresponding s_i is composed as follows:

$$s_i = x_i \cdot \Lambda + v_i - 1. \tag{9.21}$$

Note that all $s_i < \Lambda$ corresponds to the solutions where the component i is not protected.

The possible replacement frequency alternatives are ordered in vector $\boldsymbol{h} = \{h_1, h_2, ..., h_\Lambda\}$ so that $h_i < h_{i+1}$, where h_i represents the number of replacements during the operation period that corresponds to alternative i. After obtaining v_i from decoding the solution string, the number of replacement for component i can be obtained as

$$n_i = h_{v_i}. \tag{9.22}$$

Furthermore replacement interval for component i can be obtained as

$$T_i = \frac{T_c}{n_i + 1} = \frac{T_c}{h_{v_i} + 1}. \tag{9.23}$$

For each given pair of vectors (T, x) the decoding procedure first calculates l_i using (9.7) and determines the availability of each element using (9.8), after which the entire system capacity distribution can be obtained by using (9.13) and (9.14). The availability index A and the total unsupplied demand cost C_{ud} can be obtained using (9.15–9.17).

In order to let the genetic algorithm search for the solution with minimal total cost, when A is not less than the required value A^*, the solution quality (fitness) is evaluated as follows:

$$F(T,x) = \omega \cdot (A^* - A) \cdot 1(A^* - A) + C_{ud}(T,x)$$
$$+ \sum_{i=1}^{N} a_i x_i + \sum_{i=1}^{N} (T_c/T_i - 1)C_i + \sum_{i=1}^{N} l_i, \qquad (9.24)$$

where ω is a sufficiently large penalty.

For solutions that meet the requirements $A \geq A^*$, the fitness of the solution is equal to its total cost.

9.4.2 Crossover and Mutation Procedures

The cross operator for given parent strings $P1$, $P2$ and the offspring string O is defined as follows: the ith element ($1 \leq i \leq N$) of the string O is equal to the ith element of either $P1$ or $P2$ both with probability 0.5.

The mutation procedure swaps elements initially located in two randomly chosen positions.

9.5 Illustrative Example

The system considered in this example consists of eight parallel components. The lifetime of the system T_c is 120 months. The system demand W is 100.

We assume that totally $\Lambda = 6$ different replacement frequency alternatives are considered and the alternatives are $h = \{4, 9, 14, 19, 24, 29\}$. The corresponding alternatives for replacement interval are 24, 12, 8, 6, 4.8 and 4 months. The characteristics of the components are presented in Table 9.1.

For our example we assume that $q = 0.5$, $d = 30$, $\alpha = 3,000$ and the maximum protection effort allowed to be allocated on a component is $M = 50$. According to (9.18) we have $0 \leq s_i < (M + 1) \cdot \Lambda = 306$.

For a given solution string, T and x can be decoded using (9.18), (9.19) and (9.23). Thereafter (9.24) can be used to obtain the fitness function $F(T, x)$. For example, the solution string $S = \{195\ 280\ 49\ 218\ 176\ 132\ 270\ 120\}$ is decoded into $T = (6\ 4.8\ 12\ 8\ 8\ 24\ 24\ 24)$ and $x = (32\ 46\ 8\ 36\ 29\ 22\ 45\ 20)$. If we assume

Table 9.1 The characteristics of the components

i	1	2	3	4	5	6	7	8
G_i	10	10	12	12	14	15	15	20
a_i	7	8	8	9	10	10	11	12
C_i	100	100	110	110	120	140	140	150
t_i (month)	0.010	0.011	0.012	0.014	0.015	0.015	0.016	0.016
σ_i	8	8	8	7	8	6	8	7
τ_i (month)	0.012	0.014	0.015	0.016	0.017	0.017	0.018	0.018
θ_i	9	9	9	8	9	7	9	8
λ_i (4)	0.8	0.72	0.64	0.6	0.72	0.7	0.6	0.5
λ_i (4.8)	1.04	0.92	0.85	0.8	0.96	0.9	0.85	0.8
λ_i (6)	1.6	1.5	1.4	1.2	1.5	1.4	1.4	1.4
λ_i (8)	2.8	2.7	2.7	2.1	2.7	2.4	2.7	2.8
λ_i (12)	5.5	5.2	5.4	4.2	5.3	4.8	5.4	5.8
λ_i (24)	15	15	15	14	15	14	14	16

Table 9.2 Examples of solutions obtained

Constraints	T	x	A	C_{ud}	$F(T, x)$	Variation (%)
None	(24 24 24 24 24 24 24 24)	(22 17 21 14 16 14 16 15)	0.8970	2,044.6	14,206	0.014
$A^* = 0.90$	(24 24 24 24 24 24 24 24)	(24 23 23 18 20 15 23 21)	0.9000	1,973.9	14,240	0.34
$A^* = 0.95$	(6 4.8 6 6 6 8 6 6)	(44 44 49 50 47 43 47 45)	0.9500	935.64	25,941	0.57

$m = 1$ and $A^* = 0.90$, the fitness function for this solution takes the value $F(T, x) = 19,210$.

The problem is to find the optimal replacement and protection strategy (T, x) which minimizes $F(T, x)$ subject to the availability requirement $A > A^*$. As an illustration, we assume that $m = 1$. The optimal solutions obtained under different constraints are shown in Table 9.2. Each solution was obtained as the optimal one among five different runs of the GA with different randomly generated initial populations. The coefficients of variation among the values of $F(T, x)$ obtained in the five runs are also presented in Table 9.2. The low values of this coefficient evidence the good consistency of the GA.

With the increase of the reliability requirement more resources need to be put into protection actions and the components need to be replaced more frequently, thus the total cost increases. It can also be seen that C_{ud} decreases with the increase of A^*. Indeed, when the obtained system availability increases, the unsupplied demand decreases.

The maximum availability $A = 0.9592$ can be achieved when maximal possible protection and replacement frequency are applied. In this case all the components are replaced every 4 months and the protection effort on each component is 50. The corresponding total cost is 35,334.

9.6 Conclusions

This chapter considers a parallel system that can fail due to both internal failures and external impacts. It is assumed that the failure rate of each component is increasing over time and the failures between replacements are fixed by minimal repairs. The external impact frequency is assumed to be constant. The availability of an element can be increased in two ways: (1) replace the element more frequently (2) allocate more resource to protect it against external impacts. The system reliability is defined as the ability to maintain a specified performance level. In this chapter a framework is proposed to solve the optimal maintenance and protection strategy that provides the desired system reliability at minimum cost. A universal generating function technique is used to evaluate the system availability for any maintenance and protection policy. A genetic algorithm is used for the optimization.

A numerical example is shown in this paper. With the increase of the reliability requirement more resources need to be put into protection actions and the components need to be replaced more frequently, thus the total cost increases. Meanwhile with the increase of the obtained system availability, the unsupplied demand decreases. The maximum availability can be achieved when maximal possible protection and replacement frequency are applied.

References

Beichelt F, Fischer K (1980) General failure model applied to preventive maintenance policies. IEEE Trans Reliab 29(1):39–41

Beichelt F, Franken P (1983) Zuverlassigkeit und Instandhaltung. Verlag Technik, Berlin

Gen M, Cheng R (1997) Genetic algorithms and engineering design. Wiley, New York

Goldberg D (1989) Genetic algorithms in search, optimization and machine learning. Addison Wesley, Reading

Hausken K (2005) Production and conflict models versus rent seeking models. Public Choice 123(1–2):59–93

Hausken K, Levitin G (2008) Efficiency of even separation of parallel elements with variable contest intensity. Risk Anal 28(5):1477–1486

Levitin G, Lisnianski A (1999) Joint redundancy and maintenance optimization for multistate series-parallel systems. Reliab Eng Sys Saf 64(1):33–42

Lisnianski A, Levitin G (2003) Multi-state system reliability: assessment, optimization and applications. World Scientific, Singapore

Sheu SH, Chang CC (2009) An extended periodic imperfect preventive maintenance model with age-dependent failure type. IEEE Trans Reliab 58(2):397–405

Skaperdas S (1996) Contest success functions. Econ Theory 7(2):283–290

Tullock G (1980) Efficient rent-seeking. In: Buchanan JM, Tollison RD, Tullock G (eds) Toward a theory of the rent-seeking society. University Press, TX, pp 97–112

Ushakov I (1986) Universal generating function. Soviet J Comp Sys Sci 24(5):118–129

Whitley D (1989) The GENITOR algorithm and selective pressure: why rank-based allocation of reproductive trials is best. In: Schaffer D (ed) Proceedings of the 3rd international conference on genetic algorithms. Morgan Kaufmann, pp 116–121

Zhuang J, Bier VM (2007) Balancing terrorism and natural disasters-defensive strategy with endogenous attacker effort. Op Res 55(5):976–991

Chapter 10
Heuristic Optimization Techniques for Determining Optimal Reserve Structure of Power Generating Systems

Yi Ding, Lalit Goel, Peng Wang, Yuanzhang Sun, Poh Chiang Loh and Qiuwei Wu

Abstract Electric power generating systems are typical examples of multi-state systems (MSS). Sufficient reserve is critically important for maintaining generating system reliabilities. The reliability of a system can be increased by increasing the reserve capacity, noting that at the same time the reserve cost of the system will also increase. The reserve structure of a MSS should be determined based on striking a balance between the required reliability and the reserve cost. The objective of reserve management for a MSS is to schedule the reserve at the minimum system reserve cost while maintaining the required level of supply reliability to its customers. In previous research, Genetic Algorithm (GA) has been used to solve most reliability optimization problems. However, the GA is not very computationally efficient in some cases. In this chapter a new heuristic

Y. Ding (✉) · Q. Wu
Department of Electrical Engineering, Centre for Electric Technology,
Technical University of Denmark, Lyngby, Denmark
e-mail: yding@elektro.dtu.dk

Q. Wu
e-mail: qw@elektro.dtu.dk

L. Goel · P. Wang · P. C. Loh
Division of Power Engineering, School of Electrical and Electronic Engineering,
College of Engineering, Nanyang Technological University, Singapore, Singapore
e-mail: elkgoel@ntu.edu.sg

P. Wang
e-mail: epwang@ntu.edu.sg

P. C. Loh
e-mail: epcloh@ntu.edu.sg

Y. Sun
School of Electrical Engineering, Wuhan University, Wuhan, China
e-mail: yzsun@whu.edu.cn

A. Lisnianski and I. Frenkel (eds.), *Recent Advances in System Reliability*,
Springer Series in Reliability Engineering, DOI: 10.1007/978-1-4471-2207-4_10,
© Springer-Verlag London Limited 2012

optimization technique—the particle swarm optimization has been used to determine the optimal reserve structure for power generating systems, which can greatly improve the computational efficiency. The computational efficiency and accuracy of the proposed method have been compared with those of the GA technique in the illustrative example.

10.1 Introduction

Electric power generating systems are typical examples of multi-state systems (MSSs—Lisnianski and Levitin 2003) and can be represented using the corresponding equivalents (Wang et al. 2002; Wang and Billinton 2003). The Universal Generating Function (UGF) technique primarily introduced in Ushakov (1986, 1987) is the best-known method to calculate the reliability of a MSS. The paper (Levitin et al. 1998) proposes a system-structure optimization algorithm for MSSs. The algorithm proposed in Levitin (2001) is used to solve a component redundancy optimization problem for a MSS. The reserve sharing within a restructured power system through reserve agreements has been discussed in Wang and Billinton (2004). The reserve management of a generating system is a redundant resource problem and general methods have been developed in Ding et al. (2006). In order to improve the computational efficiency, in this chapter a new optimization technique based on Particle Swarm Optimization (PSO) algorithm has been used to determine the optimal reserve structure for power generating systems. The computational efficiency and accuracy of the proposed method have been compared with those of the Genetic Algorithm (GA) technique through an illustrative example.

10.2 Models and Problem Formulation

A generating system (GS) consists of generating units (GUs) with different reliabilities. A GS can operate in various states due to random failures. Different GSs are usually connected together through the supply network to share the reserve. A GS usually has the reserve contracts with other GSs to increase its reliability and to reduce the reserve cost (Ding et al. 2006). The state performance of a GS is determined by various factors such as the number of units, capacity and reliability of each unit, supply network and reserve contracts with other GSs. The reserve management of a GS is a complex optimization problem which belongs to system-structure optimization problems addressed in Ushakov (1994).

In order to determine the optimal reserve structure of a specific GS, the reliability model of the GS can be represented by a multi-state generating system (MSGS). The reliability models of all other GSs are represented by the multi-state reserve providers (MSRP). The reliability model of the supply network between

Fig. 10.1 Equivalent model of a CGS for the MSGS

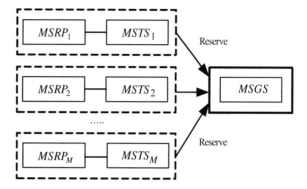

the MSGS and an MSRP is represented by the corresponding multi-state transmitting system (MSTS). The equivalent model of an interconnected generating system for the reserve optimization of the MSGS is shown in Fig. 10.1, which considers a CGS with a MSGS and M MSRPs and M MSTSs (Ding et al. 2006).

There are several factors determining the reserve contract between the MSGS and MSRP m: contractual reserve between the MSGS and MSRP m (CR_m), the contract type, unit cost of reserve capacity from MSRP m (ε_{Cm}), and unit cost of the utilized reserve provided by MSRP m (ε_{Um}) and also by the reliability of MSRP m and MSTS m. The determination of the optimal reserve structure includes the reserve contracts with MSRPs and the utilization order of MSRPs, under the minimum total reserve cost while satisfying the system reliability requirement. The total reserve cost includes the reserve capacity cost CC and the expected reserve utilization cost EC. The reserve optimization problem can be summarized as (Ding et al. 2006):

$$\text{Min} \sum_{m=1}^{M} (CC_m + EC_m), \tag{10.1}$$

$$CC_m = \varepsilon_{Cm} CR_m T, \tag{10.2}$$

$$EC_m = \sum_{n=1}^{N} \sum_{j=1}^{K} p_j \varepsilon_{Um} UR_{mj}^n T_n, \tag{10.3}$$

where UR_{mj}^n is the utilized reserve of MSRP m at load level n associated with the state j; T is the contract period; N is the number of load levels; K is the number of states; p_j is the probability of the state j.

Subject to the following constraints:

- The reserve constraints:

$$UR_{mj}^n \leq CR_m, \tag{10.4}$$

$$UR_{mj}^n \leq \min\{AR_m^{j_r}, AT_m^{j_t}\}, \tag{10.5}$$

where AR_m^{jr} is the available reserve capacity of MSRP m; AT_m^{ji} is the available transmission capacity of MSTS m.

- The reliability constraints:

$$LOLP \leq LOLP^{spec}, \tag{10.6}$$

$$EENS \leq EENS^{spec}, \tag{10.7}$$

where LOLP is the MSGS unavailability and defined as the loss of load probability. EENS is the expected energy not supplied in operation period T. $LOLP^{spec}$ and $EENS^{spec}$ are the specified reliability criteria.

- The load balance constraint:

$$AG^{jg} + \sum_{m=1}^{M} UR_{mj}^{n} = L^{n} - LL_{j}^{n}, \tag{10.8}$$

where L^n and LL_j^n are load level and load curtailment, respectively. Load curtailment LL_j^n is utilized when the total generating and reserve capacity cannot meet the load of the MSGS.

For solving the above optimization problem, the reliability indices have to be determined during the optimization process. The procedure for reliability evaluation using UGF has been shown to be very convenient for numerical realization and requires relatively small computational resources (Ushakov 1986; Lisnianski and Levitin 2003). The method of the UGF generalizes the technique that is based on using a well-known ordinary generating function, which was introduced by Ushakov (1986). The UGF technique is therefore used to evaluate the reliability indices in the reserve management problem. The UGFs for MSGS, MSRPs and MSTSs have been derived. The reliability indices for the MSGS considering the reserve agreements with MSRPs have been developed using the associated UGFs. Because of space limitations, more details about the reliability evaluation using UGF technique can be found in Ding et al. (2006).

10.3 Techniques for Solving Reliability Optimization Problems

There are many optimization methods which have been used for various reliability optimization problems (Lisnianski and Levitin 2003). The applied algorithms can be classified into two categories: heuristics and exact techniques. Most reliability optimization problems are solved by using general heuristic techniques based on artificial intelligence and stochastic techniques. Based on the classification of Lisnianski and Levitin (2003) and some recent research, the heuristics techniques include simulating annealing, ant colony, tabu search, GA and PSO.

The GA has been used to solve most reliability optimization problems. The GAs were first introduced in Holland (1975) and motivated by natural evolution phenomenon. In reliability engineering, there are many applications of GAs. A GA with the special encoding scheme that considers the structure of reserve capacity and reserve utilization order is developed for the proposed optimization problem (Ding et al. 2006).

However, the GA is not very computationally efficient in some cases. In order to improve the computational efficiency, a new optimization technique based on PSO algorithm has been used to determine the optimal reserve structure for power generating systems. The PSO was first described in Kennedy and Eberhart (1995), which is inspired from the biological behavior of bird swarms. There is some similarity between GA and PSO: a stochastic heuristic search is conducted by operating a population of solutions. However there are no evolution operators such as crossover and mutation in PSO (PSO Tutorial), which therefore is easy to be implemented. It is noticed that the information sharing mechanism between GA and PSO is totally different (PSO Tutorial): in GA the whole population of solutions moves relatively uniformly to the optimal area because solution chromosomes share information with each other; whereas in PSO only the solution parameters (gbest and pbest) send out information, which is one-way information sharing. Some recent research (PSO Tutorial; Hassan et al. 2005) concluded that the effectiveness of PSO is the same as that of GA but PSO may have better computational efficiency because the only one-way information sharing mechanism can improve the speed for searching for the best solution.

The PSO has been used to solve the optimal reactive power generation dispatch in Zhao et al. (2005). The economic dispatch problems with non-smooth cost functions have been proposed in Park et al. (2005), which has been solved by PSO. In some recent research (Zavala et al. 2005; Levitin et al. 2007), several reliability optimization problems have also been discussed by using the PSO technique.

The PSO imitates food searching behaviors of bird swarms by operating a population of solutions to look for the optimal points. The solutions in the PSO are called particles, which are determined by two parameters: velocity and position.

The position determines the fitness of a particle and velocity directs the flying of the particle (PSO Tutorial). During flying, the position of a particle is adjusted by both its experience and the experience of the bird swarm (Park et al. 2005), which is based on the best position visited by the particle and the best position visited by the swarm, respectively. The best positions visited by a particle and the swarm are presented as pbest and gbest, respectively. The velocity and the position of a particle in an updated iteration can be determined by the following equations (PSO Tutorial; Park et al. 2005):

$$V_i^{k+1} = V_i^k + c_1 \cdot \text{rand}_1 \cdot (p\text{best}_i^k - Pos_i^k) + c_2 \cdot \text{rand}_2 \cdot (g\text{best}^k - Pos_i^k), \quad (10.9)$$

$$Pos_i^{k+1} = Pos_i^k + V_i^{k+1}, \quad (10.10)$$

where V_i^k is the velocity of particle i for iteration k, Pos_i^k is the position of particle i for iteration k representing the associated fitness, c_1 and c_2 are parameters, $rand_1$ and $rand_2$ are random numbers between 0 and 1, $pbest_i^k$ is the best fitness value of particle i obtained as far, $gbest^k$ is the best fitness value of the swarm obtained as far.

The following steps have been developed to solve the optimization problem of reserve management, which are based on the PSO procedure (PSO Tutorial; Park et al. 2005):

- Step 1: Generate an initial population of solutions randomly in the search space. A particle is represented by a solution vector, which consists of the contracted reserve capacity for M MSRPs. Therefore $Pos_i^0 = (CR_1, \ldots, CR_m, \ldots, CR_M)$, where CR_m is the amount of the contracted reserve capacity of MSRP m. $V_i^0 = (v_1, \ldots, v_m, \ldots, v_M)$ represents the change of the contracted reserve capacity for M MSRPs.
- Step 2: Evaluate the fitness of each particle. The same procedures used in GA (Ding et al. 2006) are applied to obtain LOLP and EENS. Unlike GA, the reserve utilization order cannot be represented by PSO particles effectively. Therefore a heuristic algorithm is used to determine the reserve utilization order: the reserve of a MSRP with cheapest utilization cost is used first, the reserve of a MSRP with second cheapest utilization cost is used secondly, and the reserve of a MSRP with most expensive utilization cost is used finally, up to the point the load is met, or the available reserve is used up. Obviously the heuristic algorithm can minimize the reserve utilization cost in a contingency state.
- Step 3: Calculate the position and velocity for each particle in the swarm by using Eqs. (10.9) and (10.10). The updated positions of particles must be located in the search space. If a particle violates its constraints, its position is adjusted to the border point of search space.
- Step 4: If the position of a particle i is the best value so far, set it as the new $pbest_i^k$.
- Step 5: Set the $gbest^k$ as the best value of $pbest_i^k$ in the swarm.
- Step 6: Repeat the steps 2–4, until the stopping criterion is satisfied. The stopping criterion of the PSO can be the fixed number of computing cycles or that the solution of the proposed PSO reaches a set optimal value. In order to compare the performance of the GA with the PSO, the stopping criterion of the PSO is that the solution of the PSO reaches the optimal value obtained by the GA in this chapter.

10.4 System Studies

The restructured IEEE-RTS (IEEE Task Force 1979) is used to illustrate the technique. The restructured IEEE-RTS consists of six MSGSs. In this example, the optimal reserve structure of the specific MSGS, which has reserve agreements with

Fig. 10.2 Convergence characteristics of the GA for first example

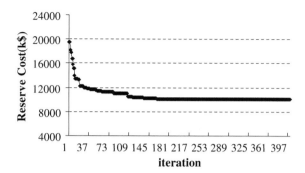

other five MSGSs, is determined using the proposed technique. The MSGS owns 6 × 50 MW hydro units, 5 × 12 MW oil units, 2 × 155 MW coal units and 2 × 400 MW nuclear units. MSRP1 owns 2 × 20 MW gas turbine units and 2 × 76 MW coal units. MSRP 2 owns 3 × 100 MW oil units. MSRP 3 owns 2 × 20 MW gas turbine units and 2 × 76 MW coal units. MSRP 4 owns 3 × 197 MW oil units. MSRP 5 owns 2 × 155 MW coal units and one 350 MW coal unit. The MSRPs are connected to the MSGS through tie lines.

All the MSRPs are firm contract providers for the MSGS. The utilization prices used in the simulation for each MSRP were set as 25, 55, 75, 60 and 20$/MWh, respectively. We also assume that capacity price for each MSRP were 2.5, 5.5, 7.5, 6 and 2$/MWh, respectively. The time frame for the contracts is set as one year. Two examples with different reliability criteria were investigated in this case. In the first example, the *LOLP* and *EENS* criteria for the MSGS were set as 0.01 and 500 MWh/year, respectively. In the second example, the reliability criteria were tightened and the *LOLP* and *EENS* for the MSGS were set as 0.0005 and 50 MWh/year, respectively.

Firstly the GA proposed in Ding et al. (2006) was used to determine the optimal reserve structure. The stopping criterion is set as performing at least 400 times of genetic cycles and there are at least five consecutive genetic cycles without improving the solution performance in a defined tolerance (Lisnianski and Levitin 2003). The population size in the GA is 50. The convergence characteristics of the proposed GA for the first and second examples are illustrated in Figs. 10.2 and 10.3, respectively.

It can be seen from Figs. 10.2 and 10.3 that the GA converges to near-optimal solutions by performing about 160 and 90 iterations for first and second examples, respectively. The execution times for performing 400 iterations in first and second examples are 7,140.3 and 7,426.4 s, respectively. The execution time for 160 iterations for first example is about 2,860 s and the execution time for 290 iterations for the second example is about 5,384 s.

Table 10.1 shows the total reserve cost and the reserve capacities between the MSGS and MSRPs for the two examples evaluated by the GA. The capacity reserve cost and expected cost of reserve utilization for each MSRP are shown in

Fig. 10.3 Convergence
characteristics of the GA
for second example

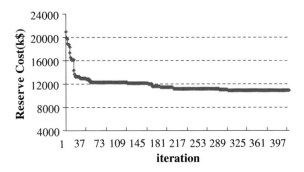

Table 10.1 The total reserve cost and reserve structure evaluated by GA

	Example 1	Example 2
Total reserve cost (k$)	10,197.0	10,896.1
CR_{MSRP1} (MW)	53.728	70.355
CR_{MSRP2} (MW)	1.789	0.0185
CR_{MSRP3} (MW)	0.298	0.0186
CR_{MSRP4} (MW)	6.495	0.0524
CR_{MSRP5} (MW)	374.82	419.02

Table 10.2 The capacity cost and expected cost of utilized reserve evaluated by GA

MSRP No.	Example 1		Example 2	
	CC_m (k$)	EC_m (k$)	CC_m (k$)	EC_m (k$)
1	1,172.6	24.7	1,536.1	18.2
2	86.0	1.02	0.89	0.0013
3	19.5	0.23	1.22	0.0017
4	340.4	4.05	2.74	0.004
5	6,549.4	1,999.0	7,321.2	2,015.7

Table 10.2. It can be seen from Tables 10.1 and 10.2 that the reserve cost increases
if we improve the reliability requirements.

Secondly the proposed PSO is used to determine the optimal reserve structure.
In order to compare the computational efficiency with the GA, the stopping
criterion is set as the solutions of the PSO to reach the optimal value obtained by
the GA. The population size in the PSO is 50.

In the first example, the execution time and the number of iterations for
obtaining the optimal values are 808.5 s and 99 iterations. In the second example,
the execution time and the number of iterations for obtaining the optimal values
are 229.4 s and 26 iterations. The total reserve cost and the reserve capacities for
the two examples evaluated by the proposed PSO are shown in Table 10.3. The
capacity reserve cost and expected cost of reserve utilization evaluated by the
proposed PSO are shown in Table 10.4.

Obviously the computational efficiency of the proposed PSO is much better
than that of the proposed GA caused because of two reasons (PSO Tutorial; Park

Table 10.3 The total reserve cost and reserve structure evaluated by PSO

	Example 1	Example 2
Total reserve cost (k$)	10,160.0	10,870.0
CR_{MSRP1} (MW)	41.777	53.844
CR_{MSRP2} (MW)	3.274	0
CR_{MSRP3} (MW)	0	0
CR_{MSRP4} (MW)	6.493	0
CR_{MSRP5} (MW)	384.81	438.48

Table 10.4 The capacity cost and expected cost of utilized reserve evaluated by PSO

MSRP No.	Example 1		Example 2	
	CC_m (k$)	EC_m (k$)	CC_m (k$)	EC_m (k$)
1	912.41	9.64	1,175.9	11.71
2	157.31	1.66	0	0
3	0	0	0	0
4	340.3	3.60	0	0
5	6,723.4	2,011.3	7,661.2	2,021.2

et al. 2005). Firstly, the PSO implements the one-way information sharing mechanism, in which only $p$best and $g$best give out information to other particles (Park et al. 2005). This kind of information sharing mechanism can make the population converge more quickly to the best solutions than the multi-way information sharing mechanism used by the GA. Secondly, the crossover and mutation operations implemented by the GA can be very time consuming, while the PSO only uses the "velocity" to update the positions of its particles.

Comparing the results shown in Tables 10.3 and 10.4 with those in Tables 10.1 and 10.2, the percentage differences for the total reserve cost evaluated by the GA and PSO are very small, which are 0.36 and 0.24% in the first and second examples, respectively. The percentage differences for the CR_{MSRP5} evaluated by the GA and the PSO are also small (2.7 and 4.6% for the two examples) because MSRP5 is the cheapest reserve provider and has the major influence on the value of the reserve cost. The percentage differences for other contracted reserve capacities evaluated by the GA and the PSO are relatively large. This is because their contributions to the reserve cost are small. Therefore even if the variations of these contracted reserve capacities are relatively large, their effects on the change of reserve cost is still small.

10.5 Conclusions

In a restructured power system, reliability is not uniform and economical considerations have become more and more important. The reserve should be determined based on striking a balance between the required reliability and the reserve cost. This chapter illustrates a practical technique for optimal reserve management of a power generating system, which is based on the particle swarm optimization method. The particle swarm optimization method can greatly improve the computational efficiency compared with the GA technique.

Acknowledgment This research was partially supported by State Key Lab. of Power System, Tsinghua University, Beijing 100084, China.

References

Ding Y, Wang P, Lisnianski A (2006) Optimal reserve management for restructured power generating systems. Reliab Eng Sys Saf 91(7):792–799

Hassan R, Cohanim B, de Weck O (2005) A comparison of particle swarm optimisation and the genetic algorithm. American Instituteof Aeronautics and Astronautics (AIAA): structures, structural dynamics and materials conference, 18–21 April 2005

Holland JH (1975) Adaptation in natural and artificial systems. The University of Michigan Press, Ann Arbor, Michigan

IEEE Task Force (1979) IEEE reliability test system. IEEE Trans Power Apparatus Sys 98: 2047–2054

Kennedy J, Eberhart RC (1995) Particle swarm optimization. In: Proceedings of IEEE international conference on neural networks, Piscataway, NJ, pp 1942–1948

Levitin G (2001) Redundancy optimization for multi-state systems with fixed resource-requirements and unreliability sources. IEEE Trans Reliab 50(1):52–59

Levitin G, Lisnianski A, Ben-Haim H, Elmakis D (1998) Redundancy optimization for series–parallel multi-state systems. IEEE Trans Reliab 47(2):165–172

Levitin G, Hu XH, Dai YS (2007) Particle swarm optimization in reliability engineering. In: Levitin G (ed) Computational intelligence in reliability engineering. Springer, Berlin, pp 83–112

Lisnianski A, Levitin G (2003) Multi-state system reliability: assessment, optimization, applications. World Scientific, Singapore

Park JB, Lee KS, Shin JR, Lee KY (2005) A particle swarm optimization for economic dispatch with nonsmooth cost functions. IEEE Trans Power Sys 20(1):34–42

Ushakov IA (1986) Universal generating function. Soviet J Com Sys Sci 24(5):118–129

Ushakov IA (1987) Optimal standby problems and a universal generating function. Soviet J Comp Sys Sci 25(4):79–82

Ushakov IA (1994) Handbook of reliability engineering. Wiley, New York

Wang P, Billinton R (2003) Reliability assessment of a restructured power system using reliability network equivalent techniques. IEE Proc Gener Transm Distrib 150(5):555–560

Wang P, Billinton R (2004) Reliability assessment of a restructured power system considering reserve agreements. IEEE Trans Power Sys 19(2):972–978

Wang P, Billinton R, Goel L (2002) Unreliability cost assessment of an electric power system using reliability network equivalent approaches. IEEE Trans Power Sys 17(3):549–556

PSO Tutorial. http://www.swarmintelligence.org/tutorials.php

Zavala AEM, Diharce ERV, Aguirre AH (2005) Particle evolutionary swarm for design reliability optimization. In: Coello Coello CA et al (eds) Evolutionary multi-criterion optimization. Springer, Berlin, pp 856–869

Zhao B, Guo CX, Cao YJ (2005) A multiagent-based particle swarm optimization approach for optimal reactive power dispatch. IEEE Trans Power Sys 20(2):1070–1078

Chapter 11
Determination of Vital Activities in Reliability Program for Multi-State System by Using House of Reliability

Shuki Dror and Kobi Tsuri

Abstract This paper presents an innovative method that enables a company to determine its vital activities in reliability program for multi-state system (MSS). The method is based on a house of reliability for translating the system's failure costs into relative importance of corresponding activities listed in the reliability program. A Mean Square Error (MSE) criterion supports the selection of vital reliability program activities. It divides a set of activities in reliability program into two groups: vital few (activities in reliability program for MSS) and trivial many. The partition minimizes the overall MSE and so, delineates two homogeneous groups. A case study is presented to illustrate the application of the developed methodology in a warfare system (a tank). The vital reliability program activities—treatment routine and spare parts storage were found to be the best activities for reducing the costs of the tank's failures.

Keywords Multi-state system · Reliability program · House of Reliability · Mean Square Error

S. Dror (✉) · K. Tsuri
Department of Industrial Engineering and Management,
Ort Braude College of Engineering,
P.O. Box 78, Karmiel, Israel
e-mail: dror@braude.ac.il

K. Tsuri
e-mail: zuria10@walla.com

A. Lisnianski and I. Frenkel (eds.), *Recent Advances in System Reliability*,
Springer Series in Reliability Engineering, DOI: 10.1007/978-1-4471-2207-4_11,
© Springer-Verlag London Limited 2012

11.1 Introduction

Reliability is an important factor in the management, planning and design of an engineering product. In order to meet the customer's requirements and achieve desired reliability level different types of activities should be performed by the system developer/manufacturer and customer. These activities have to be listed in corresponding reliability program and include inspections, spare parts storage/providing, failure analysis etc. (Dhillon 1999). On one hand, each activity from reliability program list requires some cost for its realization. On the other hand, when any activity was not performed, we should expect some losses because of system's failures. A firm should define the list of activities that could provide the desired reliability level with minimum reliability program costs. The problem of reliability program determination is solved for binary-state systems by using qualitative approach according with general guidelines (Ireson et al. 1996; Pecht 2009). For multi-state systems (MSSs) the problem is much more complex and has not even been considered in the literature till now (Liu and Kapur 2008).

This paper presents a first step in this direction. It presents the method based on which one can divide the reliability program activities for MSS into two groups. The first one is the group with vital activities and the second one is the group where the activities are essentially less important. The knowledge about vital reliability program activities is very useful and can help to reduce reliability associated costs.

The work is based on the developing a House of Reliability (HOR) for translating the MSS's failure costs into relative importance of the MSS reliability program activities. An idea to use Quality Function Deployment (QFD) in reliability engineering was primarily suggested by Yang and Kapur (1997). In their approach the QFD table was integrated with failure analysis tools such as Failure Mode and Effects Analysis and Fault Tree Analysis.

The above studies provide good reasons for developing reliability program activities providing failure costs reduction, but offers no mechanism for building and maintaining the relevance of these activities. Nevertheless, to be effective, a system's reliability program activities have to be considered in conjunction with corresponding failure costs. The current study indicates that basic guidelines are needed for selecting the reliability program activities vis-à-vis the failure costs of an individual system (Dror 2010).

In this work a HOR is building on the House of Quality Cost (HOQC) (Dror 2010). The methodology combines the following methods: the HOR method, a QFD matrix, which translates the desired improvement in failure costs into reliability program activities and ranks them by relative importance, the Analysis of Variance (ANOVA) method, which supports selection of vital reliability program activities. A case study is presented to illustrate the application of the developed methodology.

The paper proceeds as follows: Section 11.2 introduces the methodology. Section 11.3 presents a case study and Sect. 11.4 sums up and concludes the paper.

11.2 Method Description

11.2.1 House of Reliability

QFD is a method for structured product planning and development that enables a development team to clearly specify a customer's wants and needs, and then to systematically evaluate each proposed product or service capability in terms of its impact on meeting these needs (Cohen 1995). Typically, the method is described in terms of a Four-Phase Model, comprising four successive stages or matrices: an overall customer requirement planning matrix (also called the House of Quality), a final product characteristic deployment matrix, a process plan and quality control charts, and, finally, operating instructions (e.g., Akao 1990; Hauser and Clausing 1988; Wasserman 1993).

A HOQC is a QFD matrix to translate the desired improvement in failure costs (internal and external) into controllable efforts ranked according to their relative importance (prevention and appraisal costs). The output of a QFD matrix is expressed in terms of the relative importance of each column with respect to the summed up importance over all the columns. First, the absolute importance of each column is calculated. This is done by multiplying the importance of each row intersected by the respective strength of the relationship. A nonlinear scale such as (1, 3, 9) is used for highlighting the strong relationships. The values thus obtained are summed up over all the intersected rows. Then, the absolute importance of each column is normalized with respect to the total importance of all columns in the matrix.

The building sequence of the HOQC comprises the following five major components:

1. The firm's processes—causes/sources of failure (the WHATs).
2. The required improvement level of the WHATs.
3. The controllable efforts of the quality system (the HOWs).
4. The impact of each controllable effort on each process.
5. The required improvement level of the HOWs.

Based on the HOQC we developed a HOR for translating the improvement needs of a system's failure costs into relative importance of its reliability program activities.

The building sequence of the HOR comprises the following five major components:

1. The system's technologies (the WHATs)
2. The required improvement level of the WHATs—normalized failure costs (considered here as the Voice of Customer (VoC).

3. The reliability program activities of the system (the HOWs).
4. The impact of reliability program activities on each failure cost.
5. The required improvement level of the HOWs.

The calculation of the required improvement level of the HOWs is detailed below: Let $\mathbf{h} = (h_1, \ldots, h_p)$ be a vector of the required improvement level of the HOWs, $\mathbf{w} = (w_1, \ldots, w_q)$ be a vector of required improvement level of the WHATs, and $\mathbf{R}$ be an $q \times p$ matrix expresses the relationship strengths between the WHATs and the HOWs. So, $\mathbf{h} = \mathbf{w} \cdot \mathbf{R}$.

A QFD matrix is typically carried out by teams of multidisciplinary representatives from all stages of product development and manufacturing. For improving the understanding of the logical relationships between the WHATs and the HOWs of the HOR, a cross functional team is established. It might include reliability engineers, maintenance technicians as well as technical engineers and R&D representatives. Among its tasks, the team organized the process of extracting input information for HOR matrix.

11.2.2 Analysis of Variance for Selecting the Vital Few Reliability Program Activities

ANOVA is a method for decomposing the total variability in a set of observations, as measured by the sum of the squares of these observations from their average SST, into component sums of the squares that are associated with specific, defined sources of variation. In a one-way ANOVA, there are two sources of variation: the sum of the squares of the differences between the group means and the grand mean—denoted SSB—and the sum of the squares of the differences between group observations and the group mean—denoted SSE. The model for a simple ANOVA with one type of group at different levels is SST = SSB+SSE.

The Mean Square Error (MSE) is an unbiased estimator of σ^2. The Mean Square between (MSB) groups' estimates σ^2 plus a positive term that incorporates variations due to the systematic difference in group's means. The F-statistic is used to test for significant differences between the means of two or more groups. In one-way ANOVA, statistical significance is tested by comparing the F test statistic, F = MSB/MSE, to the F-distribution (Montgomery 2008).

The ANOVA method is used here for selecting the vital few reliability program activities.

Dror and Barad (2006) utilized the MSE criterion, introduced previously by Taylor (2000) as a quantitative tool for implementing the Pareto Principle. This principle was presented by Juran and De Feo (2010) as a universal principle he referred to as the "vital few and trivial many". The Pareto Principle (also known as the 80–20 rule, the law of the vital few, and the principle of factor scarcity) states that, for many events, roughly 80% of the effects come from 20% of the causes.

Here the MSE criterion supports the selection of the vital few reliability program activities (Dror and Barad 2006):

1. Arrange the normalized required improvement levels of the k components in descending order, where p_1 represents the highest improvement level needed and p_k the lowest improvement level needed, $0 \leq p_j \leq 1$, $j = 1, \ldots, k$.
2. While maintaining this order, divide the k components into two groups, A and B. Group A consists of the first m components, while group B comprises the remaining $k - m$ components. Assuming that each group includes at least one component, there are $k - 1$ possibilities for selecting an m value for dividing the items into two groups.
3. Calculate MSE(m), $m = 1, \ldots, k - 1$ using the following equation,

$$\text{MSE}(m) = \left\{ \sum_{j=1}^{m} \left(p_j - \bar{p}_A \right)^2 + \sum_{j=m+1}^{k} \left(p_j - \bar{p}_B \right)^2 \right\}, \tag{11.1}$$

where $\bar{p}_A$ and $\bar{p}_B$ are the average improvement levels in vital group A and in trivial group B, respectively.

4. Find,

$$\text{MSE}(m^*) = \underset{1 \leq m \leq k-1}{\text{Min}} \left[\text{MSE}(m) \right]. \tag{11.2}$$

11.3 Case Study

This section describes the implementation of the methodology presented in Sect. 11.2 in a warfare system (a tank) as it actives in a field environment and in operative mode.

We identify the main failures in a mechanical system (a tank) with respect to the cost of reliability program activities. The purpose of the case study is to determine the resource partition in a *reliability program* in order to reduce failure costs.

11.3.1 Warfare System

A tank is a tracked, armored fighting vehicle designed for front-line combat which combines operational mobility and tactical offensive and defensive capabilities. Firepower is normally provided by a large-caliber main gun in a rotating turret and secondary machine guns, while heavy armor and all-terrain mobility provide protection for the tank and its crew, allowing it to perform all primary tasks of the armored troops on the battlefield (Fig. 11.1).

It built of three mine systems:

- Mechanical system contains the engine, track and wheels system and active protection system.
- Turret system contains the cannon, machine guns and traverse system.
- Fire control system components.

11.3.2 Reliability Program Activities

There are several activities taking place in order to maintain high level of reliability:

- Authorization and training—The first step of building a maintenance foundation is to authorize and train manpower to carry out all reliability program activities, repair failures and preserve a high level of fitness of the tank.
- Routine treatments—Another mine element is the routine treatments of the armored vehicle. This usually determined by the manufacturer using break-down tests and failure analysis. For example: treatment determine by mileage (like 10,000 km) or every constant period of time (week, month and so forth).
- Failure analysis—Failure analysis is the process of collecting and analyzing historic data to determine the root cause of a failure. It is an important discipline that helps decide how to divide resources and prevent similar failures to repeat.
- Inspections—Inspections are being made by professionals according to a planned schedule or unplanned inspection.
- Spare part storage—Defective spare parts consumed in maintenance operations may cause repeated failures (rejects) therefore; it is a major concern to store spare parts properly. Those five elements contribute to the effectiveness and quality of the reliability program.

11.3.3 HOR Results

In order to obtain the necessary data for calculating the HOR input (the average failure cost per month (in thousands of dollars) of each type of failure), files relevant to the functional and operational aspects of the organization were analyzed, include specialist's opinion and financial reports.

The core of the HOR matrix presents the strength of the relationship between each failure cost and each reliability program activities. This allows the HOR input to be translated into the HOR output. The HOR output represents the extent of effort one has to invest in order to reduce failure costs.

Table 11.1 presents the HOR of a warfare system—a tank. The five components of the HOR are detailed below:

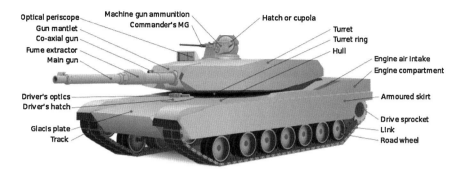

Fig. 11.1 A warfare system

1. The types of failures exist in a tank (the WHATs): electrical failure, mechanical failure, platform failure, turret failure and fire control failure.
2. Each failure as an estimated cost requires keeping the tank in proper condition. The cost composed of three main elements: spare parts cost, labor cost and support means cost (work tools, workshops, leafs). The element that affects the most is spare parts cost therefore mechanical, platform and fire control failure are rated highest. The corresponding required improvement level of the WHATs: 0.09, 0.31, 0.23, 0.08 and 0.29.
3. Reliability program activities of the system (the HOWs): treatment routine, authorization and training, failure analysis, inspections and spare parts storage.
4. The relationship strengths between the WHATs and the HOWs: electrical failures are strongly affected by authorization and trainings. There is a strong relationship between platform failure and treatment routine, and by spare parts storage. There is a strong relationship between turret failure and treatment routine.
5. The required improvement level of the reliability program activities (the HOWs): 0.42, 0.14, 0.10, 0.08, 0.26.

11.3.4 ANOVA Partitioning Results

The MSE criterion is utilized for selecting the vital reliability program activities to be selected.

There are five reliability program activities that have to be divided into two groups. The required improvement levels of the HOWs, arranged in descending order, are: 0.42, 0.26, 0.14, 0.10, 0.08.

The partitioning results—the selected vital reliability program activities—are presented in Table 11.2.

The lowest $MSE(m)$ is obtained for $MSE(2) = 0.015$. Therefore, the vital few reliability program activities are the first two on the list: treatment routine and spare parts storage.

Table 11.1 HOR of a warfare system

		HOR Input		Reliability program activities				
		Cost	Normalized	Treatment routine	Authorization and training	Failure analysis	Inspections	Spare parts storage
Failure costs	Electrical failure	30	0.09	3	9	3	3	1
	Mechanical failure	100	0.31	9	1	3	1	1
	Platform failure	75	0.23	9	0	0	3	9
	Turret failure	28	0.08	9	3	1	1	9
	Fire control failure	95	0.29	3	3	1	0	3
HOR output	Required improvement level of reliability program activities			6.71	2.25	1.56	1.35	4.07
	Normalized			0.42	0.14	0.10	0.08	0.26

Table 11.2 ANOVA for selecting the vital reliability program activities

m	The vital reliability program activities	The trivial many reliability program activities	MSE(m)
1	0.42	0.26, 0.14, 0.10, 0.08	0.020
2	0.42, 0.26	0.14, 0.10, 0.08	0.015
3	0.42, 0.26, 0.14	0.10, 0.08	0.040
4	0.42, 0.26, 0.14, 0.10	0.08	0.062

11.3.5 Implications of the Results

1. Most of the failure enquires the use of spare parts. It is crucial in every maintenance action to ensure the parts are undamaged. Spare part storage may refer to conditions, procedure or processing step at which hazards can be controlled inside the storeroom. At these activities, the defective can be prevented, eliminated or reduced to acceptable levels. Examples are: storeroom design, specific improvement procedures and preventive of reject failures.
2. Every equipment has a set of manufacturer's instructions in order to obtain an acceptable level of reliability. Those reliability program activities can be occur according to a certain mileage, engine hours or in time intervals.

A right treatment routine will ensure a high quality performance and will extend the life of the equipment.

11.4 Conclusions

In order to meet the customer's requirements and achieve desired reliability level different types of activities should be performed by the system developer/ manufacturer and customer. In this work an enhanced QFD method for selecting the vital activities of a reliability program is applied. QFD is usually applied in a wide variety of products and services and considered as a key practice of Design for Six Sigma.

Here the method was constructed to support the concept that failure costs can continue to decline over time with no corresponding increase in reliability program activities costs (Marcellus and Dada 1991; Ittner 1996).

The other existing methods provide good reasons for developing reliability program activities providing failure costs reduction, but offers no mechanism for building and maintaining the relevance of these activities. Nevertheless, to be effective, a system's reliability program activities have to be considered in conjunction with corresponding failure costs.

Our method (The HOR supported by the MSE criterion) reveals the uniqueness of the reliability program activities structure to be adopted by an individual organization. In the case study, vital reliability program activities, treatment routine and spare parts storage were found to be the best controllable efforts for reducing the tank's failure costs.

The method applied in this work proved itself capable of effectively supports the selection of vital reliability program activities. It emphasizes adopting a systemic approach for selecting the vital reliability program activities in response to failure costs.

A direction for future research is to apply the proposed method—in other areas of a MSS.

References

Akao Y (1990) Quality function deployment: integrating customer requirements into product design. Productivity Press, Cambridge

Cohen L (1995) Quality function deployment: how to make QFD work for you. Addison-Wesley, MA

Dhillon B (1999) Design reliability: fundamentals and applications. CRC Press, Boca Raton

Dror S (2010) A methodology for realignment of quality cost elements. J Model Manag 5(2): 142–157

Dror S, Barad M (2006) House of strategy (HOS)—from strategic objectives to competitive priorities. Int J Prod Res 44(18–19):3879–3895

Hauser J, Clausing D (1988) The house of quality. Harv Bus Rev 66(3):63–73

Ireson WG, Coombs C, Moss R (1996) Handbook of reliability engineering and management. McGraw-Hill, New York

Ittner CD (1996) Exploratory evidence on the behavior of quality costs. Op Res 44(1):114–130

Juran J, De Feo J (2010) Juran's quality handbook: the complete guide to performance excellence. McGraw-Hill Professional, New York

Liu YW, Kapur K (2008) New models and measures for reliability for multistate systems. In: Misra K (ed) Handbook of performability engineering. Springer, London, pp 431–446

Marcellus RL, Dada M (1991) Interactive process quality improvement. Manag Sci 37(11): 1365–1376

Montgomery D (2008) Introduction to statistical quality control. Wiley, New York

Pecht M (2009) Product reliability maintainability and supportability handbook. CRC Press, Boca Raton

Taylor W (2000) Change-point analysis: powerful new tool for defecting changes. Taylor Enterprises, Libertyville. http://www.variation.com/cpa/tech/changepoint.html

Wasserman G (1993) On how to prioritize design requirements during the QFD planning process. IIE Trans 25(3):59–65

Yang K, Kapur K (1997) Customer driven reliability: integration of QFD and robust design. In: Proceedings of annual reliability and maintainability symposium, Philadelphia, PA, USA, pp 339–345, 13–16 January 1997

Chapter 12
Multi-State Availability Modeling in Practice

Kishor S. Trivedi, Dong Seong Kim and Xiaoyan Yin

Abstract This chapter presents multi-state availability modeling in practice. We use three analytic modeling techniques; (1) continuous time Markov chains, (2) stochastic reward nets, and (3) multi-state fault trees. Two case studies are presented to show the usage of these modeling techniques: a simple system with two boards and the processors subsystem of the VAXcluster. The three modeling techniques are compared in terms of the solution accuracy and the solution time.

12.1 Introduction

There are systems which have multiple states and whose components also have multiple states. In order to capture such multi-state system availability, many techniques have been proposed. In this chapter, we provide three analytic modeling techniques which are useful in modeling multi-state availability; continuous time Markov chains (CTMC), stochastic reward nets (SRN) (Trivedi 2001), and multi-state fault trees (MFTs). Two case studies are shown including a system with two boards and the processors subsystem of the VAXcluster.

K. S. Trivedi (✉) · D. S. Kim · X. Yin
Department of Electrical and Computer Engineering, Duke University, Room 130, Hudson Hall, Durham, NC 27708 USA
e-mail: kst@ee.duke.edu

D. S. Kim
e-mail: dk76@duke.edu

X. Yin
e-mail: xy15@duke.edu

A. Lisnianski and I. Frenkel (eds.), *Recent Advances in System Reliability*,
Springer Series in Reliability Engineering, DOI: 10.1007/978-1-4471-2207-4_12,
© Springer-Verlag London Limited 2012

Fig. 12.1 The two boards
system

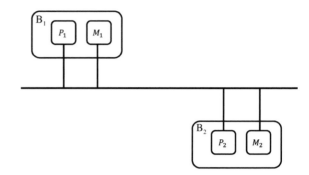

The rest of this chapter is organized as follows. Multi-state reliability modeling techniques for a simple system with two boards are described in Sect. 12.2. Multi-state availability modeling techniques for the processors subsystem of the VAXcluster are described in Sect. 12.3. The three modeling techniques are compared in terms of model solution accuracy and execution time. The chapter concludes in Sect. 12.4.

12.2 Two Boards System

Figure 12.1 shows a system with two boards (B_1 and B_2) where each board has a processor and a memory (Zang et al. 2003). The memories (M_1 and M_2) are shared by both the processors (P_1 and P_2). The processor and memory on the same board can fail separately, but statistically-dependently. Assume that the time to failure of a processor and the time to failure of a memory are exponentially distributed with rates λ_p and λ_m, respectively. The time to the common-cause failure (i.e., both a processor and the memory on the same board fail simultaneously) is exponentially distributed with rate λ_{mp}.

We consider each board as a component with four states:

- Component State 1: both P and M are down.
- Component State 2: P is functional, but M is down.
- Component State 3: M is functional but P is down.
- Component State 4: both P and M are operational.

The system states are defined as follows:

- System State 1 (S_1): either no processor or no memory is operational and hence the system is down.
- System State 2 (S_2): at least one processor and exactly one memory are operational.
- System State 3 (S_3): at least one processor and both the memories are operational.

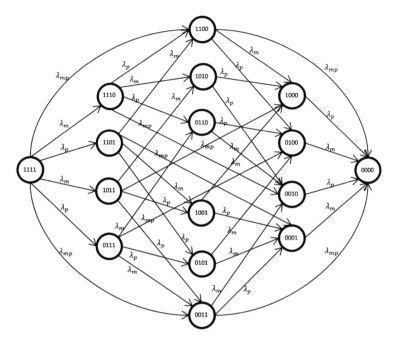

Fig. 12.2 CTMC model for the two boards system

12.2.1 Continuous Time Markov Chain Model

A complete (homogeneous) CTMC reliability model is constructed to capture the system's behavior as shown in Fig. 12.2. The states of the Markov chain are represented by $(P_1M_1P_2M_2)$, where 1 denotes up and 0 denotes down for each device.

Denote the probability that the system in state i at time t by $\pi_{S_i}(t)$. By solving the CTMC model and using reward rate assignment, we compute the expected reward rate at time t, $\pi_{S_i}(t)$ using the following formula:

$$\pi_{S_i}(t) = \sum_j r_{i,j} \cdot \pi_j(t), \tag{12.1}$$

where the reward rate assignment is:

$$
r_{1,j} = \begin{cases} 1, & j = (1010),(0101),(1000),(0100),(0010),(0001),(0000), \\ 0, & \text{otherwise}, \end{cases}
$$

$$
r_{2,j} = \begin{cases} 1, & j = (1110),(1011),(1100),(0110),(1001),(0011), \\ 0, & \text{otherwise}, \end{cases} \tag{12.2}
$$

$$
r_{3,j} = \begin{cases} 1, & j = (1111),(1101),(0111), \\ 0, & \text{otherwise}. \end{cases}
$$

Table 12.1 Results of the CTMC model for the two boards system

t (h)	$\pi_{S_1}(t)$	$\pi_{S_2}(t)$	$\pi_{S_3}(t)$
0	0	0	1
100	2.03280585e-002	1.38357902e-001	8.41314039e-001
200	7.02653699e-002	2.29253653e-001	7.00480977e-001
300	1.37122363e-001	2.84435223e-001	5.78442414e-001
∞	1	0	0

Table 12.1 shows the expected reward rate at time t, $\pi_{S_i}(t)$ of the CTMC model at different time epochs with parameters, $1/\lambda_p = 1000$ h, $1/\lambda_m = 2000$ h and $1/\lambda_{mp} = 3000$ h.

12.2.2 Stochastic Reward Net Model

Figure 12.3 shows the SRN model for the two boards system. Figure 12.3a represents the failure behavior of the processor P_1 and the memory M_1. If a token is in place $P1U$, the processor is operational, otherwise the processor is down. Figure 12.3a also presents the failure behavior of the memory M_1; if one token is in place $M1U$, the memory is operational, otherwise the memory is down. The transition $TC1F$ represents a common-cause failure. The transition $TC1F$ is enabled when there is one token in each place $P1U$ and place $M1U$. Similarly, Fig. 12.3b shows the failure behavior of the processor P_2 and the memory M_2. We can compute the probability that the system is in state i, denoted as $\pi_{S_i}(t)$, using the definition of reward functions in SHARPE (Trivedi and Sahner 2009) as shown in Eq. 12.3.

$$r_{S_1} = \begin{cases} 1, & (\#(P1U) + \#(P2U) == 0) \text{or} (\#(M1U) + \#(M2U) == 0), \\ 0, & \text{otherwise}, \end{cases}$$

$$r_{S_2} = \begin{cases} 1, & \left(\begin{array}{l} ((\#(P1U) + \#(P2U) > \ = 1) \text{and} (\#(M1U) == 1) \text{and} (\#(M2U) == 0)) \\ \text{or} ((\#(P1U) + \#(P2U) > \ = 1) \text{and} (\#(M2U) == 1) \text{and} (\#(M1U) == 0)) \end{array} \right), \\ 0, & \text{otherwise}, \end{cases}$$

$$r_{S_3} = \begin{cases} 1, & (\#(P1U) + \#(P2U) > \ = 1) \text{and} (\#(M1U) + \#(M2U) == 2), \\ 0, & \text{otherwise}. \end{cases}$$

$$(12.3)$$

Table 12.2 shows the results of $\pi_{S_i}(t)$ at different times using the SRN model. Comparing the values in Tables 12.1, 12.2, we can see that the CTMC model and the SRN model have identical results.

12.2.3 Multi-State Fault Trees Model

An alternative approach to obtain the probability of the system states is to use MFTs. By the definition in the beginning of Sect. 12.2, each board is considered as

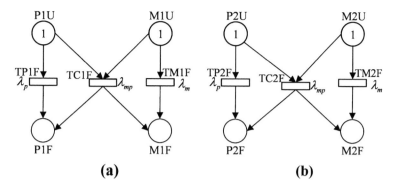

Fig. 12.3 SRN model for the two boards system

Table 12.2 Results of the SRN model for the two boards system

t (h)	$\pi_{S_1}(t)$	$\pi_{S_2}(t)$	$\pi_{S_3}(t)$
0	0	0	1
100	2.03280585e-002	1.38357902e-001	8.41314039e-001
200	7.02653699e-002	2.29253653e-001	7.00480977e-001
300	1.37122363e-001	2.84435223e-001	5.78442414e-001
∞	1	0	0

a component with four states. The failure behavior of a single component can be captured by the CTMC model in Fig. 12.4. The states of the Markov chain is represented by (*PM*), where 1 denotes up and 0 denotes down for each device.

By solving the CTMC model in Fig. 12.4, the transient probabilities for different component states $\pi_{B_{i,j}}(t)$, where $B_{i,j}$ denotes the board B_i being in state j, can be obtained. Equation 12.4 shows the formulas for $\pi_{B_{i,j}}(t)$.

$$
\pi_{B_{i,j}}(t) =
\begin{cases}
\pi_{00}(t) = \dfrac{\lambda_m\lambda_p - \lambda_{\mathrm{mp}}^2}{(\lambda_m + \lambda_{\mathrm{mp}})(\lambda_p + \lambda_{\mathrm{mp}})} e^{-(\lambda_m+\lambda_p+\lambda_{\mathrm{mp}})t} - \dfrac{\lambda_p}{\lambda_p + \lambda_{\mathrm{mp}}} e^{-\lambda_m t} \\
\qquad\quad -\dfrac{\lambda_m}{\lambda_m + \lambda_{\mathrm{mp}}} e^{-\lambda_p t} + 1, & j = 1, \\[2ex]
\pi_{10}(t) = \dfrac{\lambda_m}{\lambda_m+\lambda_{\mathrm{mp}}}\left[e^{-\lambda_p t} - e^{-(\lambda_m+\lambda_p+\lambda_{\mathrm{mp}})t} \right], & j = 2, \\[2ex]
\pi_{01}(t) = \dfrac{\lambda_p}{\lambda_p+\lambda_{\mathrm{mp}}}\left[e^{-\lambda_m t} - e^{-(\lambda_m+\lambda_p+\lambda_{\mathrm{mp}})t} \right], & j = 3, \\[2ex]
\pi_{11}(t) = e^{-(\lambda_m+\lambda_p+\lambda_{\mathrm{mp}})t}, & j = 4.
\end{cases}
$$
$$(12.4)$$

Based on the states for each component (i.e., a board), the two boards system states can be obtained using MFTs model as shown in Fig. 12.5. The definition of the system states is presented at the beginning of Sect. 12.2.

Fig. 12.4 CTMC model
for a single board

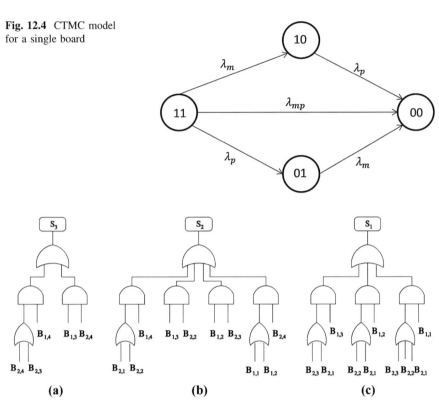

Fig. 12.5 MFTs model for the two boards system

Table 12.3 Results of the MFTs model for the two boards system

t (h)	$\pi_{S_1}(t)$	$\pi_{S_2}(t)$	$\pi_{S_3}(t)$
0	0	0	1
100	2.03223700e-002	1.38269880e-001	8.41407750e-001
200	7.02405700e-002	2.29213640e-001	7.00345800e-001
300	1.37092520e-001	2.84365240e-001	5.78342250e-001
∞	1	0	0

For any given time t, we can obtain the value of $\pi_{B_{i,j}}(t)$ by solving the CTMC model in Fig. 12.4 and Eq. 12.4, which is then assigned as the probability for event $B_{i,j}$ in the MFTs model in Fig. 12.5. By solving the MFT model, the probability for the system in state i, denoted as $\pi_{S_i}(t)$, can be obtained. Table 12.3 shows the results for the different value of t, given that $1/\lambda_p = 1,000$ h, $1/\lambda_m = 2,000$ h and $1/\lambda_{mp} = 3,000$ h. By comparing the values in Table 12.1 with those in Table 3, the multi-state fault tree model and the CTMC model (in Sect. 12.2.1) show nearly the same results. Differences are due to numerical solution errors.

Fig. 12.6 A VAXcluster
system

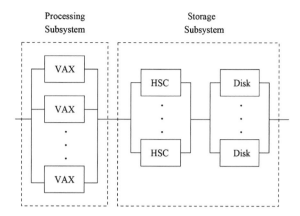

12.2.4 Model Comparison

Note that from Sects. 12.2.1, 12.2.2 and 12.2.3, similar results are obtained using three different analytic modeling techniques including CTMCs, SRNs, and MFTs. However, the states space of the CTMC model will increase exponentially as the number of boards increase. If the number of boards is n, the number of states will be 2^{2n}. If n becomes large, this approach will face a state space explosion problem. SRN model is easy to construct among these three model types, but the underlying Markov chain for the SRN model also faces state explosion problem when n becomes very large. By contrast, the multi-state fault tree model still consists of three fault trees when n increases, which is more efficient to solve than the CTMC model and the SRN model. However, the construction of such MFT model will consume more effort and are sometimes more error-prone than the CTMC model and the SRN model.

12.3 VAXcluster System

A VAXcluster, as shown in Fig. 12.6, is a closely coupled multicomputer system that consists of two or more VAX processors, one or more storage controllers (HSCs), a set of disks and a star coupler (SC). The SC can be omitted from the model since it is extremely reliable. In this Section, we concentrate on the analysis using different models for the processing subsystem in the VAXcluster (Ibe et al. 1989; Muppala et al. 1992) .

12.3.1 Continuous Time Markov Chain Model

A CTMC availability model can be developed for the processing subsystem in the VAXcluster (Ibe et al. 1989; Muppala et al. 1992) in which two types of failure and

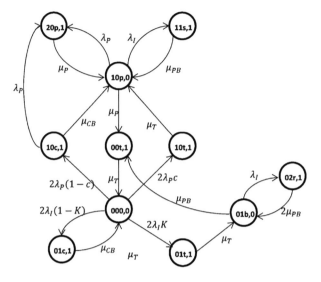

Fig. 12.7 CTMC model for the two-processor VAXcluster

a coverage factor for each failure type exist. By using a single state space model, shared repair for the processors in a cluster is taken into account. The times to failures, the repair times, and other recovery times are all assumed to be exponentially distributed. A CTMC availability model for a two-processor VAXcluster is shown in Fig. 12.7. A processor can suffer two types of failures: permanent and intermittent. A processor recovers from a permanent failure by a physical repair and from an intermittent failure by a processor reboot. These failures are further classified into covered and not covered. A covered processor failure causes a brief (in the order of seconds) cluster outage to reconfigure the failed processor out of the cluster and back into the cluster after it is fixed. Therefore, a covered failure causes a small loss in system up time. A not-covered failure causes the entire cluster to go down until it is rebooted. The parameters for the models are chosen as follows:

- Mean time to permanent failure: $1/\lambda_P = 5{,}000$ h;
- Mean time to intermittent failure: $1/\lambda_I = 2{,}000$ h;
- Mean processor repair time: $1/\mu_P = 2$ h;
- Mean processor reboot time: $1/\mu_{PB} = 6$ min;
- Mean cluster reboot time: $1/\mu_{CB} = 10$ min;
- Mean cluster reconfiguration time: $1/\mu_T = 30$ s;
- Coverage factor for permanent failure: $c = 0.9$;
- Coverage factor for intermittent failure: $K = 0.9$.

The states of the Markov chain are represented by (abc, d), where,

$a =$ number of processors down with permanent failure,

$b =$ number of processors down with intermittent failure,

$$c = \begin{cases} 0 & \text{if both processors are up,} \\ p & \text{if one processor is being repaired,} \\ b & \text{if one processor is being rebooted,} \\ c & \text{if cluster is undergoing a reboot,} \\ t & \text{if cluster is undergoing a reconfiguration,} \\ r & \text{if two processors are being rebooted,} \\ s & \text{if one is being rebooted and the other is being repaired,} \end{cases}$$

$$d = \begin{cases} 0 & \text{cluster up state,} \\ 1 & \text{cluster down state.} \end{cases}$$

The system states are defined as

- System State 1 (S_1): System is up;
- System State 2 (S_2): System is down due to cluster reboot;
- System State 3 (S_3): System is down due to cluster reconfiguration;
- System State 4 (S_4): System is down due to all processors are down (either under reboot or repair).

Denote the steady-state probability that the system is in each state as π_{S_i}, where $i = 1, 2, 3, 4$. We can solve the CTMC model to compute the steady-state probability for each state π_j and then use reward rate assignment to compute π_{S_i} based on the following formula:

$$\pi_{S_i} = \sum_j r_{i,j} \cdot \pi_j, \tag{12.5}$$

where

$$r_{1,j} = \begin{cases} 1 & j = (000,0),(10p,0),(01b0), \\ 0 & \text{otherwise,} \end{cases}$$

$$r_{2,j} = \begin{cases} 1 & j = (10c,1),(01c,1), \\ 0 & \text{otherwise,} \end{cases}$$

$$r_{3,j} = \begin{cases} 1 & j = (10t,1),(00t,1),(01t,1), \\ 0 & \text{otherwise,} \end{cases} \tag{12.6}$$

$$r_{4,j} = \begin{cases} 1 & j = (11s,1),(20p,1),(02r,1), \\ 0 & \text{otherwise.} \end{cases}$$

We computed the following results by solving the CTMC model using software package SHARPE (Trivedi and Sahner 2009) as shown in Table 12.4.

The main problem with this approach is that the size of the CTMC model grows exponentially with the number of processors in the VAXcluster system.

Table 12.4 Results of the CTMC model for the two-processor VAXcluster

	π_{S_1}	π_{S_2}	π_{S_3}	π_{S_4}
Steady-state probability	9.99955011e-001	2.33113143e-005	2.13134041e-005	3.64575688e-007

The largeness posed the following challenges; (1) the difficulty in generating the state space and (2) the capability of the software to solve the model with thousands of states for VAXcluster system with $n > 5$ processors. Using SRN model instead of CTMC can avoid the difficulty in generating the state space. However, the underlying Markov chains generated by the SRN model still face the largeness problem when n becomes large and hence exceeds the capability of the software to solve the model. The MFT model can avoid such difficulties by utilizing the features of non-state space models. The following two sections present the SRN and MFTs models, respectively.

12.3.2 Stochastic Reward Net Model

Figure 12.8 shows the SRN availability model for n-processor VAXcluster system. The number of tokens in place P_{up} represents the number of non-failed processors. The initial number of tokens in this place is n. As mentioned in Sect. 12.3.1, processors suffer two types of failures; permanent and intermittent and the failures can be covered or not. The firing of transition T_0 represents the permanent failure of one of the processors. The inhibitor arcs from the place P_{pfcov} and place $P_{pfnotcov}$ ensure that when the VAXclsuter system is undergoing a reconfiguration or reboot, no further failures occur. The firing rate of the transition T_0 is marking-dependent: Rate $(T_0) = \lambda_P \#P_{up}$, where λ_P is the permanent failure rate and $\#P_{up}$ is the number of tokens in place P_{up}. When a token appears in place P_{fail}, the immediate transitions t_{pfcov}, $t_{pfnotcov}$, and t_0 are enabled. If no token is in place P_{up}, then the immediate transition t_0 will be enabled and will be assigned a higher priority than t_{pfcov} and $t_{pfnotcov}$; this is done to ensure that for the last processor to fail and there then is no cluster reconfiguration or reboot delay. A token will be deposited in place P_{rep2} by firing immediate transition t_0. Otherwise t_{pfcov} or $t_{pfnotcov}$ will fire with probabilities c and 1-c, respectively. In the case that t_{pfcov} (covered case) fires, cluster reconfiguration (T_1) takes place to remove the failed processor from the cluster with rate μ_T and then the failed processor is repaired (T_3) with rate μ_P and another cluster reconfiguration takes place to readmitting the repaired processor into the cluster with rate μ_T. In the case that $t_{pfnotcov}$ (not-covered case) fires, the cluster is rebooted with rate μ_{CB}, the processor repair and cluster reconfiguration are followed.

The firing of transition T_6 represents intermittent failure of one of the processors. The firing rate of transition T_6 is marking-dependent: rate $(T_6) = \lambda_I \#P_{up}$, where λ_I is the failure rate. Similar to the permanent failure, if no token is in place

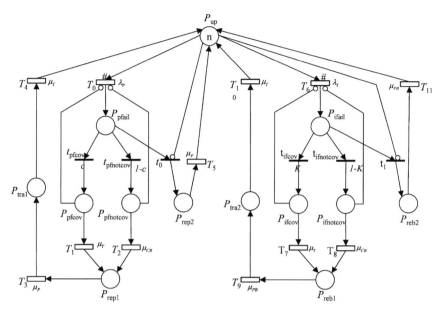

Fig. 12.8 SRN model for n-processor VAXcluster (using immediate transitions)

P_{up}, then the immediate transition t_1 will be enabled and will be assigned a higher priority than t_{ifcov} and $t_{ifnotcov}$; this is done to ensure that for the last processor to fail, there is no cluster reconfiguration or reboot delay but only process reboot of the last processor. A token will be deposited in place P_{reb2} by firing immediate transition t_1 and the processor is rebooted with rate μ_{PB}, otherwise t_{ifcov} or $t_{ifnotcov}$ will fire with probabilities K and $1-K$, respectively. In the case that t_{ifcov} (covered case) fires, cluster reconfiguration (T_7) takes place to remove the failed processor from the cluster with rate μ_T and then the failed processor is rebooted (T_9) with rate μ_{PB} and another cluster reconfiguration takes place to readmit the rebooted processor into the cluster with rate μ_T. In the case that $t_{ifnotcov}$ (not covered case) fires, the cluster is rebooted with rate μ_{CB}, the processor reboot and another cluster reconfiguration are followed.

We define reward rate functions for the system states in SHARPE (Trivedi and Sahner 2009) as follows according to the system states defined in Sect. 12.3.1.

The results of the SRN model for two-processor VAX cluster are shown in Table 12.5. Compared to the results of the CTMC model in Table 12.4, we can see that SRN model showed nearly similar results of π_{S_i}. We can easily compute the system availability for the n-processor VAXcluster using the SRN model. However, as shown in Sect. 12.3.4, the execution time of the SRN model exponentially increases when n becomes large, which makes the SRN model less efficient than the MFT model presented in the following section.

Table 12.5 Results of the SRN model for the two-processor VAXcluster

	π_{S_1}	π_{S_2}	π_{S_3}	π_{S_4}
Steady-state probability	9.99955087e-001	2.33168407e-005	2.13182409e-005	2.78074293e-007

$$r_{S_1} = \begin{cases} 1 & \left(\begin{array}{l} (\#(P_{up}) > = 1) \text{and} (\#(P_{pfcov}) + \#(P_{pfnotcov}) \\ + \#(P_{ifcov}) + \#(P_{ifnotcov}) + \#(P_{tra1}) + \#(P_{tra2}) == 0) \end{array} \right), \\ 0 & \text{otherwise,} \end{cases}$$

$$r_{S_2} = \begin{cases} 1 & ((\#(P_{pfnotcov}) > = 1) \text{or} (\#(P_{ifnotcov}) > = 1)), \\ 0 & \text{otherwise,} \end{cases}$$

$$r_{S_3} = \begin{cases} 1 & \left(\begin{array}{l} (\#(P_{pfcov}) > = 1) \text{or} (\#(P_{ifcov}) > = 1) \\ \text{or} (\#(P_{tra1}) > = 1) \text{or} (\#(P_{tra2}) > = 1) \end{array} \right), \\ 0 & \text{otherwise,} \end{cases}$$

$$r_{S_4} = \begin{cases} 1 & ((\#(P_{rep1})) + (\#(P_{rep2})) + (\#(P_{reb1})) + (\#(P_{reb2})) == n), \\ 0 & \text{otherwise.} \end{cases}$$

$$(12.7)$$

12.3.3 Multi-State Fault Trees Model

As mentioned in Sects. 12.3.1 and 12.3.2, the state space of the CTMC model and underlying Markov chains of the SRN model for n-processor VAXcluster increases exponentially with n, thereby making the availability difficult to analyze at large values of n. In this section, an approximate analysis for an n-processor VAXcluster is developed to avoid the largeness associated with state space models.

To obtain approximate system availability, the following assumptions are used:

- The behavior of each processor is modeled by a homogeneous CTMC and assume that this processor did not break the quorum rule (i.e., at least one other processor is operational). This assumption is justified by the fact that the probability of VAXcluster failure due to loss of quorum (i.e., all processors are down) is relatively low.
- Each processor has an independent repairman. This assumption is justified if the mean time to failure (MTTF) is large compared to the mean time to repair (MTTR) so that the time a faulty processor spends waiting for the repair crew to arrive is negligible.

These assumptions allow the decomposition of the n-processor VAXcluster into n independent subsystems, where each subsystem represents the behavior of one processor and can be modeled using the CTMC model shown in Fig. 12.9. Furthermore, the states of such CTMC sub-models for the individual processors are classified into the following three super-states:

- Super-state 1: the set of states in which the processor is up = {1}.
- Super-state 2: the set of states in which the processor is undergoing reboot or repair and hence the processor is down = {3, 7}.
- Super-state 3: the set of states in which the cluster is undergoing reboot = {5, 9}.
- Super-state 4: the set of states in which the cluster is undergoing reconfiguration = {2, 4, 6, 8}.

Therefore, by solving the CTMC sub-model and utilizing reward rate assignment, the steady-state probability for each super-state of a processor can be obtained. Subsequently, these super-states are considered as different states of a multi-state component (i.e., a processor) and multi-state fault trees can be constructed to compute the steady-state probability that the system is in each state, which is denoted as π_{S_i} and defined in Sect. 12.3.1. The MFTs model is illustrated in Fig. 12.10, where $P_{i,j}$ denotes that processor i is in super-state j.

The results for the MFTs model are shown in Table 12.6. Compared to the results of the complete CTMC model in Table 12.4, we can see that the MFTs model has good approximation for the exact results. Comparing to the difficulty in generating state spaces for a single complete CTMC model, we can easily extend the MFTs model to n-processor VAXcluster when n becomes large.

Based on the steady-state probabilities for different system states, the downtimes in hours per year can be computed and are shown in Fig. 12.11. Here the downtime $D(n) = U(n) \times 8760$ h/year and is expressed as Eq. 12.8.

$$D(n) = U(n) \times 8760 = \begin{cases} [1 - \pi_{S_1}(n)] \times 8760 & \text{Total downtime} \\ \pi_{S_2}(n) \times 8760 & \text{Downtime due to cluster} \\ & \text{reboot} \\ \pi_{S_3}(n) \times 8760 & \text{Downtime due to cluster} \\ & \text{reconfiguration} \\ \pi_{S_4}(n) \times 8760 & \text{Downtime due to all} \\ & \text{processors down} \end{cases}$$

(12.8)

As shown in the results, the total downtime is not monotonically decreasing with the number of processors. The reason for this behavior is that as the number of processors increases beyond 2, the primary cause of downtime is the cluster reboot and reconfiguration and the number of reboots and reconfigurations nearly linearly increase with the number of processors.

12.3.4 Model Comparison

For the n-processor VAXcluster system, the CTMC model faces largeness problem with respect to both generating state space and solving the model when n becomes large. The SRN model automates the generation of the state space and maintains

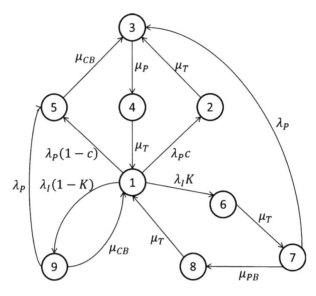

Fig. 12.9 CTMC model of a single processor

Table 12.6 Results of the MFTs model for the two-processor VAXcluster

	π_{S_1}	π_{S_2}	π_{S_3}	π_{S_4}
Steady-state probability	9.99955157e-001	2.33222980e-005	2.13232545e-005	1.97840042e-007

solution accuracy. Using the MFTs model in the top level and using the CTMC sub-model for each processor in the lower level provides an efficient approximation method to analyze the system availability. In this section, the SRN model and the MFTs model are compared with respect to both solution accuracy and efficiency. Figure 12.12 shows the downtime per year computed by these two models. The results for the CTMC model are not included due to the complexity to construct the model when n is larger than two. We can see that nearly the same downtime is obtained from these two models. However, the SRN model consumes more time to solve than the MFTs model. Figure 12.13 presents the log value of the execution time T (second) for these models.

From the above results, we can see that the execution time for the MFTs model slightly increases as n increases. By contrast, the SRN model execution time exponentially increases with n. Therefore, we can conclude that the MFTs model for VAXcluster analysis is more efficient, for both model construction and model solution when compared with the CTMC and SRN models while still maintaining the accuracy of the results.

Fig. 12.10 MFTs model
for the two-processor
VAXcluster

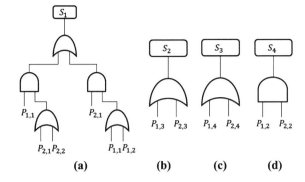

Fig. 12.11 Downtime versus
the number of processors
for the MFTs model

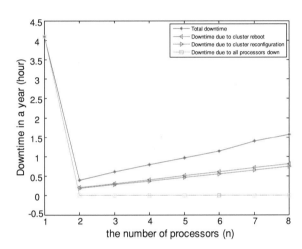

Fig. 12.12 Downtime versus
the number of processors for
different models

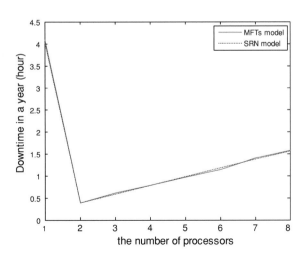

Fig. 12.13 Execution time versus the number of processors for different models

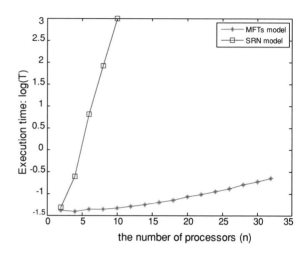

12.4 Conclusions

There are many systems which have the multi-state. We have presented the usage of three analytic modeling techniques (CTMC, SRN, and MFTs) to evaluate the reliability/availability of multi-state systems. We have illustrated the three techniques via two multi-state systems (the two boards system and the processors subsystem of the VAXcluster). We have compared three modeling techniques in terms of model accuracy and execution time. We have also shown the optimal number of processors for the processor subsystem of the VAXcluster.

References

Ibe OC, Howe RC, Trivedi KS (1989) Approximate availability analysis of VAXcluster systems. IEEE Trans Reliab 38(1):146–152

Muppala J, Sathaye A, Howe R, Trivedi KS (1992) Dependability modeling of a heterogenous VAXcluster system using stochastic reward nets. In: Avresky DR (ed) Hardware and software fault tolerance in parallel computing systems. Ellis Horwood, NJ, pp 33–59

Trivedi KS (2001) Probability and statistics with reliability, queuing and computer science applications. Wiley, New York

Trivedi KS, Sahner R (2009) SHARPE at the age of twenty two. ACM SIGMETRICS Perform Eval Rev 36(4):52–57

Zang X, Wang D, Sun H, Trivedi KS (2003) A BDD-based algorithm for analysis of multistate systems with multistate components. IEEE Trans Comp 52(12):1608–1618

Chapter 13
Recent Results in the Analysis of Redundant Systems

Mikhail Nikulin, Noureddine Saaidia and Ramzan Tahir

Abstract In Reliability, the distributions with unimodal or the so-called ∩-shape hazard functions are used quiet often. Mostly people use log-normal, log-logistic and inverse Gaussian distributions. In this paper, we consider also an application of the inverse Gaussian, log-normal and log-logistic distribution in the reliability analysis of redundant system with one main unit and $m - 1$ stand-by units. Properties of the point estimators and confidence intervals are studied.

Keywords Inverse Gaussian distribution · Log-normal distribution · Log-logistic distribution · Reliability · Sedyakin's model · Redundant system · Accelerated failure time model · Confidence interval · Failure times · Maximum likelihood estimation · Warm stand-by unit

13.1 Introduction

Following the results of Bagdonavicius et al. (2008a, 2008b, 2009, 2010) we consider the redundant system $S(1, m - 1)$ with one principal main unit operating in "hot" and $(m - 1)$ stand-by units operating in "warm" conditions. The problem

M. Nikulin (✉) · N. Saaidia · R. Tahir
IMB, Université Victor Segalen, Bordeaux 2, France
e-mail: mikhail.nikouline@u-bordeaux2.fr

N. Saaidia
e-mail: saaidianoureddine@yahoo.fr

R. Tahir
e-mail: ramzantahir7@gmail.com

N. Saaidia
Université Badji Mokhtar, Annaba, Algeria

A. Lisnianski and I. Frenkel (eds.), *Recent Advances in System Reliability*,
Springer Series in Reliability Engineering, DOI: 10.1007/978-1-4471-2207-4_13,
© Springer-Verlag London Limited 2012

is to obtain confidence intervals for the cumulative distribution funtions of the failure times of the systems of two groups of units. We assume that switching from warm to hot does not cause shock or damage to units.

Suppose that in "hot" conditions the failure time, the c.d.f., the survival function and the density of the main unit are T_1, F_1, S_1 and f_1, respectively, and in "warm" conditions the failure times, the c.d.f., the survival function and the density of the stand-by units are $T_i, (i = 2, \ldots, m), F_2, S_2$ and f_2. Using the Sedyakin principle Bagdonavicius et al. (2008a) give mathematical formulation of "fluent switch on" for the accelerated failure time (AFT) and Sedyakin's models and propose tests for verification of these models.

If a stand-by unit is switched to "hot" conditions, its distribution function is different from F_1 and F_2. The failure time of the system $S(1, m-1)$ is

$$T^{(m)} = T_1 \vee T_2 \vee \ldots \vee T_{m-1} \vee T_m = (T_1 \vee T_2 \vee \ldots \vee T_{m-1}) \vee T_m.$$

Denote by K_j the distribution function of $T^{(j)}, j = 2, \ldots, m$. We know that

$$K_j(t) = P\{T^{(j)} \leq t\} = P\{T^{(j-1)} \leq t, T_j \leq t\} = \int_0^t P\{T_j \leq t | T^{(j-1)} = y\} dK_{j-1}(y).$$

According to Bagdonavicius et al. (2008a) the "fluent switch on" hypothesis H_0 states that the conditional density of T_j given $T^{(j-1)} = y$ is

$$f_{T_j | T^{(j-1)} = y}(t) = \begin{cases} f_2(t), & \text{if } t \leq y, \\ f_1(t + ry - y), & \text{if } t > y, \end{cases}$$

for some $r > 0$. It means that the distributions of units funtioning in "warm" and "hot" conditions differ only in scale, i.e. $F_2(t) = F_1(rt)$ for all $t \geq 0$ and some $r > 0$. It means that conditionally (given $T^{(j-1)} = y$) the hypothesis H_0 corresponds to the AFT model, which implies that the cumulative distribution function of the failure time $K_j(t)$ of the system with one main unit and $j-1$ stand-by units is given by

$$K_j(t) = \int_0^t F_1(t + ry - y) dK_{j-1}(y), \quad K_1(t) = F_1(t),$$

from where it follows that the distribution function $K_m(t)$ of the system with $m-1$ stand-by units is defined recurrently using the last equation. It is often in practice that the c.d.f. of units functioning in "hot" and "warm" conditions belong to the same parametric families of distributions. Here we shall do the accent on the families of inverse Gaussian, log-logistic and log-normal distributions, having the unimodal hazard rate functions.

13.2 Parametric Point Estimators of the K_j

Bagdonavicius et al. (2008a, 2009, 2010) consider the following data:

(a) complete ordered sample $T_{11}, \ldots, T_{1n_1}$ of size n_1, T_{1i} is the failure time of units tested in "hot" conditions;
(b) complete ordered sample $T_{21}, \ldots, T_{2n_2}$ of size n_2, T_{2i} is the failure time of units tested in "warm" conditions.

Suppose that in hot conditions the c.d.f. $F_1(t; \theta)$ is absolutely continuous and depends on parameter $\theta \in \Theta \subset \mathbf{R}^k$. Let $\gamma = (r, \theta^T)^T$, the MLE's $\hat{\gamma} = (\hat{r}, \hat{\theta}^T)^T$ of the parameter γ maximizes the loglikelihood function

$$\ell(\gamma) = \sum_{i=1}^{n_1} \ln f_1(T_{1i}; \theta) + n_2 \ln r + \sum_{j=1}^{n_2} \ln f_2(rT_{2j}; \theta).$$

For $j \geq 2$, the c.d.f. $K_j(t)$ is estimated recurrently by

$$\hat{K}_j(t) = \int_0^t F_1(t + \hat{r}y - y; \hat{\theta}) d\hat{K}_{j-1}(y), \quad \hat{K}_1(t) = F_1(t; \hat{\theta}).$$

13.3 Asymptotic Confidence Interval for $K_j(t)$

Using the results from Bagdonavicius et al. (2008a, 2009), we can construct the asymptotic $(1 - \alpha)100\%$ confidence interval $(\underline{K}_j(t), \overline{K}_j(t))$ for $K_j(t)$, where

$$\underline{K}_j(t) \approx \left(1 + \frac{1 - \hat{K}_j(t)}{\hat{K}_j(t)} \exp\left\{ \frac{\hat{\sigma}_{\hat{K}_j} z_{1-\alpha/2}}{\sqrt{\hat{K}_j(t)(1 - \hat{K}_j(t))}} \right\} \right)^{-1},$$

$$\overline{K}_j(t) \approx \left(1 + \frac{1 - \hat{K}_j(t)}{\hat{K}_j(t)} \exp\left\{ -\frac{\hat{\sigma}_{\hat{K}_j} z_{1-\alpha/2}}{\sqrt{\hat{K}_j(t)(1 - \hat{K}_j(t))}} \right\} \right)^{-1},$$

with $\hat{\sigma}_{\hat{K}_j(t)}^2 = C_j^T(t, \hat{\gamma}) I^{-1}(\hat{\gamma}) C_j(t, \hat{\gamma}),$

where

$$C_j(t, \gamma) = (C_{j1}(t, \gamma), C_{j2}^T(t, \gamma))^T,$$

$$C_{j1}(t, \gamma) = \frac{\partial K_j(t)}{\partial r} = \int_0^t \frac{\partial F_1}{\partial r}(t + ry - y; \theta) dK_{j-1}(y),$$

$$C_{j2}(t, \gamma) = \left(\frac{\partial K_j(t)}{\partial \theta_l}\right)_{l=1,\dots,k}$$

$$= \left(\int_0^t \frac{\partial F_1}{\partial \theta_l}(t + ry - y; \theta) dK_{j-1}(y) + F_1(t + ry - y; \theta) d\left(\frac{\partial K_{j-1}(y)}{\partial \theta_l}\right)\right)_{l=1,\dots,k},$$

$I^{-1}(\gamma)$ is the inverse of the fisher's information matrix $I(\gamma)$ and $z_{1-\alpha/2}$ is the $(1 - \alpha/2)$-quantile of the standard normal distribution.

13.4 Inverse Gaussian Distribution

Suppose that the distribution of failure times in hot and warm conditions is inverse Gaussian (Seshadri 1994, 1998; Chhikara and Folks 1989), i.e.

$$S_1(t) = 1 - F_1(t) = \Phi\left(-\sqrt{\frac{\lambda}{t}}\left(\frac{t}{\mu} - 1\right)\right) - \exp\left(\frac{2\lambda}{\mu}\right)\Phi\left(-\sqrt{\frac{\lambda}{t}}\left(\frac{t}{\mu} + 1\right)\right),$$

where $\Phi(.)$ denotes the c.d.f. of the standard normal distribution.

The loglikelihood function is (Lemeshko et al. 2010; Nikulin and Saaidia 2009)

$$\ell(r, \mu, \lambda) = \frac{n}{2}\ln\lambda - \frac{n}{2}\ln(2\pi) - \frac{n_2}{2}\ln r + \frac{\lambda n}{\mu} - \frac{3}{2}\sum_{i=1}^{n_1}\ln(T_{1i}) - \frac{3}{2}\sum_{i=1}^{n_2}\ln(T_{2i})$$

$$- \frac{\lambda}{2\mu^2}\sum_{i=1}^{n_1}T_{1i} - \frac{\lambda}{2}\sum_{i=1}^{n_1}T_{1i}^{-1} - \frac{\lambda r}{2\mu^2}\sum_{i=1}^{n_2}T_{2i} - \frac{\lambda}{2r}\sum_{i=1}^{n_2}T_{2i}^{-1},$$

where $n = n_1 + n_2$.

In the Fig. 13.1 we represent the trajectories of the parametric estimators $\hat{F}_1, \hat{K}_2, \hat{K}_3$ and $\hat{K}_4$ for the inverse Gaussian distribution.

The score functions are

$$\frac{\partial\ell}{\partial r} = -\frac{n_2}{2r} - \frac{\lambda}{2\mu^2}\sum_{i=1}^{n_2}T_{2i} + \frac{\lambda}{2r^2}\sum_{i=1}^{n_2}T_{2i}^{-1},$$

$$\frac{\partial\ell}{\partial\mu} = \frac{\lambda}{\mu^3}\sum_{i=1}^{n_1}T_{1i} + \frac{\lambda r}{\mu^3}\sum_{i=1}^{n_2}T_{2i} - \frac{\lambda n}{\mu^2},$$

$$\frac{\partial\ell}{\partial\lambda} = n\left(\frac{1}{2\lambda} + \frac{1}{\mu}\right) - \frac{1}{2\mu^2}\sum_{i=1}^{n_1}T_{1i} - \frac{1}{2}\sum_{i=1}^{n_1}T_{1i}^{-1} - \frac{r}{2\mu^2}\sum_{i=1}^{n_2}T_{2i} - \frac{1}{2r}\sum_{i=1}^{n_2}T_{2i}^{-1}.$$

Fig. 13.1 Graphs of the trajectories of the parametric estimators $\hat{F}_1, \hat{K}_2, \hat{K}_3$ and $\hat{K}_4$ (Inverse Gaussian distribution)

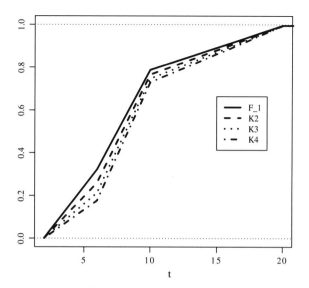

To find the estimator $\hat{\gamma} = (\hat{r}, \hat{\mu}, \hat{\lambda})^T$ one can solve the system formed by equalizing the score functions to zero.

Second partial derivatives of the loglikelihood function are

$$\frac{\partial^2 \ell}{\partial r^2} = +\frac{n_2}{2r^2} - \frac{\lambda}{r^3} \sum_{i=1}^{n_2} T_{2i}^{-1}; \qquad \frac{\partial^2 \ell}{\partial r \partial \mu} = \frac{\lambda}{\mu^3} \sum_{i=1}^{n_2} T_{2i};$$

$$\frac{\partial^2 \ell}{\partial r \partial \lambda} = -\frac{1}{2\mu^2} \sum_{i=1}^{n_2} T_{2i} + \frac{1}{2r^2} \sum_{i=1}^{n_2} T_{2i}^{-1}; \qquad \frac{\partial^2 \ell}{\partial \mu^2} = -\frac{3\lambda}{\mu^4} \sum_{i=1}^{n_1} T_{1i} - \frac{3\lambda r}{\mu^4} \sum_{i=1}^{n_2} T_{2i} + \frac{2\lambda n}{\mu^3};$$

$$\frac{\partial^2 \ell}{\partial \mu \partial \lambda} = \frac{1}{\mu^3} \sum_{i=1}^{n_1} T_{1i} + \frac{r}{\mu^3} \sum_{i=1}^{n_2} T_{2i} - \frac{n}{\mu^2}; \qquad \frac{\partial^2 \ell}{\partial \lambda^2} = -\frac{n}{2\lambda^2}.$$

So the Fisher's information matrix is

$$I(\gamma) = \begin{pmatrix} \dfrac{(\mu + 2\lambda)n_2}{2\mu r^2} & -\dfrac{\lambda n_2}{r\mu^2} & -\dfrac{n_2}{2r\lambda} \\[2ex] -\dfrac{\lambda n_2}{r\mu^2} & \dfrac{\lambda n}{\mu^3} & 0 \\[2ex] -\dfrac{n_2}{2r\lambda} & 0 & \dfrac{n}{2\lambda^2} \end{pmatrix}.$$

The inverse of the Fisher's information matrix is

$$I^{-1}(\gamma) = \begin{pmatrix} \dfrac{2nr^2\mu}{n_1n_2(\mu+2\lambda)} & \dfrac{2r\mu^2}{n_1(\mu+2\lambda)} & \dfrac{2r\mu\lambda}{n_1(\mu+2\lambda)} \\ \dfrac{2r\mu^2}{n_1(\mu+2\lambda)} & \dfrac{(n_1\mu+2n\lambda)\mu^3}{n_1n(\mu+2\lambda)\lambda} & \dfrac{2n_2\mu^2\lambda}{n_1n(\mu+2\lambda)} \\ \dfrac{2r\mu\lambda}{n_1(\mu+2\lambda)} & \dfrac{2n_2\mu^2\lambda}{n_1n(\mu+2\lambda)} & \dfrac{2(n\mu+2n_1\lambda)\lambda^2}{n_1n(\mu+2\lambda)} \end{pmatrix}.$$

For $j = 2$, the c.d.f. $K_2(t)$ is estimated by

$$\hat{K}_2(t) = \int_0^t \sqrt{\frac{\hat{\lambda}}{2\pi y^3}} \exp\left\{-\frac{\hat{\lambda}(y-\hat{\mu})^2}{2\hat{\mu}^2 y}\right\} \Phi\left(\sqrt{\frac{\hat{\lambda}}{t+\hat{r}y-y}}\left(\frac{t+\hat{r}y-y}{\hat{\mu}}-1\right)\right) dy$$

$$+ \int_0^t \sqrt{\frac{\hat{\lambda}}{2\pi y^3}} \exp\left\{-\frac{\hat{\lambda}(y-\hat{\mu})^2}{2\hat{\mu}^2 y}+\frac{2\hat{\lambda}}{\hat{\mu}}\right\} \Phi\left(-\sqrt{\frac{\hat{\lambda}}{t+\hat{r}y-y}}\left(\frac{t+\hat{r}y-y}{\hat{\mu}}+1\right)\right) dy.$$

with $\hat{\sigma}^2_{\hat{K}_2(t)} = C_2^T(t,\hat{\gamma})I^{-1}(\hat{\gamma})C_2(t,\hat{\gamma})$,

where

$$C_2(t,\gamma) = (C_{21}(t,\gamma), C_{22}(t,\gamma), C_{23}(t,\gamma))^T,$$

$$C_{21}(t,\gamma) = \int_0^t \frac{\partial F_1}{\partial r}(t+ry-y;\mu,\lambda)dF_1(y;\mu,\lambda),$$

$$C_{22}(t,\gamma) = \int_0^t \frac{\partial F_1}{\partial \mu}(t+ry-y;\mu,\lambda)dF_1(y;\mu,\lambda) + F_1(t+ry-y;\mu,\lambda)d\left(\frac{\partial F_1}{\partial \mu}(y;\mu,\lambda)\right),$$

$$C_{23}(t,\gamma) = \int_0^t \frac{\partial F_1}{\partial \lambda}(t+ry-y;\mu,\lambda)dF_1(y;\mu,\lambda) + F_1(t+ry-y;\mu,\lambda)d\left(\frac{\partial F_1}{\partial \lambda}(y;\mu,\lambda)\right).$$

To calculate these integrals, we use a numerical method.

13.4.1 Simulation Study

Let us consider the case of complete sample of size $n_1 = n_2 = 100$. Each sample is repeated 5000 times. We find by simulation the confidence levels of intervals using formulas with $1 - \alpha = 0.9$. We simulated failure times T_{1i} and T_{2j} from inverse Gaussian distribution with the parameters $T_{1i} \sim IG(\mu_1, \lambda_1)$, $T_{2j} \sim IG(\mu_2, \lambda_2)$, $\mu_1 = 8$, $\mu_2 = 4$, $\lambda_1 = 40$, $\lambda_2 = 20$.

Table 13.1 Confidence level for the c.d.f. of the failure time for samples with $n_1 = n_2 = 100$ (Inverse Gaussian distribution)

t	4	8	10	20	30	40	50
$K_2(t)$	0.05099	0.55988	0.75297	0.99049	0.99966	0.99998	0.99999
C.L. (%)	93.67	88.87	83.55	86.60	89.04	91.04	87.22

For various values of t the proportions of confidence level (C.L.) realizations covering the true value of the distribution function $K_2(t)$ are given in Table 13.1.

13.5 Log-normal Distribution

Suppose that the distribution of failure times in hot and warm conditions is log-normal, i.e.

$$S_1(t) = 1 - F_1(t) = 1 - \Phi\left(\frac{\ln t - m}{\sigma}\right).$$

In the Fig. 13.2, we represent the trajectories of the parametric estimators $\hat{F}_1, \hat{K}_2, \hat{K}_3$ and $\hat{K}_4$ for the log-normal distribution.

The loglikelihood function

$$\ell(r, m, \sigma) = -n\left(\ln \sigma + \frac{m^2}{2\sigma^2}\right) + \frac{n_2 \ln r}{\sigma^2}\left(m - \frac{\ln r}{2}\right) + \left(\frac{m}{\sigma^2} - 1\right) \sum_{i=1}^{n_1} \ln(T_{1i})$$

$$+ \left(\frac{m}{\sigma^2} - \frac{\ln r}{\sigma^2} - 1\right) \sum_{j=1}^{n_2} \ln(T_{2j}) - \frac{1}{2\sigma^2} \sum_{i=1}^{n_1} \ln^2(T_{1i}) - \frac{1}{2\sigma^2} \sum_{j=1}^{n_2} \ln^2(T_{2j}),$$

where $n = n_1 + n_2$.

The score functions are

$$\frac{\partial \ell}{\partial r} = \frac{n_2}{r\sigma^2}(m - \ln r) - \frac{1}{r\sigma^2} \sum_{j=1}^{n_2} \ln(T_{2j}),$$

$$\frac{\partial \ell}{\partial m} = -\frac{nm}{\sigma^2} + \frac{n_2 \ln r}{\sigma^2} + \frac{1}{\sigma^2} \sum_{i=1}^{n_1} \ln(T_{1i}) + \frac{1}{\sigma^2} \sum_{j=1}^{n_2} \ln(T_{2j}),$$

$$\frac{\partial \ell}{\partial \sigma} = -n\left(\frac{1}{\sigma} - \frac{m^2}{\sigma^3}\right) - \frac{n_2 \ln r}{\sigma^3}(2m - \ln r) - \frac{2m}{\sigma^3} \sum_{i=1}^{n_1} \ln(T_{1i})$$

$$+ \frac{2}{\sigma^3}(-m + \ln r) \sum_{j=1}^{n_2} \ln(T_{2j}) + \frac{1}{\sigma^3} \sum_{i=1}^{n_1} \ln^2(T_{1i}) + \frac{1}{\sigma^3} \sum_{j=1}^{n_2} \ln^2(T_{2j}).$$

To find the estimator $\hat{\gamma}$ one can solve the system formed by equalizing the score functions to zero.

Fig. 13.2 Graphs of the trajectories of the parametric estimators $\hat{F}_1, \hat{K}_2, \hat{K}_3$ and $\hat{K}_4$ (Log-normal distribution)

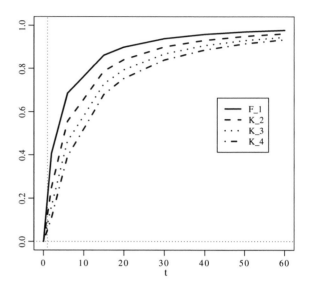

Second partial derivatives of the loglikelihood function are

$$\frac{\partial^2 \ell}{\partial r^2} = -\frac{n_2}{r^2\sigma^2}(m + \ln r + 1) + \frac{1}{r^2\sigma^2}\sum_{j=1}^{n_2}\ln(T_{2j}); \quad \frac{\partial^2 \ell}{\partial r \partial m} = \frac{n_2}{r\sigma^2};$$

$$\frac{\partial^2 \ell}{\partial r \partial \sigma} = \frac{2n_2}{r\sigma^3}(-m + \ln r) + \frac{2}{r\sigma^3}\sum_{j=1}^{n_2}\ln(T_{2j}); \quad \frac{\partial^2 \ell}{\partial m^2} = -\frac{n}{\sigma^2};$$

$$\frac{\partial^2 \ell}{\partial m \partial \sigma} = \frac{2nm}{\sigma^3} - 2n_2\frac{\ln r}{\sigma^3} - \frac{2}{\sigma^3}\sum_{i=1}^{n_1}\ln(T_{1i}) - \frac{2}{\sigma^3}\sum_{j=1}^{n_2}\ln(T_{2j});$$

$$\frac{\partial^2 \ell}{\partial \sigma^2} = -n\left(-\frac{1}{\sigma^2} + \frac{3m^2}{\sigma^4}\right) + \frac{3n_2}{\sigma^4}\left(2m\ln r - \ln^2 r\right) + \frac{6m}{\sigma^4}\sum_{i=1}^{n_1}\ln(T_{1i})$$

$$+ \frac{6}{\sigma^4}(m - \ln r)\sum_{j=1}^{n_2}\ln(T_{2j}) - \frac{3}{\sigma^4}\sum_{i=1}^{n_1}\ln^2(T_{1i}) - \frac{3}{\sigma^4}\sum_{j=1}^{n_2}\ln^2(T_{2j}).$$

So the Fisher's information matrix is

$$I(\gamma) = \begin{pmatrix} \dfrac{n_2}{r^2\sigma^2} & -\dfrac{n_2}{r\sigma^2} & 0 \\[2ex] -\dfrac{n_2}{r\sigma^2} & \dfrac{n}{\sigma^2} & 0 \\[2ex] 0 & 0 & \dfrac{2n}{\sigma^2} \end{pmatrix}.$$

The inverse of the Fisher's information matrix is

$$
I^{-1}(\gamma) =
\begin{pmatrix}
\dfrac{nr^2\sigma^2}{n_1 n_2} & \dfrac{r\sigma^2}{n_1} & 0 \\[2mm]
\dfrac{r\sigma^2}{n_1} & \dfrac{\sigma^2}{n_1} & 0 \\[2mm]
0 & 0 & \dfrac{\sigma^2}{2n}
\end{pmatrix}.
$$

Taking $j = 2$, the c.d.f. $K_2(t)$ is estimated by

$$
\hat{K}_2(t) = \frac{1}{\sqrt{2\pi}\hat{\sigma}} \int_0^t \frac{1}{y} \Phi\left(\frac{\ln(t+\hat{r}y-y)-\hat{m}}{\hat{\sigma}}\right) \exp\left\{-\frac{\ln y - \hat{m}}{2\hat{\sigma}^2}\right\} dy.
$$

$$
C_2(t,\gamma) = (C_{21}(t,\gamma), C_{22}(t,\gamma), C_{23}(t,\gamma))^T,
$$

$$
C_{21}(t,\gamma) = \int_0^t \frac{\partial F_1}{\partial r}(t+ry-y,m,\sigma)\,dF_1(y,m,\sigma),
$$

$$
C_{22}(t,\gamma) = \int_0^t \frac{\partial F_1}{\partial m}(t+ry-y,m,\sigma)\,dF_1(y,m,\sigma) + F_1(t+ry-y,m,\sigma)\,d\left(\frac{\partial F_1}{\partial m}(y,m,\sigma)\right),
$$

$$
C_{23}(t,\gamma) = \int_0^t \frac{\partial F_1}{\partial \sigma}(t+ry-y,m,\sigma)\,dF_1(y,\mu,\lambda) + F_1(t+ry-y,m,\sigma)\,d\left(\frac{\partial F_1}{\partial \sigma}(y,m,\sigma)\right).
$$

The partial derivatives are

$$
\frac{\partial F_1}{\partial r}(t+ry-y,m,\sigma) = \frac{y}{\sigma(t+ry-y)}\,\varphi\left(\frac{\ln(t+ry-y)-m}{\sigma}\right),
$$

$$
\frac{\partial F_1}{\partial m}(t+ry-y;m,\sigma) = -\frac{1}{\sigma}\,\varphi\left(\frac{\ln(t+ry-y)-m}{\sigma}\right),
$$

$$
\frac{\partial F_1}{\partial \sigma}(t+ry-y;m,\sigma) = -\frac{\ln(t+ry-y)-m}{\sigma^2}\,\varphi\left(\frac{\ln(t+ry-y)-m}{\sigma}\right),
$$

Table 13.2 Confidence level for the c.d.f. of the failure time for samples with $n_1 = n_2 = 100$. (Log-normal distribution)

t	10	50	100	200	300
$K_2(t)$	0.7119	0.9556	0.9855	0.9961	0.9983
C.L. (%)	82.25	88.87	83.55	86.60	89.04

we can write also

$$C_{21}(t,\gamma) = \frac{1}{\sqrt{2\pi}} \int_0^t \frac{1}{\sigma^2(t+ry-y)} \exp\left\{-\frac{(\ln y - m)^2}{2\sigma^2}\right\} \varphi\left(\frac{\ln(t+ry-y)-m}{\sigma}\right) dy,$$

$$C_{22}(t,\gamma) = -\frac{1}{\sigma^2\sqrt{2\pi}} \int_0^t \frac{1}{y} \exp\left\{-\frac{(\ln y - m)^2}{2\sigma^2}\right\} \varphi\left(\frac{\ln(t+ry-y)-m}{\sigma}\right) dy$$

$$+ \frac{1}{\sigma^3} \int_0^t \Phi\left(\frac{\ln(t+ry-y)-m}{\sigma}\right) \left(\frac{\ln y - m}{y}\right) \varphi\left(\frac{\ln y - m}{\sigma}\right) dy,$$

$$C_{23}(t,\gamma) = -\frac{1}{\sigma^3\sqrt{2\pi}} \int_0^t \frac{\ln(t+ry-y)-m}{\sigma^2} \exp\left\{-\frac{(\ln y - m)^2}{2\sigma^2}\right\} \varphi\left(\frac{\ln(t+ry-y)-m}{\sigma}\right) dy$$

$$+ \frac{1}{\sigma^2} \int_0^t \frac{1}{y} \Phi\left(\frac{\ln(t+ry-y)-m}{\sigma}\right) \left(\frac{(\ln y - m)^2}{\sigma^2} - 1\right) \varphi\left(\frac{\ln y - m}{\sigma}\right) dy,$$

where $\varphi(.)$ is the density function of the standard normal distribution.

13.5.1 Simulation Study

Let us consider the case of complete sample of size $n_1 = n_2 = 100$. Each sample is repeated 5000 times. We find by simulation the confidence levels of intervals using formulas with $1 - \alpha = 0.9$. We simulated failure times T_{1i} and T_{2j} from log-normal distribution with the parameters $T_{1i} \sim LN(m_1, \sigma_1)$, $T_{2j} \sim LN(m_2, \sigma_2)$, $m_1 = 1$, $m_2 = -0.3862944$, $\sigma_1 = \sigma_2 = 1.5$.

For various values of t the proportions of confidence interval (C.L.) realizations covering the true value of the distribution function $K_2(t)$ are given in Table 13.2.

13.6 Log-logistic Distribution

This distribution has been well studied by Bagdonavicius et al. (2010). Suppose that the distribution of failure times in hot and warm conditions is log-logistic, i.e.

$$S_1(t) = 1 - F_1(t) = \frac{1}{1 + \left(\frac{t}{\mu}\right)^v}.$$

The loglikelihood function is given by

$$\ell(r, \mu, v) = n \ln v - nv \ln \mu + n_2 v \ln r + (v - 1)\left(\sum_{i=1}^{n_1} \ln(T_{1i}) + \sum_{j=1}^{n_2} \ln(T_{2j})\right)$$

$$- 2 \sum_{i=1}^{n_1} \ln\left(1 + \left(\frac{T_{1i}}{\mu}\right)^v\right) - 2 \sum_{j=1}^{n_2} \ln\left(1 + \left(\frac{rT_{2j}}{\mu}\right)^v\right)$$

where $n = n_1 + n_2$.

The Fisher's information matrix is

$$I(r, \mu, v) = \begin{pmatrix} \dfrac{n_2 v^2}{3r^2} & -\dfrac{n_2 v^2}{3r\mu} & 0 \\[2mm] -\dfrac{n_2 v^2}{3r\mu} & \dfrac{n_2 v^2}{3\mu^2} & 0 \\[2mm] 0 & 0 & \dfrac{n\{3 + 2\Gamma''(2) - 2[\Gamma'(2)]^2\}}{3v^2} \end{pmatrix}.$$

The inverse of the Fisher's information matrix is

$$I^{-1}(r, \mu, v) = \begin{pmatrix} \dfrac{3nr^2}{n_1 n_2 v^2} & \dfrac{3r\mu}{n_1 v^2} & 0 \\[2mm] \dfrac{3rv}{n_1 v^2} & \dfrac{3\mu^2}{n_1 v^2} & 0 \\[2mm] 0 & 0 & \dfrac{3v^2}{n\{3 + 2\Gamma''(2) - 2[\Gamma'(2)]^2\}} \end{pmatrix},$$

where $\Gamma'(\cdot)$ and $\Gamma''(\cdot)$ are respectively the first and second derivatives of the gamma function.

Taking $j = 2$, the c.d.f. $K_2(t)$ is estimated by

$$\hat{K}_2(t) = 1 - \frac{1}{1 + \left(\frac{t}{\hat{\mu}}\right)^{\hat{v}}} - \frac{\hat{v}}{\hat{\mu}} \int_0^t \left(\frac{y}{\hat{\mu}}\right)^{\hat{v}-1} \frac{1}{1 + \left(\frac{t + \hat{r}y - y}{\hat{\mu}}\right)^{\hat{v}}} \frac{1}{\left(1 + \left(\frac{y}{\hat{\mu}}\right)^{\hat{v}}\right)^2} \, dy.$$

See Bagdonavicius et al. (2010) for more details.

Table 13.3 Confidence level for the c.d.f. of the failure time for samples with $n_1 = n_2 = 100$. (Log-logistic distribution)

t	50	100	200	300	400	500
$K_2(t)$	0.016	0.138	0.517	0.743	0.851	0.905
C.L. (%)	89.0	88.8	90.4	89.6	89.5	90.5

13.6.1 Simulation Study

Let us consider the case of complete sample of size $n_1 = n_2 = 100$. Each sample is repeated 2000 times. We find by simulation the confidence levels of intervals using formulas with $1 - \alpha = 0.9$. We simulated failure times T_{1i} and T_{2j} from log-logistic distribution with the parameters $T_{1i} \sim LL(\alpha_1, \beta_1)$, $T_{2j} \sim LL(\alpha_2, \beta_2)$ and $\alpha_1 = \alpha_2 = 2$, $\beta_1 = 100$, $\beta_2 = 300$.

For various values of t the proportions of confidence interval (C.L.) realizations covering the true value of the distribution function $K_2(t)$ are given in Table 13.3.

13.7 Conclusion

In the analysis of redundant systems we considered the practical approach for the family of distributions whose hazard rate functions are $\cap$-shape or unimodal. Globally the numerical simulation gives very good results. Note that the calculation of functions K_j and the vector C_j takes a lot of time for inverse Gaussian distribution compared to log-normal and log-logistic distributions.

References

Bagdonavicius V, Masiulaityle I, Nikulin M (2008a) Statistical analysis of redundant system with "warm" stand-by units Stochastics. Int J Probab Stochastical Processes 80(2–3):115–128

Bagdonavicius V, Masiulaityle I, Nikulin M (2008b) Statistical analysis of redundant system with one stand-by unit. In: Huber C, Limnios N, Messbah M, Nikulin M (eds) Mathematical methods in survival analysis reliability and quality of life. ISTE & Wiley, London, pp 189–202

Bagdonavicius V, Masiulaityle I, Nikulin M (2009) Asymptotic properties of redundant systems reliability estimators. In: Nikulin M, Limnios N, Balakrishnan N, Kahle W, Huber C (eds) Advances in degradation modeling: applications to reliability, survival analysis and finance. Springer, pp 293–310

Bagdonavicius V, Masiulaityle I, Nikulin M (2010) Parametric estimation of redundant system reliability from censored data. In: Rykov V, Balakrishnan N, Nikulin M (eds) Mathematical and statistical models and methods in reliability. Springer, pp 177–191

Chhikara RS, Folks JL (1989) The inverse Gaussian distribution. Marcel Dekker, New York

Lemeshko BY, Lemeshko SB, Akushkina KA, Nikulin M, Saaidia N (2010) Inverse Gaussian model and its applications in reliability and survival analysis. In: Rykov VV, Balakrishnan N,

Nikulin MS (eds) Mathematical and statistical models and methods in Reliability. Birkhäuser, Boston, pp 293–315

Nikulin MS, Saaidia N (2009) Inverse Gaussian family and its applications in reliability. Study by simlulation. In: Proceedings of the 6th St. Petersburg Workshop on Simulation, St. Petersburg, June 28–July 4. VVM comm. Ltd. St. Petersburg, Russia, pp 657–661

Seshadri V (1994) The inverse Gaussian distribution: a case study in exponential families. Oxford University Press, New York

Seshadri V (1998) The inverse Gaussian distribution: statistical theory and applications. Springer, New York

Chapter 14
Multiobjective Reliability Allocation in Multi-State Systems: Decision Making by Visualization and Analysis of Pareto Fronts and Sets

Enrico Zio and Roberta Bazzo

Abstract Reliability-based design, operation and maintenance of multi-state systems lead to multiobjective (multicriteria) optimization problems whose solutions are represented in terms of Pareto Fronts and Sets. Among these solutions, the decision maker must choose the ones which best satisfy his\her preferences on the objectives of the problem. Visualization and analysis of the Pareto Fronts and Sets can help decision makers in this task. In this view, a recently introduced graphical representation, called Level Diagrams, is here used in support of the analysis of Pareto Fronts and Sets aimed at reducing the number of non-dominated solutions to be considered by the decision maker. Each objective and design parameter is represented on separate "synchronized" diagrams which position the Pareto front points according to their proximity to ideal preference points and on the basis of this representation a two-step front reduction procedure is proposed. An application to a redundancy allocation problem of the literature concerning a multi-state system is used to illustrate the analysis.

Keywords Multi-state system redundancy allocation · Reliability · Availability · Multiobjective decision making · Pareto Front · Pareto Set · Level Diagrams

E. Zio (✉) · R. Bazzo
Dipartimento di Energia, Politecnico di Milano, Milano, Italy
e-mail: enrico.zio@ecp.fr; enrico.zio@polimi.it

E. Zio
Chair "Systems Science and Energetic Challenge",
European Foundation for New Energy—EDF,
Ecole Centrale Paris-Supelec, Paris, France

R. Bazzo
e-mail: roberta.bazzo@alice.it

A. Lisnianski and I. Frenkel (eds.), *Recent Advances in System Reliability*,
Springer Series in Reliability Engineering, DOI: 10.1007/978-1-4471-2207-4_14,
© Springer-Verlag London Limited 2012

14.1 Introduction

Solutions to multiobjective problems in reliability-based multi-state system design, operation and maintenance are not unique as they have to satisfy simultaneously various conflicting objectives. The set of solutions for consideration of the decision maker (DM) forms the Pareto Set[1] in the space of the decision variables, and the Pareto Front[2] in the space of the objectives.

In practice, the solution to a multiobjective optimization problem is found by a search algorithm as a discrete approximation of the Pareto Front, from which the DM has to select one, or more, solutions of preference. For this task to be successful, the DM should be confronted with a relatively small number of solutions representative of feasible alternatives on the Pareto Front.

In this view, visualization techniques can represent valuable tools for analyzing the multidimensional Pareto Front and Set. For two-dimensional (and at times for three-dimensional) problems, it is usually possible to make an accurate graphical analysis of the Pareto Front and Set, but this becomes soon impractical for higher dimensions. Some of the most common techniques proposed for multidimensional visualization are (ATKOSoft 1997):

- *Scatter diagrams* The visualization consists of an array of scatter diagrams arranged in the form of an $s \times s$ matrix, where s is the number of objectives. Each objective function defines one row and one column of the matrix. The complexity of the representation increases notably with the number of dimensions.
- *Parallel coordinates* A multidimensional point is plotted in a two-dimensional graph, one for the objective functions and one for the decision variables. Each dimension of the original data is transformed into a coordinate in the two-dimensional plot. This is a very compact way of presenting multidimensional information, but it soon loses clarity with large sets of data and the analysis becomes difficult.
- *Interactive decision maps* (Lotov et al. 2004) Two-dimensional projections of the objective function space are used to display decision maps of contour lines parameterized by a third, color-associated objective; the process of parameterizing the two-dimensional contours can become time-consuming and cumbersome when many objectives are involved.

Recently, Level Diagrams[3] have been introduced for visualizing multidimensional Pareto Fronts and Sets (Blasco et al. 2008); they can be employed a priori,

---

[1] The so-called Pareto Set is the set containing all non-dominated solutions resulting from multiobjective optimization problems. Multiobjective optimization problems generally do not have a unique solution, but rather a set of solutions, which dominate all others but among them none is better. Dominance and non-dominance is determined by pair-wise vector comparisons of the multiobjective values corresponding to the pair of solutions under comparison.

[2] The Pareto Front is the set of values of the objective functions in correspondence to the solutions of the Pareto Set.

[3] Level Diagrams are a recently introduced technique for visualizing Pareto Front and Sets, which will be further explained in the chapter

interactively or a posteriori of the optimization process to help the DM defining his or her preferences during the solution selection phase. The visualization is based on a metric distance from an ideal solution which optimizes all objectives simultaneously; a solution coloring procedure can be adopted to visualize the DM preferences.

In the present work, Level Diagrams are drawn for the Pareto Front and Set of a multi-state system redundancy allocation problem of the literature which involves three conflicting objectives: system availability to be maximized, system cost and weight to be minimized (Messac and Wilson 1998). The decision variables define the system configurations, each variable indicating the number of components of a particular type allocated in the configuration to provide redundancy.

The Level Diagrams analysis is then exploited for establishing a two-step procedure of reduction of the number of possible solutions represented, which renders it easier for the DM to apply his or her preferences.

The remainder of the paper is organized as follows. Section 14.2 recalls the main ideas behind the Level Diagrams representation and the related procedure (Blasco et al. 2008). In Sect. 14.3, the multi-state system redundancy allocation problem of interest is formulated (Messac and Wilson 1998). Section 14.4 reports the analysis of the Pareto Front and Set using Level Diagrams, with the proposed procedure of front reduction. Finally, some conclusions are drawn in Sect. 14.5.

14.2 Level Diagrams Representation of Pareto Fronts and Sets

14.2.1 Pareto Front and Set

Without loss of generality, let us consider a multiobjective minimization problem

$$\min_{\theta \in D} J(\theta), \tag{14.1}$$

where $\theta = [\theta_1, \ldots, \theta_p] \in D$ is the vector of decision variables, D is the decision space and $J(\theta) = [J_1(\theta), \ldots, J_s(\theta)]$ is the vector of objective functions.

Generally, there is a set of solutions to (14.1) dominating all others but among them none is superior. Dominance is determined by comparing the values of the multiple objectives in correspondence to pairs of solutions: solution θ^1 dominates solution θ^2, if

$$\forall i \in \{1, \ldots, s\}, \ J_i(\theta^1) \leq J_i(\theta^2) \ \wedge \ \exists k \in \{1, \ldots, s\} : J_k(\theta^1) < J_k(\theta^2) \tag{14.2}$$

The set Θ_p of non-dominated solutions is called the Pareto Set; the vector $J(\theta)$ of the values of the objective functions in correspondence to the solutions $\theta \in \Theta_p$, defines the Pareto Front.

14.2.2 Level Diagrams

From the discrete set of Pareto solutions obtained by a multiobjective optimization algorithm, the DM is called to choose the best solution according to his or her preferences. This task can be quite difficult for high dimensional problems and must be aided by effectively visualizing the results of the multiobjective optimization.

Level Diagrams represent an interesting visualization tool which allows classifying the Pareto solutions according to their distance from the ideal solution, i.e., the one which optimizes all objectives simultaneously (Blasco et al. 2008).

Assuming, in general, that l objectives are to be minimized and m maximized, with $s = l + m$, each objective is normalized with respect to its minimum or maximum values on the Pareto Front:

$$\overline{J}_i(\theta) = \frac{J_i(\theta) - J_i^{\min}}{J_i^{\max} - J_i^{\min}}, \quad i = 1, \ldots, l \tag{14.3}$$

and

$$\overline{J}_i(\theta) = \frac{J_i^{\max} - J_i(\theta)}{J_i^{\max} - J_i^{\min}}, \quad i = 1, \ldots, m \tag{14.4}$$

so that now,

$$0 \le \overline{J}_i(\theta) \le 1, \quad i = 1, \ldots, s. \tag{14.5}$$

The value $\overline{J}_i(\theta) = 0$ means that the solution θ gives the best value for the ith objective, whereas the value $\overline{J}_i(\theta) = 1$ corresponds to the solution θ giving the worst value for the ith objective.

The distance to the ideal point

$$\theta^* : \overline{J}_i(\theta^*) = 0 \ \forall \ i = 1, \ldots, s \tag{14.6}$$

can be computed with different norms, giving different views on the characteristics of the Pareto Front (Blasco et al. 2008). The norm considered here is the 1-norm:

$$\left\|\overline{J}(\theta)\right\|_1 = \sum_{i=1}^{s} \left|\overline{J}_i(\theta)\right| \text{ with } 0 \le \left\|\overline{J}(\theta)\right\|_1 \le s. \tag{14.7}$$

The Level Diagrams are drawn as follows: each objective (J_i) and each decision variable (θ_j) is plotted separately; the X axis corresponds to the objective or the decision variable in physical units of measurement, while the Y axis corresponds to the value $\left\|\overline{J}(\theta)\right\|_1$ for all the graphs; this means that all the plots are synchronized with respect to the Y axis, i.e., all the information for a single point of the Pareto Set will be plotted at the same level of the Y axis.

14.3 The Multi-State System Redundancy Allocation Problem

The redundancy allocation problem (RAP) considered here regards the selection of the system redundancy configuration that maximizes availability and minimizes cost and weight. The system is decomposed in a number of subsystems, for each of which there are multiple component choices to be allocated in redundancy (Kulturel-Konak et al. 2003; Blasco et al. 2008).

The RAP case study here considered is taken from (Messac and Wilson 1998). The system is made of $u = 5$ units (subsystems) connected in series; redundancy can be provided to each unit by selecting components from m_i types available in the market, $i = 1, \ldots, 5$. Each component can be in two states: functioning at nominal capacity or failed, at zero capacity. The collective performance of these binary components leads to a multi-state system behavior. The types of components available are characterized by their availability, nominal capacity, cost and weight in arbitrary units (Table 14.1). Without loss of generality, component capacities are measured in terms of percentages of the maximum system demand. The different demand levels for a given period are given in Table 14.2.

The three objectives in mathematical terms are:

$$
\begin{aligned}
\text{Availability:} \quad & \max_{\theta} J_1(\theta) = \max_{\theta} \left[\prod_{i=1}^{u} A_i(\theta_i) \right], \\
\text{Cost:} \quad & \min_{\theta} J_2(\theta) = \min_{\theta} \left[\sum_{i=1}^{u} \sum_{j=1}^{m_i} c_{ij} \theta_{ij} \right], \\
\text{Weight:} \quad & \min_{\theta} J_3(\theta) = \min_{\theta} \left[\sum_{i=1}^{u} \sum_{j=1}^{m_i} w_{ij} \theta_{ij} \right], \\
\text{subject to:} \quad & 1 \le \sum_{j=1}^{m_i} \theta_{ij} \le n_{\max,i}, \quad \forall\, i = 1, \ldots, u,
\end{aligned}
\tag{14.8}
$$

where A_i—availability of the ith subsystem, θ—vector of the indexes of the system configuration, θ_{ij}—component of vector θ, i.e. index of the ith component in jth subsystem, u—number of subsystems, m_i—number of components in the ith subsystem, $n_{\max,i}$—user-defined maximum number of redundant components that can be placed in subsystem i, n_{ij}—number of components of type j used in subsystem i, c_{ij}, w_{ij}—cost and weight of the jth type of component in subsystem i.

The problem has been solved in (Taboada et al. 2007) using the MOMS-GA algorithm coupled with the Universal Moment Generating Function (UMGF) approach (Ushakov 1986; Levitin and Lisnianski 2001; Levitin 2005) for computing the system availability (8). The resulting Pareto Front of 118 points is shown in Fig. 14.1 in the objective functions space.

Table 14.1 Characteristics of the components available on the market

Subsystem	Component type	Availability	Capacity (%)	Cost	Weight
1	1	0.980	120	0.590	35.4
	2	0.977	100	0.535	34.9
	3	0.982	85	0.470	34.1
	4	0.978	85	0.420	33.9
	5	0.983	48	0.400	34.2
	6	0.920	31	0.180	34.3
	7	0.984	26	0.220	32.6
2	1	0.995	100	0.205	26.5
	2	0.996	92	0.189	22.4
	3	0.997	53	0.091	20.3
	4	0.997	28	0.056	21.7
	5	0.998	21	0.042	25.2
3	1	0.971	100	7.525	42.1
	2	0.973	60	4.720	41.7
	3	0.971	40	3.590	40.8
	4	0.976	20	2.420	39.6
4	1	0.977	115	0.180	25.4
	2	0.978	100	0.160	23.9
	3	0.978	91	0.150	24.7
	4	0.983	72	0.121	24.6
	5	0.981	72	0.102	23.6
	6	0.971	72	0.096	26.2
	7	0.983	55	0.071	25.5
	8	0.983	25	0.049	22.6
	9	0.977	25	0.044	24.8
5	1	0.984	128	0.986	15.4
	2	0.983	100	0.825	15.3
	3	0.987	60	0.490	14.9
	4	0.981	51	0.475	15.0

Table 14.2 Demand levels

Demand (%)	100	80	50	20
Duration (h)	4203	788	1228	2536
Duration (%)	0.48	0.09	0.14	0.29

14.4 Visualization With the Level Diagram

14.4.1 Objective Functions

The Level Diagrams of the Pareto Front of Fig. 14.1 are shown in Fig. 14.2 for the system availability, cost and weight.

In the Level Diagrams of all the three objectives, one observes an initial decreasing trend of the norm value up to a minimum and then an increasing trend;

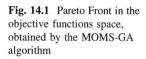

Fig. 14.1 Pareto Front in the objective functions space, obtained by the MOMS-GA algorithm

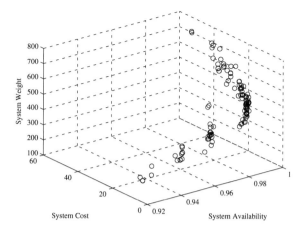

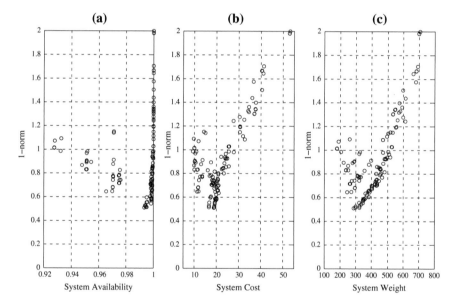

Fig. 14.2 Level Diagram representation for availability objective, J1 (*a-left*), cost objective, J2 (*b-center*), weight objective, J3 (*c-right*)

this is reasonable, due to the conflict among the objectives. To understand this behaviour, let us consider for example Fig. 14.2a related to the availability objective: system design configurations of low availability (far from the optimal value) are characterized by small costs and weights (close to the optimal values) and as a result, the values of the 1-norm (14.7) are large; for values of the system availability increasing towards the optimal value, the norms decrease up to the minimum value, optimal with respect to all three objectives, after which the increase of cost and

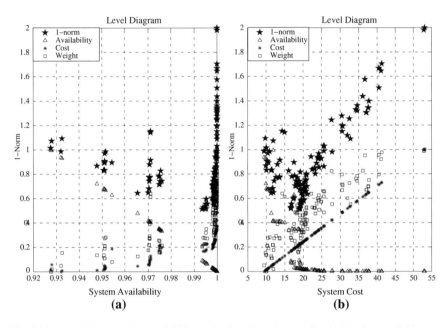

Fig. 14.3 Level Diagram for availability objective J1 and normalized objectives (*a-left*), cost objective J2 and normalized objectives (*b-right*)

weight is such that the norms increase as the point in the Pareto Front moves away from the ideal one, optimal with respect to all three objectives.

The Level Diagrams representation allows also a qualitative evaluation of the sensitivity of the results. Looking again at Fig. 14.2 for the system availability objective, one can observe that the absolute value of the slope of the ascending part is larger than in Fig. 14.2b and c for the cost and the weight objectives, respectively: for example, system availabilities above 0.99 are achieved by rapidly moving away from the ideal point of minimum norm, due to significant cost and weight increases.

An interesting feature is the presence of data clusters in the 1-norm Level Diagram of availability (Fig. 14.2). This is due to the fact that whereas the cost and weight objectives grow "regularly" with increasing availability, the values that this latter objective can take are discontinuous, depending on the discrete redundant configurations which can be devised with different costs and weights but the same availability values. This becomes clear if one considers Fig. 14.3 in which the normalized values (14.3) or (14.4) of the three objectives (14.8) are also plotted. Note for example how at a given value of system availability (Fig. 14.3a) corresponds to very different values of weight, creating the vertical alignment that can be seen in the Level Diagrams of Figs. 14.2 and 14.3. Something similar occurs in Fig. 14.3b, where at cost values around 20 corresponds various values of weight, with consequent vertical alignment of the 1-norm values. These vertical alignments are caused by solutions which have approximately the same value of the objective represented in the Level Diagram, but different values of the other objectives.

The representation of the Pareto Front by means of Level Diagrams has identified the presence of system design configurations of similar values of availability but different cost and weight values. The observations thereby obtained suggest a Pareto Front reduction through the suppression of the configurations with higher values of 1-norm among those with similar values of one objective, availability. In Sect. 14.4.3, the results of this reduction will be shown to provide a less "crowded" Pareto Front, easier to be analyzed by the DM for his or her preferences evaluation on the way to a final decision.

14.4.2 Decision Variables

In the case study considered, the decision variables define the system configurations in terms of the subsystem of redundant components. The configuration is contained in a vector of 29 discrete decision variables; each variable indicates the number of components of that particular type present in the configuration.

Figure 14.4 shows the Level Diagrams relative to the 7 decision variables characterizing the redundancy configuration of the first subsystem. Note how some of the low capacity components (e.g., 6 and 7) are not selected in any of the Pareto Set configurations. On the contrary, the components of large capacity (the highest in the list of Table 14.1) are typically used, possibly in redundancy to ensure good values of availability. This is due to the definition of the multi-state system availability, which considers not only the components' availabilities but also their feeding capacities (Levitin and Lisnianski 2001).

In Fig. 14.4, one also notes that to some decision variables values are associated a wide range of 1-norm values: for example, there is a large number of solutions in the Pareto Set with $\theta_{11} = 1$ which are associated to different values of the 1-norm on the Level Diagrams, i.e., different distances from the optimal point, ideal with respect to all the three objectives (top-left graph). Solutions with $\theta_{11} = 1$ are thus quite ubiquitous in the space of the objective functions (Fig. 14.1) and the characteristics of the Pareto front may not be substantially affected if it were reduced to only the solutions with $\theta_{11} = 1$.

14.4.3 Pareto Front and Set Reduction

The Pareto Front and Set generated by a search algorithm often result in a crowded set of dominant solutions gradually different among them; on the contrary, less crowded Pareto Front and Set would be beneficial to the DM who must analyze the different solutions and identify the preferred ones.

One way of proceeding to a reduction of the solutions contained in the Pareto Front and Set is that of focusing on the values of one objective function (in this case availability), and keeping only few representative solutions of the clusters of

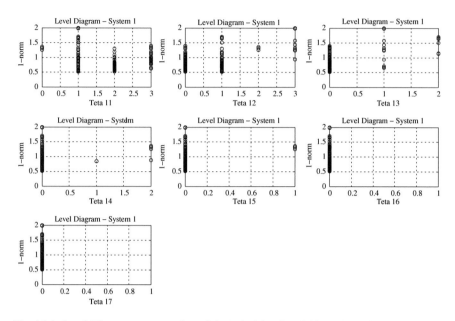

Fig. 14.4 Level Diagram representation of the 7 decisional variables values of subsystem 1

solutions with approximately equal value of that objective function (availability) and different values of the other objective functions (cost and weight). The solutions retained as representative of a cluster at a given value of availability are those with lowest costs and weights. The clusters along the Pareto Front are then reduced to a few representative points different in terms of availability and optimal in terms of low costs and weights.

To systematically perform the Pareto Front reduction, a criterion must be established to define when two solutions on the Level Diagrams can be considered vertically aligned in a cluster of equal value of an objective (availability). To this purpose, let us consider a vector J_1 of length n_j containing all the values of the availability objective of the Pareto Front sorted in ascending order, let $i_{\min}$ be the position in the vector J_1 of the optimal, ideal point (optimal with respect to all the three objectives), i.e., the solution with the lowest values of the 1-norm (14.7) and let $norm1$ be the vector containing the values of the 1-norm (14.7) for each solution on the Pareto Front. Since the differences of the availability values of two successively ranked solutions, i.e., $J_1(i+1) - J_1(i)$ are very different in the two branches of the Front (Figs. 14.2 and 14.3), two distinct criteria are needed to decide whether two solutions can be considered vertically aligned in a cluster on the left and right of the minimum value of the norm. Then, in the vertical cluster alignment, solutions with a higher value of the 1-norm, i.e., with a worse overall system performance, are discarded. In this work, the solution $\theta(i+1)$ corresponding to $J_1(i+1)$ and to $norm1(i+1)$, belongs to a vertical cluster alignment if the previous solution $\theta(i)$ is such that:

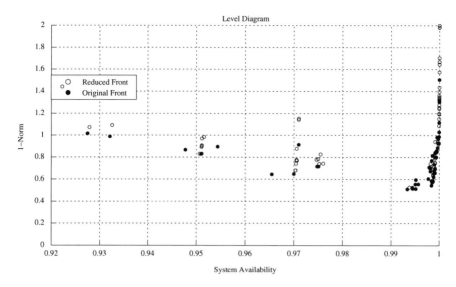

Fig. 14.5 1-norm Level Diagram representation of the reduced Pareto Front for the availability objective, J_1

$$J_1(i+1) - J_1(i) \leq 10^{-3} \quad \text{for } 1 \leq i \leq i_{\min} \quad (14.9)$$

i.e., on the right of the ideal point, while

$$J_1(i+1) - J_1(i) \leq 10^{-5} \quad \text{for } i_{\min} < i \leq n_j \quad (14.10)$$

i.e., on the left of the ideal point. Then, the solution $\theta(i)$ is discarded if

$$norm1(i+1) < norm1(i); \quad (14.11)$$

vice versa, the solution $\theta(i+1)$ is discarded if

$$norm1(i+1) > norm1(i). \quad (14.12)$$

The results of the reduction performed on the Pareto Front under analysis are plotted in Fig. 14.5: the Pareto Front is reduced from the original 118 points to 52, with the clusters indeed reduced to individual best points, i.e., those of lowest 1-norm at basically equal availability values.

In synthesis, the examination of the Pareto Front portrayed by Level Diagrams has allowed reduction by approximately 56% of the solutions, before imposing any preference or constraint.

A further reduction of the solutions contained in the Pareto Front and Set can be achieved by focusing on the values of selected decision variables. The solutions in the Pareto Front and Set can be reduced to those whose selected decision variables have given values, while maintaining the Front shape and characteristics. The

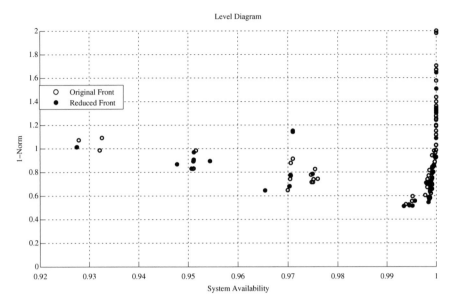

Fig. 14.6 1-norm Level Diagram representation of the reduced Pareto Front obtained by decision variables selection for the availability objective, J_1

underlying idea has been briefly introduced in Sect. 14.4.2, with reference to the decision variable θ_{11} at its value of 1 (top-left graph of Fig. 14.4), but in general one may consider more than one decision variable.

The decision variables Level Diagrams can be used to guide the selection of the decision variables and their values to be subjected to reduction, because they visualize the different 1-norm values corresponding to the decision variable values, thus showing the extent to which the solutions characterized by a given value of a decision variable cover the variety of solutions on the Pareto Front. With reference to Fig. 14.4 for subsystem 1, the decision variables and their values worth to be considered for guiding the reduction, because offering a wide coverage of different Pareto Front solutions, are: $\theta_{11} = 1$ (top-left graph), $\theta_{12} = 1$ (top-center graph), $\theta_{14} = 0$ (middle-left graph), $\theta_{15} = 0$ (middle-center graph); looking at the Level Diagrams related to the other subsystems (not shown here), also $\theta_{44} = 0$, $\theta_{45} = 0$ and $\theta_{54} = 0$ would be worth consideration for driving the reduction. In the final selection of the decision variables and values for the reduction of the Pareto Front and Set solutions, particular attention must be given to preserving as much as possible the representation of the region closest to the minimum 1-norm value in the Level Diagrams because it contains the best solutions, of minimum distance from the ideal point. Applying these considerations to the original Pareto Front and Set reduces the solutions to only 36. Figure 14.6 shows, for example, the reduced 1-norm Level Diagram of the availability objective: note, however, that although the shape is preserved, the clusters of solutions are not reduced to the best

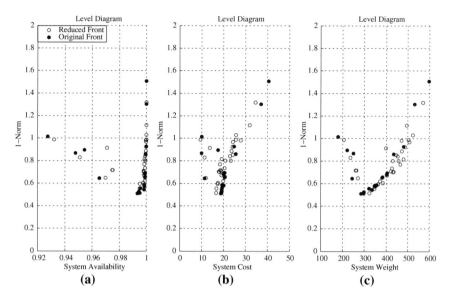

Fig. 14.7 1-norm Level Diagram representation of the reduced Pareto Front obtained by the two-steps reduction, for the availability objective, J1 (*a-left*), the cost objective, J2 (*b-center*) and the weight objective, J3 (*c-right*)

performing solutions, i.e., the minimum 1-norm solutions, as in the previous reduction procedure.

This suggests applying the two reduction procedures in succession: the result is a Pareto Front and Set of only 19 solutions. Figure 14.7 shows the representation of the Pareto Front in terms of the Level Diagrams of the three objective functions.

Looking for example at Fig. 14.7a, one can note that the representation of the solutions in the region of the low 1-norm values is well preserved. Experiments performed by changing the selected decision variables and values have show that the resulting reduced Pareto Front and Set are quite sensitive to these choices, in particular with respect to preserving the region of greatest interest, i.e., of low 1-norm values.

14.5 Conclusion

Level Diagrams used in the analysis of the Pareto Front and Set of multiobjective optimization problems can drive their reduction by removal of less efficient solutions, while maintaining the Front characteristics. In particular, the Pareto Set analysis by Level Diagrams can be used for the reduction of the Pareto Front by selecting solutions characterized by preferred values of selected decision variables.

In this work, two Pareto Front and Set reduction methods have been proposed for application in succession; a multi-state system redundancy allocation case study of the literature has shown their capabilities of leading to a significant reduction of the solutions to be considered by the DM.

Acknowledgments The authors are thankful to Professor David Coit of Rutgers University for providing the Pareto Front and Set data. This work has been funded by the Foundation pour une Culture de Securité Industrielle of Toulouse, France, A0-2009-04.

References

ATKOSoft (1997) Survey of visualization methods and software tools. http://europa.eu.int/en/comm/eurostat/research/supcom.96/30/result/a/visualisation_methods.pdf

Blasco X, Herrero JM, Sanchis J, Martínez M (2008) A new graphical visualization of n-dimensional Pareto Front for decision-making in multiobjective optimization. Int J Inform Sci 178(20):3908–3924

Kulturel-Konak S, Smith A, Coit DW (2003) Efficiently solving the redundancy allocation problem using tabu search. IEEE Trans 35(6):515–526

Levitin G (2005) Universal generating function in reliability analysis and optimization. Springer, London

Levitin G, Lisnianski A (2001) A new approach to solving problems of multi-state system reliability optimization. Qual Reliab Eng Int 17(2):93–104

Lotov AV, Bushenkov VA, Kamenev GK (2004) Interactive decision maps approximation and visualization of Pareto frontier. Kluwer, Boston

Messac A, Wilson BH (1998) Physical programming for computational control. AIAA J 36(2):219–226

Taboada H, Espiritu J, Coit D (2007) MOMS-GA: a multiobjective multi-state genetic algorithm for system reliability optimization design problems. IEEE Trans Reliab 57(1):182–191

Ushakov I (1986) A universal generating function. Soviet J Comp Sys Sci 24(5):37–49

Chapter 15
Optimal Incomplete Maintenance in Multi-State Systems

Waltraud Kahle

Abstract In this research, we are concerned with the modeling of optimal maintenance actions in multi-state systems. We consider an incomplete mainte- nance model, that is, the impact of a maintenance action is not minimal (as bad as old) and not perfect (as good as new) but lies between these boundary cases. Further, we assume that the costs of maintenance depend on the degree of repair. After a failure the system is repaired minimally. Cost optimal maintenance policies for some cost functions are considered.

15.1 Introduction

The concept of multi-state systems is widely used to describe the performance of technical systems which are subjected during their lifetime to aging and degra- dation. For an overview of ideas in MMS reliability theory we refer to (Lisnianski and Levetin 2003). In (Lisnianski and Frenkel 2009) reliability measures for multi- state systems such as mean accumulated performance deficiency, mean number of failures, and others, are computed using a non-homogeneous Markov reward model.

We consider a multi-state system with n states. Further, a time scale is intro- duced, and we assume that in time 1 the system is in state one, in time 2 it is in state two and so on. The time scale is considered because of two reasons: first, the notation becomes more simple, and, secondly, there can be used results from the

W. Kahle (✉)
Fakultaet fuer Mathematik, Institute of Mathematical Stochastics,
Otto-von-Guericke-Universitaet Magdeburg, Postfach 4120,
39016 Magdeburg, Germany
e-mail: waltraud.kahle@ovgu.de

A. Lisnianski and I. Frenkel (eds.), *Recent Advances in System Reliability*,
Springer Series in Reliability Engineering, DOI: 10.1007/978-1-4471-2207-4_15,
© Springer-Verlag London Limited 2012

theory of incomplete repairs for systems with discrete lifetime distribution. The results can be generalized to any time scale.

In each state the system can fail with some probability. The failure rate in state i, $i = 1, \ldots, n$ is assumed to be increasing. On failure, the system is minimally repaired. Additionally, maintenance decisions are regularly carried out. The distance between preventive maintenance is τ. We assume that such actions impact the failure intensity. Specifically we assume that maintenance actions served to adjust the virtual age of the system, that is, after a preventive maintenance the system is not as good as new (and, of course, not as bad as old). For multi-state systems that means that the maintenance resets the system in a lower (younger) state. Let v be the state after preventive maintenance. Note that in the special case of $v = 0$ a block replacement is carried out.

In Sect. 15.2 we describe three possible failure rates for the multi-state system. These are failure rates of a system with discrete time-to-failure distribution. Some properties of such failure rates are given. Then, we describe the concept of virtual age and degree of repair introduced by Kijima (1989) and Kijima et al. (1988). In difference to the classical model, however, we assume that after a failure the system is repaired minimally. Only preventive maintenance actions reset the virtual age of the system, it means that after such a maintenance the system switches to a lower (younger) state. In the last Sect. 15.4 several possible cost functions are introduced and optimal maintenance strategies for the introduced failure rates are given.

15.2 Discrete Lifetime Distributions and Their Failure Rate

There are a number of possibilities for consideration of discrete lifetime distributions. In this paper we consider three of them:

- the discrete uniform distribution,
- the (shifted and truncated) Poisson distribution, and
- the (truncated) discrete Weibull distribution.

15.2.1 The Discrete Uniform Distribution

Let the states of the system be $1, \ldots, n$. At each state the system fails with probability $1/n$. Let T be the random time to failure. The failure rate is given by

$$h(t) = \frac{P(T = t)}{P(T \geq t)} = \frac{1}{n + 1 - t}, \quad t = 0, 1, \ldots, n,$$

and the cumulative hazard is

$$H(t) = -\log(1 - F(t)) = -\log\left(\frac{n-t}{n}\right), \quad t = 0, 1, \ldots, n,$$

where $F(t)$ is the cumulative distribution function. It is easy to see that the discrete uniform distribution has an increasing hazard rate.

15.2.2 The Poisson Distribution

Let the lifetime T be shifted as per Poisson distribution that is, $T = X + 1$ where

$$P(X = t) = \frac{\lambda^t}{t!}e^{-\lambda}, \quad t = 0, 1, \ldots.$$

The failure rate is given by

$$h(t) = \frac{P(T = t)}{P(T \geq t)} = \frac{\frac{\lambda^t}{t!}}{\sum_{k=t}^{\infty}\frac{\lambda^k}{k!}}, \quad t = 0, 1, \ldots.$$

Since $\sum_{k=t+1}^{\infty}\frac{\lambda^k}{k!} = \sum_{k=t}^{\infty}\frac{\lambda^{k+1}}{(k+1)!} \leq \frac{\lambda}{t+1}\sum_{k=t}^{\infty}\frac{\lambda^k}{k!}$ we get $\frac{\lambda^t/t!}{\sum_{k=t}^{\infty}\lambda^k/k!} \leq \frac{\lambda^{t+1}/(t+1)!}{\sum_{k=t+1}^{\infty}\lambda^k/k!}$ that is, the Poisson distribution has an increasing failure rate. We use the shifted version of this distribution, that is, the first failure may appear in state $t = 1$. The cumulative hazard is given by

$$H(t) = -\log(1 - F(t)) = -\log\left(\sum_{k=t+1}^{\infty}\frac{\lambda^{k-1}}{(k-1)!}e^{-\lambda}\right), \quad t = 1, 2, \ldots.$$

If we assume again that the system has n states, then a truncated Poisson distribution should be considered, that is, the probabilities are

$$P(T = t) = A \cdot \frac{\lambda^{(t-1)}}{(t-1)!}e^{-\lambda}, \quad t = 1, \ldots, n$$

where $A = \sum_{k=n+1}^{\infty}\frac{\lambda^{(t-1)}}{(t-1)!}e^{-\lambda}$.

15.2.3 The Discrete Weibull Distribution

There are different *discrete Weibull distributions*, as example introduced by Padgett and Spurrier (1985) or the Nakagawa-Osaki model (Nakagawa and Osaki 1975). In this paper, we use the following model. Let X be Weibull distributed with density function

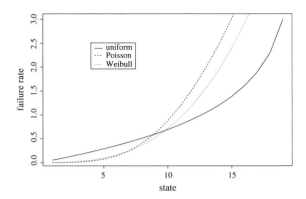

Fig. 15.1 Three cumulative hazard functions

$$f(x) = \frac{\beta}{\alpha}\left(\frac{x}{\alpha}\right)^{\beta-1}\exp\left(-\left(\frac{t}{\alpha}\right)^{\beta}\right), \quad x \geq 0.$$

Now, for the discrete version we put the probability mass of the interval $(t-1, t]$ into the point t that is,

$$P(T = t) = \int\limits_{s=t-1}^{t} f(s)ds = \exp\left(-\left(\frac{t-1}{\alpha}\right)^{\beta}\right) - \exp\left(-\left(\frac{t}{\alpha}\right)^{\beta}\right), \quad t = 1, 2, \ldots.$$

$$(15.1)$$

If $\beta = 1$, then we get the geometric distribution

$$P(T = t) = \left(\exp\left(-\frac{1}{\alpha}\right)\right)^{t-1} - \left(\exp\left(-\frac{1}{\alpha}\right)\right)^{t-1} \cdot \exp\left(-\frac{1}{\alpha}\right) = p^{t-1}(1-p).$$

The cumulative hazard is the same as for the continuous Weibull distribution

$$H(t) = -\log(1 - F(t)) = \exp\left(-\left(\frac{t}{\alpha}\right)^{\beta}\right).$$

In Fig. 15.1 three hazard functions are plotted. We assume 20 possible states for the system. If the time to failure is uniformly distributed, then it has the expectation 10. The parameters of both other distributions are chosen so that the expectation is 10, too: $\lambda = 10$ for the Poisson distribution and $\beta = 3$, $\alpha = 11.2$ for the Weibull distribution.

Remark Note that for discrete failure time distributions the failure rate is a probability unlike for continuous distributions. Further, the cumulative hazard is

$$H(t) = -\ln R(t) = -\ln(1 - F(t)) \neq \sum_{i=1}^{t} h(t),$$

where $F(t)$ and $R(t)$ are distribution function and survival function, respectively. It is convex if $h(t)$ is increasing. For an other definition of failure rates for discrete distributions we refer to (Xie et al. 2002) and the references given in their paper.

15.3 Kijima Type Repairs

First, we describe the classical Kijima repair model. Consider the impact of repairs. A system (machine) starts working with an initial prescribed failure rate $\lambda_1(t) = \lambda(t)$. Let t_1 denote the random time of the first sojourn. At this time t_1 the item will be repaired with the degree ξ_1. When the system is minimally repaired then the degree is equal to one, and if the repair makes the system as good as new then this degree is zero. The virtual age of the system at the time t_1, following the repair, is $v_1 = \xi_1 t_1$, implying the age of the system is reduced by maintenance actions. The distribution of the time until the next sojourn then has failure intensity $\lambda_2(t) = \lambda(t - t_1 + v_1)$. Assume now that t_k is the time of the kth $(k \geq 1)$ sojourn and that ξ_k is the degree of repair at that time. We assume that $0 \leq \xi_k \leq 1$, for $k \geq 1$ (Kijma 1989; Kijma et al. 1988).

After repair the failure intensity during the $(k + 1)$th sojourn is determined by

$$\lambda_{k+1}(t) = \lambda(t - t_k + v_k), \quad t_k \leq t < t_{k+1}, k \geq 0,$$

where the virtual age v_k is for Kijima's Type II imperfect repair model

$$v_k = \xi_k(v_{k-1} + (t_k - t_{k-1})),$$

that is, the repair resets the intensity of failure proportional to the virtual age.

Kijima's Type I imperfect repair model suggests that upon failure, the repair undertaken could serve to reset the intensity only as far back as the virtual age at the start of working after the last failure that is:

$$v_k = t_{k-1} + \xi_k(t_k - t_{k-1}).$$

The process defined by $v(t, \xi_k, k = 1, 2, \ldots) = t - t_k + v_k$, $t_k \leq t < t_{k+1}$, $k \geq 0$ is called the *virtual age process* (Last and Szekli 1998).

In this paper we assume that, after a failure, a minimal repair is undertaken. This means that the failure rate, and also the state, of the system is the same as just before failure. Additionally, preventive maintenance action are regularly carried out. Let us assume that v is the state after such a maintenance and that the next maintenance is undertaken τ states later. Every time when the system reaches the state $v + \tau$ a maintenance resets it to state v. If the system has n states there are the following possibilities:

- $\tau = n - 1; v = 1,$
- $\tau = n - 2; v = 1$ or $v = 2,$
- $\tau = n - 3; v = 1, v = 2$ or $v = 3,$
- $\tau = 1; v = 1, \ldots, v = n - 1.$

In terms of the Kijima model we have a degree of repair $\xi = v/\tau$ for the Kijima type I model and $\xi = v/(v + \tau)$ for the Kijima type II model.

Note that in both cases a perfect ($\xi = 0$) maintenance resets the system to state 0 and a minimal repair ($\xi = 1$) leads to $v = \tau$ in case of Kijima type I and to $\tau = 0$ in Kijima type II models. We exclude these two extreme cases. Our aim is now to find the states v and $v + \tau$ which minimizes the complete costs of failure and maintenance.

15.4 Cost Optimal Maintenance

Now we consider optimal maintenance policies with respect to costs. Again, each failure is removed by a minimal repair and at fixed times τ, 2τ, preventive maintenance actions are undertaken. In classical policies the maintenance renews the system, then we have a block replacement. We do not make the assumption that after maintenance the item is as good as new. Every maintenance action leads to an (constant) age v. In Nakagawa (2002) an incomplete maintenance model was considered under the following assumptions:

- the age after the kth pm falls to $a_k t$,
- $N - 1$ pm's at x_i, $i = 1, \ldots, N - 1$, renewal at x_N,
- minimal repairs at failures,
- each incomplete repair has the same costs.

The problem was to find the times of pm's x_i, $i = 1, \ldots$, and the number N of preventive maintenances with minimal costs per time.

We think that the assumption *each incomplete repair has the same costs* is very restrictive and defines a cost function which describes the costs of repair actions according to the degree of repair. Let c_F and c_M be the cost of a failure and the cost of a maintenance, respectively. We do not make the assumption that after main-tenance the item is as good as new. Every maintenance action leads to an age v. Then the costs per time unit are given by

$$C(v, \tau) = \frac{c_F \cdot (H(\tau + v) - H(v)) + c_M(v, \tau)}{\tau} . \qquad (15.2)$$

This function should be minimized with respect to v and τ.

Fig. 15.2 Total costs of
failure and repair

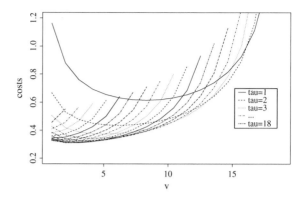

15.4.1 Costs Proportional to the Removed Hazard

The first idea is that the costs are proportional to the cumulative hazard which is
removed be the maintenance action that is,

$$c_M(v, \tau) \sim H(\tau + v) - H(v) .$$

In this case the optimization problem has a trivial solution:

1. Since for distributions with increasing failure rates the cumulative hazard $H(t)$
   is convex, $C(v, \tau)$ is increasing in v. Consequently, the pm should be a renewal.
2. If the pm renews the system, then the essential part of (15.2) is $\frac{H(\tau)}{\tau}$, which is
   increasing in τ. For continuous lifetimes the problem does not have a solution.
   For discrete lifetimes the optimal policy is a renewal at each time.

15.4.2 Costs Proportional to the Impact of Repair

We consider a very simple cost function which depends only on the state after
preventive maintenance:

$$c_M(v, \tau) = c_I \frac{1}{v^\delta} ,$$

where δ is a parameter which makes this cost function more flexible.

In Fig. 15.2 the resulting costs are plotted for a uniform distribution with n states,
the ratio of costs $c_F/c_I = 3$ and $\delta = 0.5$. Each line belongs to a different τ. The points
are connected with lines for a better visibility. It can be seen that the cost function has
a unique minimum. The optimal states before and after maintenance $v + \tau$ and v for
all three hazard rates and different parameters are given in Table 15.1. Again, the
parameters of both other distributions are chosen so that the expectation is 10, too:
$\lambda = 10$ for the Poisson distribution and $\beta = 3$, $\alpha = 11.2$ for the Weibull distribution.

Table 15.1 Optimal values for v and τ

		Uniform	Poisson	Weibull
$c_F/c_I = 3$	$\delta = .5$	$v = 2$	$v = 2$	$v = 2$
		$\tau = 8$	$\tau = 4$	$\tau = 5$
	$\delta = 1$	$v = 3$	$v = 2$	$v = 2$
		$\tau = 6$	$\tau = 4$	$\tau = 3$
	$\delta = 2$	$v = 4$	$v = 3$	$v = 3$
		$\tau = 3$	$\tau = 2$	$\tau = 2$
$c_F/c_I = 2$	$\delta = .5$	$v = 3$	$v = 2$	$v = 2$
		$\tau = 8$	$\tau = 5$	$\tau = 5$
	$\delta = 1$	$v = 4$	$v = 2$	$v = 2$
		$\tau = 6$	$\tau = 4$	$\tau = 5$
	$\delta = 2$	$v = 4$	$v = 3$	$v = 3$
		$\tau = 3$	$\tau = 2$	$\tau = 2$
$c_F/c_I = 1$	$\delta = .5$	$v = 3$	$v = 2$	$v = 2$
		$\tau = 10$	$\tau = 6$	$\tau = 7$
	$\delta = 1$	$v = 4$	$v = 3$	$v = 3$
		$\tau = 8$	$\tau = 4$	$\tau = 5$
	$\delta = 2$	$v = 4$	$v = 3$	$v = 3$
		$\tau = 5$	$\tau = 3$	$\tau = 3$

15.4.3 Costs Proportional to the Degree of Repair

The next function, which can be good interpreted, is

$$c_R \cdot \left(1 - \left(\frac{v}{\tau + v}\right)^\delta\right),$$

where c_R is the costs of a renewal. This function has the following properties:

- If the system is renewed ($v = 0$) then we get c_R.
- If $\tau = 0$ there are no costs.
- If $\delta < 1$ then the costs of a relatively small repair are closed to that of a renewal, if $\delta > 1$ then the costs of a large repair are relatively small.

Note that $v/(\tau + v)$ is ξ, where ξ is the degree of repair in Kijima II model. Then the costs per time unit are given by

$$C(v, \tau) = \frac{c_F \cdot (H(\tau + v) - H(v)) + c_M \cdot \left(1 - \left(\frac{v}{\tau+v}\right)^\delta\right)}{\tau}. \tag{15.3}$$

In this case we can find a unique optimum for both, v and τ, too.

As before, in Fig. 15.3 the resulting costs are plotted for a uniform distribution with n states, the ratio of costs $c_F/c_I = 3$ and $\delta = 0.5$. Each line belongs to a different τ. The points are connected with lines for a better visibility. It can be seen

Fig. 15.3 Total costs of failure and repair

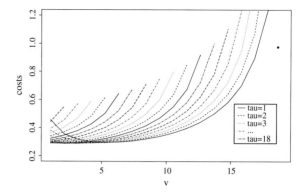

Table 15.2 Optimal values for v and τ

		Uniform	Poisson	Weibull
$c_F/c_I = 3$	$\delta = .5$	$v = 2$	$v = 2$	$v = 1$
		$\tau = 6$	$\tau = 2$	$\tau = 4$
	$\delta = 1$	$v = 1$	$v = 1$	$v = 1$
		$\tau = 9$	$\tau = 5$	$\tau = 5$
	$\delta = 2$	$v = 1$	$v = 1$	$v = 1$
		$\tau = 10$	$v = 2$	$\tau = 6$
$c_F/c_I = 2$	$\delta = .5$	$v = 3$	$v = 2$	$v = 2$
		$\tau = 6$	$\tau = 3$	$\tau = 3$
	$\delta = 1$	$v = 1$	$v = 1$	$v = 1$
		$\tau = 10$	$v = 2$	$\tau = 6$
	$\delta = 2$	$v = 1$	$v = 1$	$v = 1$
		$\tau = 10$	$\tau = 6$	$\tau = 6$
$c_F/c_I = 1$	$\delta = .5$	$v = 4$	$v = 2$	$v = 2$
		$\tau = 7$	$\tau = 4$	$v = 2$
	$\delta = 1$	$v = 1$	$v = 1$	$v = 1$
		$\tau = 13$	$\tau = 7$	$\tau = 8$
	$\delta = 2$	$v = 1$	$v = 1$	$v = 1$
		$\tau = 13$	$\tau = 7$	$\tau = 8n$

that the cost function has a unique minimum. The optimal states before and after maintenance $v + \tau$ and v for all three hazard rates and different parameters are given in Table 15.2, where the parameters are the same as in Table 15.1.

The numerical result lead to the following conclusions:

1. If the costs of maintenance depend on the degree of repair and $\delta \leq 1$, then the optimal state after maintenance is state 1. This means that the optimal policy is a renewal because the costs of some incomplete maintenance are closed to those of a renewal.

2. If the hazard function is Weibull or Poisson then the results are quit similar, where, for the uniform distribution, we get different optimal solutions. The reason is the different shape of the cumulative hazard.
3. If the hazard function is that of an uniform distribution, then the optimal distance between maintenances is larger than for the other two hazard rates.

15.5 Conclusion

We have considered failure rate optimal maintenance under the assumption that the maintenance action has an impact between the two extreme cases, minimal repair and renewal. For finding cost optimal maintenance it was necessary to define a cost function which describes the costs of repair actions according to the degree or impact of repair. There are many other possible cost functions, which can be considered. Further, we can consider models where the repair after a failure is not minimal, but imperfect, too.

References

Kijima M (1989) Some results for repairable systems with general repair. J Appl Probab 26:89–102

Kijima M, Morimura H, Suzuki Y (1988) Periodical replacement problem without assuming minimal repair. Eur J Op Res 37:194–203

Last G, Szekli R (1998) Stochastic comparison of repairable systems by coupling. J Appl Probab 35(2):348–370

Lisnianski A, Frenkel I (2009) Non-homogeneous Markov reward model for aging multi-state system under minimal repair. Int J Perform Eng 5(4):303–312

Lisnianski A, Levetin G (2003) Multi-state system reliability: assessment,optimization and applications. World Scientific, Singapore

Nakagawa T (2002) Imperfect preventive maintenance models. In: Osaki S (ed) Stochastic models in reliability and maintenance. Springer, Berlin, pp 125–143

Nakagawa G, Osaki S (1975) The discrete Weibull distribution. IEEE Trans Reliab R-24(5):300–301

Padgett WJ, Spurrier JD (1985) Discrete failure models. IEEE Trans Reliab R-34(3):253–256

Xie M, Gaudoin O, Bracquemond C (2002) Redefining failure rate function for discrete distributions. Int J Reliab, Qual Saf Eng 9(3):275–285

Chapter 16
Nonparametric Estimation of Marginal Temporal Functionals in a Multi-State Model

Somnath Datta and A. Nicole Ferguson

Abstract Multi-state models are generalizations of traditional survival analysis and reliability studies. They are common in medical and engineering applications where a subject (say a patient or a machine) is moving through a succession of states (each representing a stage of disease progression or the condition of a machine) with time. In addition, several key questions in event history analysis and multi-variate survival analysis can be formulated in terms of a staged system making the use of multi-state models extremely broad. While the use of parametric and semiparametric models to various transitions are the most common approaches to study multi-state data, this overview paper deals entirely with nonparametric methods. In addition, we limit our exposition to estimation questions related to marginal models rather than conditional (e.g., regression) models. We review a number of methods from the recent past dealing with estimation of hazards, transition and state occupation probabilities; state entry, exit and waiting time distributions and also discuss some ongoing and future research problems on these topics. Various forms of censoring that occur in the collection of multi-state data are discussed including right and interval censoring.

Keywords Multi-state models · Survival · Multi-stage · Censoring · Current status · Interval censoring · State occupation probabilities

S. Datta (✉) · A. N. Ferguson
Department of Bioinformatics and Biostatistics,
University of Louisville, Louisville, KY 40202, USA
e-mail: somnath.datta@louisville.edu

A. N. Ferguson
e-mail: nicole.ferguson@louisville.edu

A. Lisnianski and I. Frenkel (eds.), *Recent Advances in System Reliability*,
Springer Series in Reliability Engineering, DOI: 10.1007/978-1-4471-2207-4_16,
© Springer-Verlag London Limited 2012

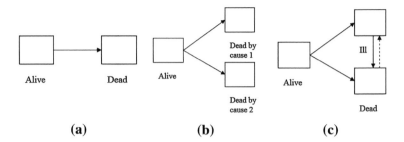

Fig. 16.1 Flowgraphs of multi-state models. **a** survival, **b** competing risk, **c** illness–death

16.1 Introduction

Multi-state models are a type of multi-variate survival data which provide a framework for describing a complex system where individuals make a transition through a series of distinct states. This framework, which is often represented with a directed graph, illustrates the different states (or events) that individuals may experience, as well as the possible transitions between states. Transitions between states may be reversible or irreversible while states can be either absorbing (meaning further transitions cannot occur) or transient. Multi-state models have a range of applications including event history data, epidemiology, clinical trials where individuals progress through the different stages of a disease such as cancer and AIDS, and in systems engineering where a machine may experience various systems conditions with age.

Standard survival analysis models measure the time span from some time origin (e.g., birth) until the occurrence of the event of interest (e.g., death). This corresponds to the simplest multi-state model, the two-state model with one transient root state (alive) and one absorbing state (dead). This could be expanded to include several absorbing states corresponding to different causes of death and is called the competing risk or multiple decrement model. Another simple example of a multi-state model, which allows for a branching event, is the so-called illness–death model. In this model, individuals start in the well state. Some individuals subsequently move to the illness state and the rest of the individuals eventually experience death without ever visiting the illness state. In the irreversible version of the model all such individuals eventually move to the "dead" state without any possible recovery from the illness while in the reversible version, an individual in the illness state may recover and thus make a transition back to the well state. All these simple models are represented by directed graphs or flowgraphs (Huzurbazar 2005) in Fig. 16.1. Multi-state models can offer various degrees of complexities where individuals can pass through multiple transient states before entering a number of possible absorbing states.

There are several key questions which arise in studying multi-state models. What is the probability that a subject is in a specific state j at a time t? What is the hazard (rate) at which a subject in a given state j at time t transitions to a future stage j'?

What is the distribution of the time spent (waited) in a state j? More formally, these questions ask what are the state occupation probabilities, the state transition intensities (or transitional hazards), and the state waiting time distributions, respectively. Distribution functions for the state entry and exit times are also of interest. Estimators of these quantities have been proposed in the recent past under a variety of parametric and nonparametric assumptions as well as structural assumptions on the system (such as, progressive, Markov, semi-Markov etc.). In this paper we restrict ourselves to the nonparametric methods. Moreover, we concern ourselves with the estimation questions in a marginal model and not a conditional (e.g., regression) model. Thus, we do not discuss the semiparametric models in this paper. Generally speaking, results for the survival setup (e.g., a two-state progressive model) are widely available in the literature and are not discussed in this paper.

In the standard survival analysis setting, especially with right censored data, the nonparametric likelihood type methods have been the usual choice. As for example, the classical Kaplan–Meier estimator for the survival function can be obtained as a nonparametric maximum likelihood estimator. It is possible to apply this technique to certain multi-state models such as a Markov or a semi-Markov which simplifies the likelihood formulation (Aalen 1978; Aalen and Johansen 1978; Frydman 1992; Satten and Sternberg 1999; etc.). However, in the absence of such additional structural assumptions the likelihood of an event may depend on all past events (state occupation) and event times. Thus, a likelihood approach in general is not feasible. In addition, there are additional challenges brought on by inherent incompleteness in the observed data due to various forms of censoring. As we shall see, a combination of nonparametric functional estimation techniques, mostly various forms of averaging or smoothing are needed to form the estimators in multi-state models.

The following general notations will be used throughout the paper. A multi-state process is a stochastic process $\mathbb{S} = \{S(t) : t \geq 0\}$, where t denotes time and $S(t)$ denotes the state occupied at time t. We can think of $S(t-) = \lim_{s \to t-} S(s)$ as the state occupied just before time t. We assume a finite state space $\mathbf{X} = \{0, 1, \ldots, M\}$. Under the marginal model, we will assume that the multi-state processes for n individuals $\mathbb{S}_i = \{S_i(t) : t \geq 0\}$, $1 \leq i \leq n$, are independent and identical (i.i.d., hereafter) realizations of $\mathbb{S}$.

For many applications, it is reasonable to assume that the system is progressive in which case the directed graph will have a tree structure and we will denote the root node by 0. For a given state j, $p_j(t) = \Pr\{S(t) = j\}$ is the state occupation probability of state j as a function of time. In a multi-state model representation of the standard survival analysis setup, we let state $0 = $ "alive" and state $1 = $ "dead". Then $p_0(t)$ is the survival function and $p_1(t)$ is the distribution function of the failure time. For simplicity of exposition, throughout the paper, we will assume that the process has at most one jump in an infinitesimal time interval $[t, t + dt)$ leading to (marginal) hazard rates of transitions from states j and j', $\alpha_{jj'}(t) = \lim_{dt \to 0} \Pr\{S(s) = j'$, for some $s \in [t, t + dt) | S(t-) = j\}$, and integrated (or cumulative) hazard rates $A_{jj'}(t) = \int_0^t \alpha_{jj'}(s)ds$. Similarly, the (marginal) rates of entry to and exit out of state j are given by $\alpha_{\cdot j}(t) = \lim_{dt \to 0} \Pr\{S(s) = j, \text{ for some } s \in [t, t + dt) | S(t-) \neq j\}$ and $\alpha_{j \cdot}(t) = \lim_{dt \to 0} \Pr\{S(s) \neq j, \text{ for some } s \in [t, t + dt) | S(t-) = j\}$. For defining the state waiting times, we need to impose the restriction that a given state j can be entered at most once. For handling situations with repeated events, one would therefore add additional states to the system such as first entry, second entry, and so on; this would mean that we can keep track of the occurrence of multiple entries to a given state. In this case, we can define the state entry, exit, and waiting (sojourn) times by $U^j = \inf\{t : S(t) = j\}$ and $V^j = \sup\{t : t > U^j, S(t) \neq j\}$, $W^j = V^j - U^j$, when $U^j < \infty$. Note that by convention, $U^j < \infty$, if state j is never entered and $V^j = \infty$, if either state j is never entered or j is an absorbing state (in which case it is never left). The (marginal) state entry, exit, and waiting time distributions will be denoted by $F^j(t) = \Pr\{U^j \leq t | U^j < \infty\}$, $G^j(t) = \Pr\{V^j \leq t | U^j < \infty\}$ and $H^j(t) = \Pr\{W^j \leq t | U^j < \infty\}$, respectively.

The rest of the paper is organized as follows. The following section of the paper introduces various estimation methodologies to handle right censored multi-state data. Right censoring is perhaps the prevalent form of censoring in time to event studies. Section 16.3 considers more severe forms of censoring when individuals are not constantly monitored. The paper ends with a discussion (Sect. 16.4).

16.2 Estimation Under Right Censoring

There are a number of reasons why right censoring is often, if not always, present in time to event data including multi-state models. Generally, studies have a finite duration and the event of interest may not take place during the study interval leading to right censoring of the event. More generally, in a multi-state model framework, an individual may still be at a transient state at the end of the study or follow-up time which means that there are potential future transitions whose exact times will be unknown. Mathematically speaking, a multi-state process $\mathbb{S}$ that is right censored by a censoring variable C is given by the stochastic process, $\mathbb{S}^c = \{S(t \wedge C) : t \geq 0\}$. Basically, it means that we observe all the transition times and the state occupation up to time C and nothing beyond that. Thus, the right censored data will be i.i.d. realizations of $\mathbb{S}^c$ given by $\mathbb{S}_1^c, \ldots, \mathbb{S}_n^c$ together with $\delta_i C_i, 1 = 1, \ldots, n$, where δ_i is the indicator of the event that the last transition time for the ith individual is censored. The most common assumption on the censoring times is that they are i.i.d. and are independent of the original multi-state processes $\mathbb{S}_1, \ldots, \mathbb{S}_n$. This is the so-called "random censoring" assumption and will be assumed for the following two sub sections.

16.2.1 Nelson–Aalen Estimators

The Nelson–Aalen estimators (Aalen 1978; Andersen et al. 1993) are obtained on the basis of rate calculations. Using the independent censoring assumption, one can establish that the observed rates of transitions between states in a censored experiment is the same as that in an uncensored experiment. The former rate can be empirically estimated based on available data and it leads to the Nelson–Aalen estimators of integrated (marginal) transition hazards. More formally, let for states j and j', $N_{jj'}$ and $N_{jj'}^c$ be counting processes with jumps given by $\Delta N_{jj'}(t) = \sum_{i=1}^n I\{S_i(t-) = j,\ S_i(t) = j'\}$ and $\Delta N_{jj'}^c(t) = \sum_{i=1}^n I\{S_i(t-) = j,\ S_i(t) = j',\ C_i \geq t\}$, respectively, recording the transition counts from states j to j' in the uncensored and censored experiments, respectively. Also, let $Y_j(t) = \sum_{i=1}^n I(S_i(t-) = j)$ and $Y_j^c(t) = \sum_{i=1}^n I(S_i(t-) = j,\ C_i \geq t)$ be the number of individuals at state j just before time t in the uncensored and censored experiments, respectively. Then as, $n \to \infty$, the two instantaneous rates $dN_{jj'}(t)/Y_j(t)$ and $dN_{jj'}^c(t)/Y_j^c(t)$ converge (in probability) to $\alpha_{jj'}(t)dt$ and $\alpha_{jj'}^c(t)dt$, respectively, where $\alpha_{jj'}$ is defined earlier and

$$
\begin{aligned}
\alpha_{jj'}^c(t) &= \lim_{dt \to 0} \frac{\Pr\{S(s) = j', \text{for some } s \in [t, t+dt), C \geq t\}}{\Pr\{S(t-) = j, C \geq t\}} \\
&= \lim_{dt \to 0} \frac{\Pr\{S(s) = j', \text{for some } s \in [t, t+dt)\}\Pr\{C \geq t\}}{\Pr\{S(t-) = j\}\Pr\{C \geq t\}},
\end{aligned}
$$

using independence of $\mathbb{S}$ and C. The last expression, however, equals $\alpha_{jj'}(t)$. In other words, the two hazard rates $\alpha_{jj'}(t)$ and $\alpha_{jj'}^c(t)$ are equal at all time points t. Therefore the integrated hazard rate $A_{jj'}(t)$ in the marginal model can be estimated by the integrated empirical hazard rate from the right censored multi-state data leading to the Nelson–Aalen estimator

$$
\hat{A}_{jj'}(t) = \int_0^t I\left(Y_j^c(s) > 0\right) \frac{dN_{jj'}^c(s)}{Y_j^c(s)}. \tag{16.1}
$$

Since this estimator is a step function, in order to obtain a legitimate estimator of the hazard rate $\alpha_{jj'}$, one needs to apply kernel smoothing to it. To that end, let K be a symmetric kernel (e.g., a symmetric density function) and let $0 < h = h(n) \downarrow 0$ be a bandwidth sequence. Then a non-parametric estimator of the marginal hazard rate of transition from state j to j' is given by

$$
\hat{\alpha}_{jj'}(t) = \frac{1}{h} \int K\left(\frac{t-s}{h}\right) d\hat{A}_{jj'}(s).
$$

16.2.2 Aalen–Johansen Estimators

For a Markov multi-state process, the transition probabilities $p_{jj'}(s,t) = \Pr\{S(t) = j'|S(s) = j\}$ can be computed by product integration of the marginal hazard function $\mathbf{P}(s,t) = \Pi_{(s,t]}(I + \mathrm{d}\mathbf{A}(u))$, where $\mathbf{P}(s,t)$ is a matrix with (j,j')th entry $p_{jj'}(s,t)$ and $\mathbf{A}$ is a matrix with (j,j')th entry $A_{jj'}$, if $j' \neq j$, and $= -\sum_{k \neq j} A_{jk}$, if $j' = j$. This leads to the construction of the Aalen–Johansen estimator (Aalen and Johansen 1978) of transition probabilities of a Markov multi-state model obtained by substituting the Nelson–Aalen estimators of $\mathbf{A}$ into this formula

$$\hat{\mathbf{P}}(s,t) = \Pi_{(s,t]}\big(I + \mathrm{d}\hat{\mathbf{A}}(u)\big). \tag{16.2}$$

For multi-state models with only one transient state, such as classical survival analysis and the competing risk model, the assumption of Markovity holds trivially and thus the Aalen–Johansen estimators are valid. In particular, for the survival setting it is just the Kaplan–Meier estimator. As mentioned earlier, the Aalen–Johansen estimator can also be obtained as a non-parametric maximum likelihood estimator under the Markov assumption. Valid estimators for the three state progressive non-Markov illness–death model are proposed by Meira-Machado et al. (2006). Nonparametric estimators of transition probabilities for general multi-state models without the Markovity assumption are not currently available.

One can set the initial time $s = 0$ in the Aalen–Johansen estimator and combine it with the initial state occupation to obtain the following natural estimators of state occupation probabilities $p_j(t) = \Pr\{S(t) = j\}$,

$$\hat{p}_j(t) = n^{-1}\sum_{k=0}^{M}\hat{p}_{kj}(0,t)Y_k(0+). \tag{16.3}$$

Interestingly, Datta and Satten (2001) noted that this estimator remains valid (e.g., consistent) even without the Markov assumption; also, see Gliden (2002) for a different proof of the same result. In other words, the Markov assumption which is often unverified but routinely assumed is not really needed if one is only interested in estimation of state occupation probabilities as a function of time. Unfortunately, this fact remains to be relatively unknown amongst practitioners even till date.

16.2.3 Datta–Satten Estimators

Datta and Satten (2002) extended the Nelson–Aalen and Aalen–Johansen estimators to situations where the censoring random variable is not necessarily independent of the multi-state process, but rather only conditionally independent given an observable time varying covariate $\mathbf{Z} = \{Z(t) : t \geq 0\}$. In their treatment, they estimate the two processes $N_{jj'}$ and Y_j separately using the principle of inverse probability of censoring weights (Koul et al. 1981; Robins and Rotnitzky 1992; Robins 1993; Satten et al. 2001) rather than their ratio. The estimates are not equal to the censored data versions defined earlier; however, under the model of independent censoring these estimated processes are proportional to the respective censored data processes defined before and so the Datta–Satten estimators under the independent censoring hazard assumption reduce to Nelson–Aalen and Aalen–Johansen estimators. State occupation probability estimators in an illness–death model using a different reweighting scheme to handle a specific type of dependent censoring was considered in Datta et al. (2000a) but the present treatment is more general.

In general, to construct these estimators, a model for the censoring hazard

$$\lambda_c\left(t|\overline{Z}(t)\right) = \lim_{dt \to 0} \Pr\{C \in [t, t + dt) | C \geq t, Z(s), 0 \leq s \leq t, \mathbb{S}\}$$
$$= \lim_{dt \to 0} \Pr\{C \in [t, t + dt) | C \geq t, Z(s), 0 \leq s \leq t\}$$

given the time-dependent covariates Z is needed to obtain an estimate $\hat{K}(t) = \exp\left\{-\hat{\Lambda}_c(t|\overline{Z}(t)\right\}$. In particular, Datta and Satten advocated the use of Aalen's linear hazards model (Aalen 1980) for this purpose. Using the reweighting principle (see Datta and Satten 2002, for a formal argument) one can construct the following estimators of the complete data counting and at risk processes

$$\Delta\hat{N}_{jj'}(t) = \sum_{i=1}^{n} \frac{I\{S_i(t-) = j, S_i(t) = j', C_i \geq t\}}{\hat{K}_i(t-)} \tag{16.4}$$

and

$$\hat{Y}_j(t) = \sum_{i=1}^{n} \frac{I(S_i(t-) = j, C_i \geq t)}{\hat{K}_i(t-)}. \tag{16.5}$$

Substituting these expressions (16.4, 16.5) into the formula (16.1) in places of $\hat{N}_{jj'}^c$ and $\hat{Y}_j^c$ we obtain the Dat ta–Satten estimators of integrated (marginal) transition hazards. Using the Datta–Satten estimator of $\hat{A}$ and the at risk set $\hat{Y}_j$ in formulas (16.2) and (16.3) we get in turn the Datta–Satten estimators of transition probabilities (for Markov systems) and state occupation probabilities (for possibly non-Markov systems) under dependent censoring. See Cook et al. (2009) for an application of the Datta–Satten estimator to bone cancer data.

For some applications, state entry and exit time distribution functions are of interest. The estimators of state occupation probabilities constructed above can be used to estimate these distributions by state pooling as follows. For this purpose, we assume that the model can be expanded into a progressive tree-like structure with a root node 0 so that each state can be entered and exited at most once. For cyclic models (such as a reversible illness–death and recurrent events data), each entry of a given state needs to be interpreted as a new state. After the state occupation probabilities of this expanded system are calculated they can be pooled (e.g., summed) to obtain estimators in the original system.

Let $\mathbb{S}^j$ denote the collection of all states $j' \neq j$ such that state j appears on the path connecting states 0 and j'. In other words, $\mathbb{S}^j$ is the collection of all states which proceed state j. Then estimators of entry and exit time distributions of state j are given by

$$\hat{F}^j(t) = \frac{\sum_{k \in \{j\} \cup \mathbb{S}^j} \hat{p}_k(t)}{\sum_{k \in \{j\} \cup \mathbb{S}^j} \hat{p}_k(\infty)} \quad \text{and} \quad \hat{G}^j(t) = \frac{\sum_{k \in \mathbb{S}^j} \hat{p}_k(t)}{\sum_{k \in \{j\} \cup \mathbb{S}^j} \hat{p}_k(\infty)}. \tag{16.6}$$

We end this section with an introduction of the Satten and Datta (2002) estimators of state waiting time distributions that are valid without the Markovity assumption. Furthermore, these estimators use reweighting based on censoring hazard and thus available covariate information can be incorporated that might be related to the censoring mechanism. The form of the reweighting reflects the fact that waiting time distributions are measured since state entry and not in calendar times. Once again, assume that a transient state j can be entered at most once.

Let $\hat{K}_i(t) = \exp\left\{-\hat{\Lambda}_c(t|\overline{Z}_i(t)\right\}$ be as before. Then, the estimated counting processes for waiting times in a given state j is a jump process with jump size equal to

$$\Delta \hat{N}_j^W(t) = \sum_{i=1}^n \frac{I\{W_i^j = t, C_i \geq V_i^j\}}{\hat{K}_i(V_i^j-)}$$

which can be computed based on the available right censored data since if $C_i \geq V_i^j$, then the state j waiting time W_i^j is available. The inverse weighting factor is essentially the estimated conditional probability of the event $\{C_i \geq V_i^j\}$, given $\{V_i^j, W_i^j\}$. Next, the size of the "at risk" set of state j waiting times is estimated by

$$\hat{Y}_j^W(t) = \sum_{i=1}^n \frac{I\{W_i^j \geq t, C_i \geq t + U_i^j\}}{\hat{K}_i\left((t + U_i^j)-\right)}.$$

Note that, once again, this quantity can be computed based on the available data and in particular, even if the exit time is right censored. Finally, a nonparametric estimator of state j waiting time distribution is obtained by a Kaplan–Meier-type product limit formula using these two sets

$$\hat{H}^j(t) = 1 - \prod_{s \leq t}\left(1 - \frac{d\hat{N}_j^W(ds)}{\hat{Y}_j^W(s)}\right).$$

These estimators are valid even when the censoring is not independent. Other versions of nonparametric estimators of state waiting times for certain types of multi-state models under independent censoring assumption were obtained in Wang and Wells (1998) and Wang (2003).

16.2.4 The Pepe Estimator and Its Extensions

Pepe (1991) suggested using the difference of two Kaplan–Meier estimators to estimate the state occupation probability of a transient state in a four-state leukemia progression model. Another nonparametric estimator of the state occupation probability was proposed by Datta et al. (2000b) and involved using a "fractional size at risk set" and a reweighting approach in the three-state irreversible illness–death model. The fractional weights representing the probabilities of traversing a future path in a more general multi-state model with a tree structure were considered by Datta and Satten (2000). These weights can be combined with right censored entry and exit times to calculate marginal estimators of state entry and exit time distributions. A Pepe-type subtraction estimator can also be constructed using these in a general multi-state model with a tree structure.

Suppose we have a progressive model that can be expanded into a rooted directed tree with the root node 0. Let $N^c_{\cdot j}$ and $N^c_{j\cdot}$ be the counting processes of observed entry and exits to state j with jumps given by

$$\Delta N^c_{\cdot j}(t) = \sum_{i=1}^{n} I(S_i(t) = j, \ S_i(t-) = j_*, C_i \geq t) \qquad (16.7)$$

and

$$\Delta N^c_{j\cdot}(t) = \sum_{i=1}^{n} I\big(S_i(t) \in \mathbb{S}^j, \ S_i(t-) = j, C_i \geq t\big), \qquad (16.8)$$

where j_* is the state that precedes j in the path from 0 to j. The corresponding "numbers at risk" processes at time t in the censored experiment are given by

$$Y^c_{\cdot j}(t) = \sum_{i=1}^{n} I\big(C_i \geq t, \ S_i(t-) \in \{j \cup \mathbb{S}^j\}^c, \ S_i(u) = j, \ \text{for some } u \geq 0\big)$$

and

$$Y^c_{j\cdot}(t) = \sum_{i=1}^{n} I\big(C_i \geq t, \ S_i(t-) \in \{\mathbb{S}^j\}^c, \ S_i(u) = j, \ \text{for some } u \geq 0\big).$$

However, these later two processes cannot be evaluated from the observed data if certain individuals are censored at a state, say $S(C)$, from which eventual passage through state j is possible but not guaranteed. Such individuals should contribute a fractional count ϕ_j to the "at risk sets" which represents their probability of passing through state j in future had there been no censoring. This idea leads to the following "fractional at risk sets"

$$\hat{Y}_{\cdot j}^c(t) = \sum_{i=1}^n \hat{\phi}_{ij} I\left(C_i \ge t, \ S_i(t-) \in \left\{j \cup \mathbb{S}^j\right\}^c\right) \qquad (16.9)$$

and

$$\hat{Y}_{j\cdot}^c(t) = \sum_{i=1}^n \hat{\phi}_{ij} I\left(C_i \ge t, \ S_i(t-) \in \left\{\mathbb{S}^j\right\}^c\right) \qquad (16.10)$$

that can be computed from the available data. Here $\hat{\phi}_{ij}$ is an estimate of $\Pr\{S_i(u) = j, \text{ for some } u \ge 0 \,|\, C_i, \mathbb{S}_i^c\}$. These fractional weights are recursively calculated based on the distance of the state $S(C_i)$ from the root node 0 (Datta and Satten 2000).

For the time being, we drop the index i to keep the notation simple. First consider the case when $S(C) = 0$ and j can be reached from 0 in one step. Let N_0^c. be the counting process of transitions out of state 0 defined as above. Then, $\hat{\phi}_j$ can be calculated using the Aalen–Johansen transition probability estimates in a competing risk model

$$\hat{\phi}_j = \int_{(C,\infty)} \left\{\prod_{(C,u)} \left(1 - \frac{dN_{0\cdot}^c(v)}{\hat{Y}_{0\cdot}^c(v)}\right)\right\} \frac{dN_{0j}^c(u)}{\hat{Y}_{0\cdot}^c(u)}. \qquad (16.11)$$

since one can view $\hat{\phi}_j$ as an eventual occupation probability $\hat{P}_{0j}(C,\infty)$ of stage j in a collapsed network (where all future states beyond j are equated with j and so on).

Next, let $S(C) = k \ (\ne 0)$ and j can be reached from 0 in two steps with k as the intermediate step. Observe that $\hat{\phi}_k$ can be calculated by the above formula (16.11), so $\hat{Y}_{k\cdot}^c(t)$ is now well-defined. Now, define $\hat{\phi}_j$ by the above formula (16.11) with 0's replaced by k throughout.

Finally, for the general case when j can be reached from 0 in m steps with $S(C) = k$ on the path to j. Let $k = j_1 \to j_2 \to \cdots \to j_{m'} = j$ be the path from k to j for some $m' \le m$. Assume, that by induction, we have calculated $\hat{\phi}_{\tilde{j}}$ whenever 0 and $\tilde{j}$ are separated by less than m step. Note that in this case $\hat{Y}_{ji\cdot}$, for $l < m'$, are all well-defined and hence are

$$\hat{\psi}_l = \int_{(C,\infty)} \left\{\prod_{(C,u)} \left(1 - \frac{dN_{j_l\cdot}^c(v)}{\hat{Y}_{j_l\cdot}^c(v)}\right)\right\} \frac{dN_{j_l j_{l+1}}^c(u)}{\hat{Y}_{j_l\cdot}^c(u)}$$

for $l = 1, \dots, m' - 1$. Finally, let $\hat{\phi}_j = \prod_{l=1}^{m'-1} \hat{\psi}_l$.

The counting and fractional size at risk processes given by (16.7–16.10) can be used to compute alternative estimators of the state entry and exit time distributions using the product-limit formulas

$$\hat{F}^j(t) = 1 - \prod_{s \le t}\left(1 - \frac{dN^c_{\cdot j}(s)}{\hat{Y}^c_{\cdot j}(s)}\right) \text{ and } \hat{G}^j(t) = 1 - \prod_{s \le t}\left(1 - \frac{dN^c_{j \cdot}(s)}{\hat{Y}^c_{j \cdot}(s)}\right).$$

(16.12)

Unlike (16.6), these estimators are guaranteed to be monotonic. These estimators, in turn, can be combined to obtain an estimator of the state occupation probabilities which are extensions of Pepe's (1991) estimators to more complex multi-state models

$$\hat{p}_j(t) = \left(n^{-1}\sum_{i=1}^{n}\hat{\varphi}_{ij}\right)\{\hat{F}^j(t) - \hat{G}^j(t)\}.$$

(16.13)

However, since these are based on a subtraction formula, unlike the Aalen–Johansen (or the Datta–Satten) estimators (16.3), these estimators may sometimes assume negative values which is not desirable of probability estimates.

Martingale representations of all the estimators reviewed in this section are available although these could be quite complex, especially, when dependent censoring is present. Bootstrap resampling is an attractive alternative to large sample calculations for these estimators leading to variance estimates and point-wise confidence intervals.

16.3 Estimation Under Current Status and Interval Censored Data

Marginal nonparametric estimation for multi-state current status data was undertaken in Datta and Sundaram (2006), Datta et al. (2009) and Lan and Datta (2010); the special case of competing risk models was investigated by Jewell et al. (2003) and Groeneboom et al. (2008). Of course, there is a sizable literature for the case of interval censored survival data including additional complications such as truncation (Huber et al. 2009) which we do not review here.

As before, for an individual i and a time $t \ge 0$, $S_i(t)$ denotes the state individual i is in at time t; C_i denotes the random time at which the individual i gets inspected. The censoring times and the state occupation processes $\{C_i, S_i(t), t \ge 0\}$ for the individuals are assumed to be independent and identically distributed. For simplicity of development, we will make the assumption of random censoring, which means C_i is independent of $\mathbb{S}_i = \{S_i(t) : t \ge 0\}$. We further assume that all transition and censoring times are continuous and that the allowable transitions give rise to a rooted directed tree structure, in which every state $j \in \mathbb{S}$ can be reached from an initial state

0 (the root node) by a unique path $\pi(j) : 0 = s_1 \rightarrow s_2 \cdots \rightarrow s_{j+1} = j$. The observed data consist of $\{C_i, S_i(C_i)\}$ for $i = 1, \ldots, n$.

16.3.1 Estimators of State Occupation Probabilities

Consider two states j and j'. Let $U^{jj'}$ denote the (unobserved) transition time of an individual from state j to j' (define it to be ∞, if this transition is not made by the individual). Let $N_{jj'}(t)$ denote the usual counting process counting the number of j to j' transitions in $[0, t]$ with the complete data. By the laws of large numbers, assume that $n^{-1}N_{jj'}(t) \xrightarrow{P} P\{U^{jj'} \leq t\} = n_{jj'}(t)$.

Consider the indicator function $I(U^{jj'} \leq C)$ of the event that the j to j' transition has taken place by time C. Then, for any $t \geq 0$,

$$E\left(I\left(U^{jj'} \leq C\right)|C = t\right) = \Pr\left\{U^{jj'} \leq t\right\}.$$

Therefore, $n^{-1}\hat{N}_{jj'}(\cdot)$ can be obtained by a nonparametric regression estimator of $I(U^{jj'} \leq C)$ given C. Since $\Pr\{U^{jj'} \leq t\}$ is monotonic in t, $n^{-1}\hat{N}_{jj'}(\cdot)$ can be constructed by an isotonic regression of $I(U^{jj'} \leq C)$ on C, based on the pairs $\left(C_i, I\left(U_i^{jj'} \leq C_i\right)\right)$.

Next, note that $P_j(t-) = \Pr\{S(t-) = j\}$ is the (in probability) limit of $n^{-1}Y_j(t)$, where $Y_j(t)$ denotes the size of the "at risk" set of transitions out of stage j with the complete data. However, unlike the counting process of transition counts, the Y_j process does not have to be monotonic for a transient state j. Therefore, one can use kernel smoothing rather than isotonic regression to estimate this process leading to

$$\hat{Y}_j(t) := \frac{\sum_{i=1}^n I(S_i(C_i) = j)K_h(C_i - t)}{n^{-1}\sum_{i=1}^n K_h(C_i - t)},$$

where K is a density kernel, $h = h(n)$ is a bandwidth sequence, and $K_h(\cdot) = h^{-1}K(\cdot/h)$.

With the above estimators in place, the class of state occupation probabilities will be computed as in Sect. 16.2 using the relationship (16.3), where $\hat{P}(0, t) = \prod_{(0,t]}(I + d\hat{A}(u))$; however, the integrated conditional transition hazards are now calculated using $\hat{N}$ and $\hat{Y}$ defined in this section

$$\hat{A}_{jj'}(t) = \begin{cases} \int_0^t J_j(u)\hat{Y}_j(u)^{-1}d\hat{N}_{jj'}(u) & j \neq j' \\ -\sum_{j' \neq j} \hat{A}_{jj'}(t) & j = j', \end{cases}$$

where $J_j(u) = I\left(\hat{Y}_j(u) > 0\right)$.

16.3.2 State Entry and Exit Time Distributions

A similar approach as in the previous section can be followed to obtain these. However, for current status data, the basic ingredients, namely, the counting processes of transition counts and the size of the at risk sets in and out of a given state j are computed using a different machinery.

Since the indicators $I\left(U_i^j \leq C_i\right)$ and $I\left(V_i^j \leq C_i\right)$ are calculable from the available current status information (along with the topological knowledge of the system) one could regress them (say, by isotonic regression) to obtain $\hat{N}_{\cdot j}(\cdot)$ and $\hat{N}_{j\cdot}(\cdot)$. More precisely, $n^{-1}\hat{N}_{j\cdot}(\cdot)$ is a step function for taking values $n^{-1}\hat{N}_{j\cdot}\left(C_{(i)}\right) = R_i$, that minimizes the sum of squares $\sum_{i=1}^{n}\left\{R_i - I\left(U_{[i]}^j \leq C_{(i)}\right)\right\}^2$ subject to $R_1 \leq \cdots \leq R_n$, where $[i]$ denotes the index corresponding to the ith largest C; $n^{-1}\hat{N}_{\cdot j}(\cdot)$ is computed the same way with U's replaced by V's. These can be obtained using the well-known pooled adjacent violator algorithm (Barlow et al. 1972).

The "size at risk" sets will be computed by antitonic regression but with fractional weights representing the probability of ever making it to state j. Thus, $n^{-1}\hat{Y}_{\cdot j}(\cdot)$ is a step function taking values $n^{-1}\hat{Y}_{\cdot j}\left(C_{(i)}\right) = R_i$, that minimize the sum of squares $\sum_{i=1}^{n}\left\{R_i - \hat{\varphi}_{[i]j}I\left(U_{[i]j} \geq C_{(i)}\right)\right\}^2$ subject to $R_1 \geq \cdots \geq R_n$; $n^{-1}\hat{Y}_{j\cdot}(\cdot)$ is computed the same way with U's replaced by V's.

The fractional weights are successively (recursively) calculated from the root node to the distant states as before $\hat{\varphi}_j = \prod_l \hat{\psi}_l$, where

$$\hat{\psi}_l = \int_{(C,\infty)}\left\{\prod_{(C,u)}\left(1 - \frac{d\hat{N}_{j_l\cdot}(v)}{\hat{Y}_{j_l\cdot}(v)}\right)\right\}\frac{d\hat{N}_{j_l j_{l+1}}(u)}{\hat{Y}_{j_l\cdot}(u)}.$$

In the above formula, $\hat{N}_{j_l j_{l+1}}$ are calculated by isotonic regression of the pairs $\left(C_i, I\left(U_i^j \leq C_i\right)\right), 1 \leq i \leq n$. Estimators of F^j, G^j and the Pepe-type alternative estimator of p_j can be formed using these processes by formulas (16.12) and (16.13), where we use $\hat{N}_{\cdot j}$ and $\hat{N}_{j\cdot}$ instead of $N_{\cdot j}^c$ and $N_{j\cdot}^c$, respectively; similarly, $\hat{Y}_{\cdot j}$ and $\hat{Y}_{j\cdot}$ are used in places of $\hat{Y}_{\cdot j}^c$ and $\hat{Y}_{j\cdot}^c$, respectively.

16.3.3 State Waiting Time Distributions

The calculation of state waiting time distributions with current status data poses additional difficulty since we cannot directly regress the indicators of events involving the waiting times because the state entry times are also unknown. Some progress can be made with additional structural assumptions. As for example, under the Markov assumption (Datta et al. 2009), we could obtain the following identity

$$H^j(t) = 1 - \int_0^\infty \prod_{u < s \le u+t} \left(1 - \mathrm{d}\Lambda_{j.}(s)\right) \mathrm{d}F^j(u), \; t \ge 0,$$

where $\Lambda_{j.}$ is integrate transition hazard out of state j. Using this and the quantities defined earlier we obtain a non-parametric regression estimator of the state waiting time survival function

$$H^j(t) = 1 - \int_0^\infty \left\{ \prod_{u < s \le u+t} \left(1 - \frac{\mathrm{d}\hat{N}_{j.}(s)}{\hat{Y}_j(s)}\right) \right\} \mathrm{d}\hat{F}^j(u), \; t \ge 0.$$

16.3.4 Interval Censoring

Multi-state interval censored data are perhaps more common than multi-state current status data, although in the literature such data are often rounded or approximated and turned into right censored data. Such data arise due to lack of constant monitoring of the multi-state process. A fully nonparametric approach to this problem is not currently available in the literature unless one is considering some very special models, e.g., competing risk (Hudgens et al. 2001) or Markov illness–death (Frydman 1995).

Individuals progressing through a multi-state system are inspected multiple times and their status at those times is recorded. Often only intervals ending with those inspection times where a state change has taken place are kept. Just like the current status data, the exact transition times are not observed but known to have taken place in an interval. The data can be represented by $\{C_{ik}, S_i(C_{ik})\}$ for $1 \le k \le n_i$; $i = 1, \ldots, n$, where n_i denotes the number of inspection times retained in the available data for the ith individual and at each such time C_{ik}, the corresponding state information is given by $S_i(C_{ik})$. In general, the richness of interval censored data will be greater than that of current status data. However, efficient use of possibly non-independent information coming from the same individual poses additional methodological challenge. One can ignore this dependence and apply the same methodology as described above to obtain valid (e.g., consistent) but inefficient nonparametric estimators of various marginal quantities. Efficient use of this dependence requires weighted regression techniques involving calculation and estimation of variance–covariance matrix of various event indicators. Such estimators are currently under construction and will be reported elsewhere.

16.4 Discussion

Generally speaking, while parametric (and semiparametric) methods produce relatively precise inference for various model characteristics and the effects of covariates under the correct model, their performance under incorrect model assumptions is questionable. This is one compelling reason why a fully non-parametric approach is preferable although such a formulation is often difficult with time to event data. A large sample size may be necessary to derive the full utility of nonparametric methods; in addition, in dealing with time to event data, one faces additional difficulty and loss of information due to various forms of censoring. The situation with multi-state models that generalize the traditional survival setup is even more challenging. Nevertheless, only nonparametric answers represent truly empirical (or evidence based) calculations. They can at least serve as a guideline to the shape of the various marginal aspects of the system even if a semiparametric or parametric calculation is ultimately performed. Doksum and Yandell (1982) made similar points with compelling comparative illustrations of nonparametric calculations versus semiparametric calculations using the well-known Stanford heart transplant data. We hope that this paper serves as an overview of nonparametric approaches to study certain marginal temporal characteristics of a broad class of multi-state models. There is scope of future work in these areas including bivariate estimation such as that of transition probabilities without the Markov assumption and the joint distribution estimation of two waiting times. Estimation of related functionals such as measures of association are also of interest. Estimation of sojourn time distribution under current status and intervals censored data in non-Markov models remain an open problem as well.

Acknowledgments This research was supported in parts by grants from the United States National Science Foundation (DMS-0706965) and National Security Agency (H98230-11-1-0168).

References

Aalen OO (1978) Nonparametric inference for a family of counting processes. Ann Stat 6:701–726
Aalen OO (1980) A model for nonparametric regression analysis of counting processes. In: Klonecki W, Kozek A, Rosiski J (eds) Lecture notes on mathematical statistics and probability. Springer, New York
Aalen OO, Johansen S (1978) An empirical transition matrix for nonhomogeneous Markov chains based on censored observations. Scandinav J Stat 5:141–150
Andersen PK, Borgan Ø, Gill RD, Keiding N (1993) Statistical models based on counting processes. Springer, New York
Barlow RE, Bartholomew JM, Bremne JM, Brunk HD (1972) Statistical inference under order restrictions. Wiley, New York
Cook RJ, Lawless JF, Lakhal-Chaieb L, Lee K (2009) Robust estimation of mean functions and treatment effects for recurrent events under event-dependent censoring and termination: application to skeletal complications in cancer metastatic to bone. J Am Stat Ass 104:60–75

Datta S, Satten GA (2000) Estimating future stage entry and occupation probabilities in a multi-state model based on randomly right-censored data. Stat Prob Lett 50:89–95

Datta S, Satten GA (2001) Validity of the Aalen–Johansen estimators of stage occupation probabilities and integrated transition hazards for non-Markov models. Stat Prob Lett 55: 403–411

Datta S, Satten GA (2002) Estimation of integrated transition hazards and stage occupation probabilities for non-Markov systems under stage dependent censoring. Biometrics 58: 792–802

Datta S, Sundaram R (2006) Nonparametric estimation of state occupation probabilities in a multi-state model with current status data. Biometrics 62:829–837

Datta S, Satten GA, Datta S (2000a) Nonparametric estimation for the three-stage irreversible illness–death model. Biometrics 56:841–847

Datta S, Satten GA, Datta S (2000b) Estimation of stage occupation probabilities in multi-state models. In: Balakrishnan N (ed) Advances on theoretical and methodological aspects of probability and statistics. Gordon and Breach, New York

Datta S, Lan L, Sundaram R (2009) Nonparametric estimation of waiting time distributions in a Markov model based on current status data. J Stat Plan Infer 139:2885–2897

Doksum KA, Yandell BS (1982) Properties of regression estimates based on censored survival data. In: Bickel PJ, Doksum KA, Hodges, JL (eds) Festschrif for Erich Lehmann. Wadsworth, Belmont

Frydman H (1992) A non-parametric estimation procedure for a periodically observed three state Markov process, with application to AIDS. J Royal Stat Soc B 54:853–866

Frydman H (1995) Nonparametric estimation of a Markov 'illness–death' process from interval-censored observations, with application to diabetes survival data. Biometrika 82:773–789

Glidden D (2002) Robust inference for event probabilities with non-Markov data. Biometrics 58:361–368

Groeneboom P, Maathuis MH, Wellner JA (2008) Current status data with competing risks: limiting distribution of the MLE. Ann Stat 36:1064–1089

Huber C, Solev V, Vonta F (2009) Interval censored and truncated data: Rate of convergence of NPMLE of the density. J Stat Plan Infer 139:1734–1749

Hudgens MG, Satten GA, IM LonginiJR (2001) Nonparametric maximum likelihood estimation for competing risk survival data subject to interval censoring and truncation. Biometrics 57:74–80

Huzurbazar Aparna V (2005) Flowgraph models for multi-state time-to-event data. Wiley, Hoboken

Jewell NP, van der Laan MJ, Henneman T (2003) Nonparametric estimation from current status data with competing risks. Biometrika 90:183–197

Koul H, SusarlaV Van, Ryzin J (1981) Regression analysis of randomly right censored data. Ann Stat 9:1276–1288

Lan L, Datta S (2010) Nonparametric estimation of state occupation, entry and exit times with multi-state current status data. Stat Meth Med Res 19:47–165

Meira-Machado L, de Uña-Alvarez J, Cadarso-Su'arez C (2006) Nonparametric estimation of transition probabilities in a non-Markov illness–death model. Lifetime Data Anal 12: 325–344

Pepe MS (1991) Inference for events with dependent risks in multiple end point studies. J Am Stat Ass 86:770–778

Robins JM (1993) Information recovery and bias adjustment in proportional hazards regression analysis of randomized trials using surrogate markers. In: Proceedings of American statistical association—biopharmaceutical section, pp 24–33

Robins JM, Rotnitzky A (1992) Recovery of information and adjustment for dependent censoring using surrogate markers. In: Jewell N, Dietz K, Farewell V (eds) AIDS epidemiology—methodological issues. Birkhauser, Boston

Satten GA, Datta S (2002) Marginal estimation for multistage models: waiting time distributions and competing risk analyses. Stat Med 21:3–19

Satten GA, Sternberg MR (1999) Fitting semi-Markov models to interval-censored data with unknown initiation times. Biometrics 55:507–513

Satten GA, Datta S, Robins J (2001) Estimating the marginal survival function in the presence of time-dependent covariates. Stat Prob Lett 54:397–403

Wang W (2003) Nonparametric estimation of the sojourn time distributions for a multipath model. J R Stat Soc B 65:921–935

Wang W, Wells MT (1998) Nonparametric estimation of successive duration times under dependent censoring. Biometrika 85:561–572

Chapter 17
Frailty or Transformation Models in Survival Analysis and Reliability

Filia Vonta

Abstract Frailty models are generalizations of the well-known Cox model (Cox, J Roy Stat Soc B 34:187–202, 1972), introduced by Vaupel et al. (Demography 16:439–454, 1979) which are included in a bigger class of models called transformation models. They have received considerable attention over the past couple of decades, especially for the analysis of medical and reliability data that display heterogeneity, which cannot be sufficiently explained by the Cox model. More specifically, the frailty parameter is a random effect term that acts multiplicatively on the hazard intensity function of the Cox model. In this paper we present older and recent results on frailty and transformation models in the parametric and semiparametric setting and for various observational schemes. We deal with efficient estimation of parameters in the uncensored case, right censored case and interval censored and truncated data case.

17.1 Introduction

Frailty models are generalizations of the well-known Cox model (Cox 1972), introduced by Vaupel et al. (1979). They have received considerable attention over the past couple of decades, especially for the analysis of medical or reliability data that display heterogeneity, which cannot be sufficiently explained by the Cox model. The hazard intensity function of a frailty model is that of the Cox model multiplied by a random effect term which follows a distribution appropriate for a positive

F. Vonta (✉)
Department of Mathematics, School of Applied Mathematical
and Physical Sciences, National Technical University of Athens,
9 Iroon Polytechneiou Str., Zografou Campus, 15780 Athens, Greece
e-mail: vonta@math.ntua.gr

A. Lisnianski and I. Frenkel (eds.), *Recent Advances in System Reliability*,
Springer Series in Reliability Engineering, DOI: 10.1007/978-1-4471-2207-4_17,

random variable. This random effect could represent misspecified or omitted covariates. Different frailty distributions give rise to different frailty models. More specifically, a frailty model is defined through the conditional survival function

$$S(t|z, \eta) = \exp\left(-\eta e^{\beta^{tr} z} \Lambda(t)\right), \tag{17.1}$$

where from now on tr denotes transpose, $\beta \in R^d$ is the parameter of interest, η an unobservable positive 'frailty' parameter, z a d-dimensional vector of covariates and Λ the baseline cumulative hazard function. By integrating out the unknown frailty parameter we obtain the model

$$S(t|z) = \int_0^\infty \exp\left(-x e^{\beta^{tr} z} \Lambda(t)\right) dF_\eta(x) = \exp\left(-G\left(e^{\beta^{tr} z} \Lambda(t)\right)\right), \tag{17.2}$$

where $G(y) = -\ln(\int_0^\infty \exp(-xy) dF_\eta(x))$ and F_η is the cdf of the frailty η.

This function G is the $-\ln$ of the Laplace transform of the distribution function of the frailty. These frailty models are in fact special cases of the transformation models (17.2) which are valid for a larger class of G functions which may not be defined through a Laplace transform but nevertheless satisfy appropriate regularity conditions. The class of transformation models defined by Cheng et al. (1995):

$$g(S(t|z)) = h(t) + \beta^{tr} z,$$

where g is known and h unknown, is equivalent to the class of models (17.2) for $g(x) \equiv \log(G^{-1}(-\log x))$ and $h(t) \equiv \log \Lambda(t)$. The case $G(x) = x$ reduces the model (17.2) to the Cox model. The well-known Clayton–Cuzick model (Clayton and Cuzick 1985, 1986) is obtained when η is taken to be distributed with a Gamma distribution with mean 1 and variance b. The function G takes in this case the form $G(x) = b^{-1} \ln(1 + bx)$.

Various frailty distributions are discussed in Hougaard (1984, 1986 and references therein) such as the Inverse Gaussian or the positive stable frailty. Other examples include the log-normal frailty (McGilchrist and Aisbett 1991), the power variance frailty (Aalen 1988), the uniform frailty (Lee and Klein 1988) and the threshold frailty (Lindley and Singpurwalla 1986).

Frailty or more generally transformation models are very effective in describing dependency within a cluster of observations or heterogeneity between clusters. In some cases however we are dealing with univariate frailty models in which each individual/item has his/its own frailty. It is expected that people/items with high frailty values are more at risk for morbidity or mortality or failure.

Slud and Vonta (2004) consider the class of nonproportional hazards models (17.2) and prove for right-censored data and known G function, the large-sample consistency of the NPMLEs of the baseline continuous cumulative hazard function and the regression parameter. Slud and Vonta (2005) proposed a new strategy for obtaining large-sample efficient estimators of finite-dimensional regression parameters within semiparametric statistical models, a very important example of which is model (17.2). For the latter model and known G function, they provided semiparametric information bounds in a more tractable form than previously available. Parner (1998) deals with the asymptotic theory for maximum likelihood estimation of model (17.2) with clusters of size greater than or equal to 2 and shared frailty parameter η that follows the gamma distribution with unknown variance. Kosorok et al. (2004) for the class of transformation models (17.2) with appropriate conditions on the function G showed uniform consistency and semi-parametric efficiency of estimators of the regression parameter, the nuisance parameter and the frailty parameter. The case of misspecification of the model was also treated. Huber-Carol and Vonta (2004) for the case of interval censored and truncated data, examined identifiability and estimation of regression as well as nuisance parameters within model (17.2).

In this paper we review some of our past work in the area of frailty or trans-formation models and we also present some work in progress. In the first part of Sect. 17.2 we deal with efficient estimation of the structural and nuisance parameters in model (17.2) for the uncensored parametric case. The nuisance parameter is assumed to be a vector of arbitrarily high dimension. The approach we propose (Vonta 1996) is based on the use of jointly implicitly defined esti-mators of the structural and nuisance parameters which depend on a continuous kernel function K to be chosen by the statistician.

In the next part, model (17.2) is written as a generalized linear model and we proceed (Vonta and Karagrigoriou 2007) to use information criteria, Akaike's Information Criterion (AIC) and its equivalent, to select the important covariates within model (17.2). We discuss theoretical properties of the model selection criteria and propose a multiple imputations technique in order to handle censored observations and illustrate our method through real data.

In Sect. 17.3 we deal with the semiparametric case where the nuisance parameter is an infinite-dimensional parameter. We derive semiparametric infor-mation bounds (Slud and Vonta 2005) for the regression parameter in transfor-mation models (17.2) and provide an example based on a Gamma distributed frailty. Finally, we refer to work in progress towards semiparametric efficiency for the regression parameter within model (17.2) for the case of interval censored and truncated data, utilizing results of Huber-Carol et al. (2006, 2009).

17.2 Parametric Case

17.2.1 Estimation

In this section we give an overview of the approach proposed in Vonta (1996). Let (T_i, Z_i), $i = 1, \ldots, N$ be i.i.d. random pairs of variables where the distribution of T_i given $Z_i = z$ is

$$F(t, \mu, \beta|z) = 1 - e^{-G\left(e^{\beta^{tr}z}\Lambda(t,\mu)\right)}, \tag{17.3}$$

where $\beta \in R^d$ and the nuisance parameter μ is a vector of dimension ρ. Note that the baseline cumulative hazard function is assumed known up to a ρ dimensional nuisance parameterμ, that is, we model the cumulative hazard function parametrically and that we do not assume censoring. The function G is a known strictly increasing, concave function with $G(0) = 0$ and $G(\infty) = \infty$, and Λ is a continuous increasing function of t with $\Lambda(0, \mu) = 0$ and $\Lambda(\infty, \mu) = \infty$.

Consider the following joint implicit definition of the estimators $\tilde{\beta}$ and $\tilde{\mu}$ of β and μ:

$$\frac{1}{N}\sum_{j=1}^{N} e^{\tilde{\beta}^{tr} z_j}\psi(\tilde{\beta}, z_j) = \frac{1}{N}\sum_{j=1}^{N} e^{\tilde{\beta}^{tr} z_j}K\left(z_j, \Lambda\left(t_j, \tilde{\mu}\right)\right), \tag{17.4}$$

$$\tilde{\mu} = \arg\max_\mu \log L\left(t, z, \mu, \tilde{\beta}\right), \tag{17.5}$$

where L is the likelihood function and ψ is assumed to be a continuously differentiable vector-function, defined by

$$\psi(\beta, z) = \int_0^\infty K(z, \Lambda(t, \mu))dF(t, \theta|z), \tag{17.6}$$

where $\theta = (\mu, \beta)^{tr}$. It is easy to verify that the above integral is a function of the structural parameter β alone. The kernel function K is a fixed continuous d-dimensional vector-function to be chosen by the statistician. The optimal choice of K will be given below. The motivation for the estimating Eq. 17.4 comes from the semiparametric Cox model case and from the fact that when the optimal kernel function is used in this case, the estimator of the parameter of interest emerging from Eq. 17.4 coincides with the maximum partial likelihood estimator of β. This prompted us to use Eq. 17.4 to obtain good estimators in the more general frailty or transformation model case.

Theorem 1 *The pair of estimators* $\left(\tilde{\mu}, \tilde{\beta}\right)^{tr} = \tilde{\theta}$ *is the unique solution of the system of Eqs. 17.4 and 17.5, in a ball of radius $\epsilon_\star$ around the true point $\theta^0 = \left(\mu^0, \beta^0\right)^{tr}$ that does not depend on N and is consistent for θ^0, under regularity conditions given in Vonta (1996).*

Define the function $\Phi : X \times \Theta \to R^{\rho+d}$ as follows:

$$\Phi(T_i, Z_i, \theta) = \begin{pmatrix} \nabla_\mu \log f(T_i, Z_i, \theta) \\ e^{\tilde{\beta}^{tr} z_j}(K(Z_i, \Lambda(T_i, \mu)) - \psi(\beta, Z_i)) \end{pmatrix} \tag{17.7}$$

and the function

$$\Phi_0(\theta) = E_{\theta^0}(\Phi(T_1, Z_1, \theta)). \tag{17.8}$$

Lemma 1 *Under regularity conditions* (Vonta 1996), *the asymptotic distribution of* $\sqrt{N}\left(\tilde{\theta} - \theta^0\right)$ *is normal with mean* $\underline{0}$ *and covariance matrix* $\mathbf{A}^{-1}\Sigma\left(\mathbf{A}^{-1}\right)^{\text{tr}}$, *where* $\mathbf{A}$ *is the derivative of the function* Φ_0 *with respect to* θ *taken at the true point* θ^0 *and* Σ *is the covariance matrix of the function* $\Phi\left(t, z, \theta^0\right)$.

Theorem 2 *Consider the nonproportional hazards model* (17.3). *Under regularity conditions, which include an identifiability condition given in* Vonta (1996), *if* $\bar{\beta}$ *is the estimator of the structural parameter* β *defined in* (17.4) *and* $\hat{\beta}$ *is the maximum likelihood estimator (MLE) of* β, *then*

i.   *the matrix* $a^{22} - I^{22}$ *is nonnegative definite, where* I^{22} *and* a^{22} *are the asymptotic covariance matrices of* $\hat{\beta}$ *and* $\tilde{\beta}$ *respectively;*

ii.   *the estimator* $\tilde{\beta}$ *is asymptotically efficient as compared to the MLE* $\hat{\beta}$, *namely,* $a^{22} = I^{22}$ *if*

$$K(z, x) = -\left(-G'\left(e^{\left(\beta^0\right)^{\text{tr}}z}x\right) + \frac{G''\left(e^{\left(\beta^0\right)^{\text{tr}}z}x\right)}{G'\left(e^{\left(\beta^0\right)^{\text{tr}}z}x\right)}\right)xz. \tag{17.9}$$

A similar theorem holds for the estimator $\tilde{\mu}$ of the nuisance parameter μ which results in the same optimal form of the kernel K. Observe that the optimal function K depends on the true parameter β^0 which therefore needs to be estimated by a preliminary estimator $\tilde{s}$. Then we have the following theorem.

Theorem 3 *The pair of estimators* $\left(\tilde{\mu}, \tilde{\beta}\right)^{\text{tr}} = \tilde{\theta}$, *where* $\tilde{\mu}$ *is defined in* (17.5) *and* $\hat{\beta}$ *is defined by*

$$\frac{1}{N}\sum_{j=1}^{N}e^{\tilde{\beta}^{\text{tr}}z_j}\psi\left(\tilde{\beta}, z_j\right) = \frac{1}{N}\sum_{j=1}^{N}e^{\tilde{\beta}^{\text{tr}}z_j}K\left(z_j, \Lambda\left(t_j, \tilde{\mu}\right), \tilde{s}\right) \tag{17.10}$$

is the unique solution of the Eqs. 17.5 *and* 17.10, *in a ball of radius* $\epsilon_{\star}$ *around the true point* $\theta^0 = \left(\mu^0, \beta^0\right)^{\text{tr}}$ *that does not depend on* N *and is consistent for this point, under regularity conditions and if* $\tilde{s}$ *is strongly consistent for* β^0.

Lemma 2 *Under regularity conditions* (Vonta 1996), *the estimators* $\tilde{\mu}$ *and* $\tilde{\beta}$ *defined in* (17.5) *and* (17.10), *respectively are asymptotically normally distributed, namely,*

$$\sqrt{N}\left(\tilde{\theta} - \theta_0\right) \xrightarrow{L} N\left(\underline{0}, \mathbf{A}^{-1}\Sigma\left(\mathbf{A}^{-1}\right)^{\text{tr}}\right)$$

and therefore retain their full efficiency as compared to the MLE's $\hat{\mu}$ *and* $\hat{\beta}$ *in the case of the optimal function* K *where the true parameter* β^0 *has been replaced by a strongly consistent preliminary estimator* $\tilde{s}$.

In Vonta (1996) we propose a preliminary estimator $\tilde{s}$ which is consistent for β_0.

In Vonta (2004) we propose an easily implemented algorithmic procedure in order to obtain in practice the efficient estimators $\tilde{\beta}$ and $\tilde{\mu}$ defined in (17.4) and (17.5). The behaviour of the algorithm has been illustrated through simulated and real data. A unique feature of the algorithm is that it is a hybrid between an estimating equation (in the parameter β) and restricted maximum likelihood (once β is fixed) for the parameter μ. It is important to point out that the proposed estimators and subsequently the proposed algorithm, for the optimal choice of the kernel function K, constitute a novel way to obtain MLE for frailty models when the nuisance parameter is finite-dimensional. Censoring in the algorithm is being treated through a multiple imputations technique.

The estimating Eq. 17.4 has yet to be generalized to the censored data case. This problem remains open.

17.2.2 Model Selection

In this section a multiple imputations method is proposed (Vonta and Karagrigoriou 2007) for model selection in survival models for censored data. Censoring is successfully being handled through a multiple imputations technique (Little and Rubin 1987). Note that we do not assume here the estimating method of the previous section.

The frailty or transformation model defined in the previous section and in particular in the parametric setting given in (17.3), can be written after some simple algebra, as a transformation model of the form $g(S(t|z)) = h(t) + \beta^{\mathrm{tr}}z$ (Eq. 1.3 in Cheng et al. 1995), namely,

$$g(S(t|z)) \doteq \log\{G^{-1}(-\log(S(t|z)))\} = h^*(t, \mu) + \beta^{\mathrm{tr}}z,$$

where $h^*(t, \mu) = \log(\Lambda(t, \mu))$. We assume further that the function $h^*(t, \mu)$ could be written after some simplifications or possibly transformations as $h(t) + \kappa(\mu)$ where h is a function of t alone and κ a function of μ alone. Then, the above model could equivalently be written as a linear model

$$Y \equiv h(T) = -\kappa(\mu) - \beta^{\mathrm{tr}}Z + \varepsilon, \tag{17.11}$$

or equivalently as $Y = -\beta_*^{\mathrm{tr}}Z + \varepsilon$, where ε has a continuous cdf F_ε with mean μ_ε and variance σ_ε^2. If F_ε is the extreme value distribution then (17.11) corresponds to the proportional hazards model while if F_ε is the standard logistic distribution then (17.11) corresponds to the proportional odds model. If e^ε follows a Pareto distribution, then (17.11) corresponds to the Clayton and Cuzick frailty model, that is, the semiparametric Pareto model. If e^ε follows a $e^{N(0,1)}$ distribution, then the model (17.11) corresponds to the generalized Box-Cox model.

Assume that theoretically, the true process of the observations is of infinite order (the true model consists of an infinite number of variables) and we attempt to approximate the infinite order by a finite one, among a set of candidate models, so that the mean squared error (MSE) of prediction is the smallest possible. Note that this innovative assumption is due to Shibata (1980) who was the first to introduce this particular concept of asymptotic efficiency. For Shibata's criterion S_n (Shibata 1980), defined through the formula, $S_n(j) = (n + 2j)\hat{\sigma}^2(j)$ where $n\hat{\sigma}^2(j) = ||Y - \hat{\beta}_*^{tr}(j)Z||^2$ with the norm being the Euclidean norm, we have the following theorem.

Theorem 4 *Assume model* (17.11) *where* $\{\varepsilon_i\}, i = 0, 1, 2, \ldots$ *the sequence of errors, has the distribution function* F_ε *with mean* μ_ε, *variance* σ_ε^2, *and* $E(\varepsilon_i)^4 < \infty$. *Then, under regularity conditions given in* Vonta and Karagrigoriou (2007), *the order selected by* S_n, *namely* $\hat{\jmath}$, *is asymptotically efficient in the sense of* Shibata (1980).

In other words, $\hat{\jmath}$, the order selected by S_n, is optimal in the sense that the sum of squared errors for prediction attains the lower bound. Note that the above result holds for each member of the family of AIC-type criteria, namely AIC (Akaike 1974), Shibata's Criterion (S_n), Final Prediction Error Criterion (FPE), Criterion Autoregressive Transfer function (CAT), AIC Corrected Criterion (AIC$_c$) and Extended Information Criterion (EIC).

Remark 1 One should also mention that, to our knowledge, the conditions of the above theorem are the weakest possible conditions for Shibata's asymptotic efficiency. These conditions however are satisfied by most of the known distributions including those that most often emerge in survival analysis or reliability, like the extreme value, the logistic, the normal and the Pareto.

Let us assume that our data follow a parametric frailty or transformation model of the class defined in (17.3). We propose the following algorithm for the selection of the significant variables in the given data set as well as the estimation of their corresponding coefficients. We start with some preliminary estimators of β and μ, which we denote by $\hat{\beta}^{(1)}$ and $\hat{\mu}^{(1)}$, obtained by some standard method of analysis. It is then straightforward to impute complete data in place of the censored values with the current set of estimated parameters. Using these estimates we randomly generate as many failure times as the censored observations in the data set, given that these failure times are greater than the observed censoring times. Let c_i denote the ith observed censoring time among the censored individuals. A failure time T_i for this individual is imputed using the conditional survival model

$$\Pr(T_i > t | z_i, c_i, T_i > c_i) = e^{-G\left(e^{\beta^{tr}z_i}\Lambda(t,\mu)\right) + G\left(e^{\beta^{tr}z_i}\Lambda(c_i,\mu)\right)} \qquad (17.12)$$

for all $t > c_i$. Using multiple imputations, we repeat the same procedure independently M times $(M \geq 2)$, with the observed censoring times c_i fixed in all imputations. The M independently completed data sets are analyzed by applying to them the AIC procedure and results are combined to obtain $\hat{\beta}^{(2)}$ and $\hat{\mu}^{(2)}$. At each iteration we repeat the same steps for imputing survival times under relation

(17.12) using the current estimates of β and μ, until convergence takes place. The last iteration of this algorithmic procedure provides the final selection of the significant variables of the model as well as their corresponding estimated coefficients.

The way to combine the results of the M completed data sets at step i of the algorithm is the following. We apply the AIC to each one of the M independently completed data sets in order to obtain $\hat{\beta}_{(m)}$ and $\hat{\mu}_{(m)}$ for $m = 1, \ldots, M$ (if a certain variable is not selected its corresponding coefficient estimate is regarded as 0) and then we define $\hat{\beta}^{(i)} = \sum_{m=1}^{M} \hat{\beta}_{(m)}^{(i)}/M$ and $\hat{\mu}^{(i)} = \sum_{m=1}^{M} \hat{\mu}_{(m)}^{(i)}/M$. The algorithm ends when the difference between $\hat{\beta}^{(i)}$ and $\hat{\mu}^{(i)}$ for two consecutive steps of the algorithm (say j and $j + 1$) is less than 10^{-6}. At this point the final $\hat{\beta}$ and $\hat{\mu}$ are obtained. At the same time we are also faced with M (possibly different) variable selections. However, based on those selections we can easily identify the actually significant set of variables, on one hand as those variables with the largest frequencies among the M selections and on the other hand by conducting tests as to which coefficients are actually different from 0 (based on $\hat{\beta}$ and its empirical variance) at a 5% level of significance. From our experience we conclude that an M equal to 3 or 5 is not enough to obtain good estimates but a value of M equal to 10 or 20 is sufficient to provide good estimators at an acceptable computational cost.

17.2.2.1 Real Data Example

In this section, we analyze the Primary Biliary Cirrhosis Data (PBC) (Fleming and Harrington 1991) which include 17 covariates. The censoring is quite heavy (113 observations or 52.31%). It has been shown in the literature that this set of data is adequately described by the Cox model so we have considered $G(x) = x$ (Cox 1972).

The AIC applied to the regression exponential model selects 8 variables in the final model which are the age, albumin and bilirubin, presence of edema, urine copper, SGOT, prothrombin time in seconds and histologic stage of disease. Through the AIC which was applied to an exponential regression model where the censored observations were completed through the multiple imputations technique, we obtained, at the final stage of the algorithm, the exact same covariates. The algorithm needed 25 iterations until convergence and the M used was 20. The estimators of the coefficients and the standard errors (or empirical standard errors) are very comparable. The results are presented in Table 17.1.

The intercept was estimated as 7.072 from the exponential model and 6.550 from the exponential-imputations model.

For this set of data we also report in Table 17.2 the results obtained when the Weibull model is applied. In this case, the intercept obtained by the Weibull model is 5.000 while that obtained by the Weibull-imputations model is 4.963. The corresponding values for the scale parameter of the Weibull model are 0.615 and 0.596, respectively.

Table 17.1 Analysis of PBC data for exponential baseline hazard

	Exponential-imputations		Exponential	
	Estimator	Standard error	Estimator	Standard error
$\hat{a}_1$	−0.026	0.006	−0.024	0.010
$\hat{a}_2$	0.414	0.187	0.438	0.260
$\hat{a}_3$	−0.047	0.018	−0.057	0.018
$\hat{a}_4$	−0.002	0.000	−0.002	0.000
$\hat{a}_5$	−0.753	0.295	−0.626	0.333
$\hat{a}_6$	−0.201	0.068	−0.144	0.063
$\hat{a}_7$	−0.002	0.001	−0.003	0.001
$\hat{a}_8$	−0.306	0.080	−0.375	0.140

In this set of data where censoring is quite heavy and the completion of the data set by the imputation technique is more challenging, the AIC selects during the M repetitions of the imputation method, between eight and eleven covariates. However, as it was said before, the covariates finally reported are those that have successfully passed the relevant test of significance at the 0.05 level of significance. It is also reconfirmed that the reported covariates are those that appear with the largest frequencies among the M repetitions of the method. The analysis of the PBC data clearly shows that the proposed imputation method accompanied by the implementation of the AIC criterion can successfully handle the heavy censoring and therefore constitutes a powerful alternative to the traditional methods of analysis in survival modelling.

Based on the analysis of simulated data we can also report that the imputations technique diminishes the tendency of AIC for overestimation even under misspecification. The results favour the imputations technique for large n but the method should be used with caution in case of small n and heavy censoring.

17.3 Semiparametric Case

In this section we consider the case where no parametric form for the baseline hazard is assumed, so that the baseline hazard is an infinite-dimensional nuisance parameter. We focus on semiparametric efficiency bounds, as opposed to methods of estimation. We review the results of Slud and Vonta (2005) regarding semiparametric information bounds for the regression parameter in the transformation model (17.2), and provide an example based on a Gamma distributed frailty. Finally, we refer to work in progress towards semiparametric efficiency bounds for the regression parameter in model (17.2) for the case of interval censored and truncated data.

In preparation for the main result, we present an information bound result for general semiparametric statistical models. Assume that a sample of independent identically distributed (i.i.d.) random variables $X_1, X_2, \ldots, X_n$ is observed and

Table 17.2 Analysis of PBC data for Weibull baseline hazard

	Weibull-imputations		Weibull	
	Estimator	Standard error	Estimator	Standard error
$\hat{a}_1$	−0.015	0.003	−0.017	0.006
$\hat{a}_2$	0.392	0.113	0.393	0.164
$\hat{a}_3$	−0.046	0.010	−0.049	0.011
$\hat{a}_4$	−0.002	0.000	−0.002	0.000
$\hat{a}_5$	−0.589	0.179	−0.555	0.209
$\hat{a}_6$	−0.159	0.040	−0.144	0.063
$\hat{a}_7$	−0.002	0.000	−0.002	0.001
$\hat{a}_8$	−0.257	0.047	−0.272	0.090

assumed to follow a marginal probability law $\mu_0 = P_{(\beta^0, \lambda^0)}$ where $\beta^0 \in U \subset R^d$, $\lambda^0 \in V \subset L^0(R, v)$ (Borel-measurable functions), where U is a fixed open set; V is a fixed set of positive measurable functions; and the σ-finite measure v is fixed on R. Assume also that there is a family $\{P_{(\beta, \lambda)}, (\beta, \lambda) \in U \times V\}$ of Borel probability measures on R, such that $P_{(\beta, \lambda)} \ll \mu_0$. Let the densities $f_X(\cdot, \beta, \lambda) \equiv dP_{(\beta, \lambda)}/d\mu_0$.

With $a^{\otimes 2} \equiv a\,a^{\mathrm{tr}}$ for $a \in R^d$, let us define the following bilinear forms which constitute blocks of the information matrix

$$A_{\beta^0, \lambda^0}(a, a) = \int a^{\mathrm{tr}} \left(\nabla_\beta \log f_X\left(x, \beta^0, \lambda^0\right)\right)^{\otimes 2} a\, d\mu_0(x) \tag{17.13}$$

$$B_{\beta^0, \lambda^0}(a, \gamma) = \int \left(a^{\mathrm{tr}}\nabla_\beta \log f_X\left(x, \beta^0, \lambda^0\right)\right) D_\lambda \log f_X(x, \beta^0, \lambda^0)(\gamma)\, d\mu_0(x), \tag{17.14}$$

$$C_{\beta^0, \lambda^0}(\gamma, \gamma) = \int \left(D_\lambda \log f_X\left(x, \beta^0, \lambda^0\right)(\gamma)\right)^2 d\mu_0(x), \tag{17.15}$$

where $D_\lambda \phi(\lambda)(\gamma)$ stands for the *Gâteaux* directional derivative of the function $\phi : V \to R$ in the direction $\gamma \in G_0$. For further regularity conditions see Slud and Vonta (2005).

Theorem 5 *Under regularity conditions (Slud and Vonta 2005) the information bound or least favourable information matrix I_β^0 (Bickel et al. 1993, p. 23), for the parameter of interest β in the presence of the nuisance parameter λ, is given as*

$$a^{\mathrm{tr}}I_\beta^0 a = A_{\beta^0, \lambda^0}(a, a) + B_{\beta^0, \lambda^0}(a, \gamma), \tag{17.16}$$

where A and B are defined in (17.13) and (17.14) respectively and $\gamma = \nabla_\beta \lambda_\beta|_{\beta=\beta_0}$ is the least favourable direction γ. The least favourable direction corresponds, and is being found through, the least favourable parametric submodel (β, λ_β) of the general semiparametric model $P_{(\beta, \lambda)}$.

17.3.1 Right Censored Data

The semiparametric problems which motivated this work concern transformation and frailty models in survival analysis or reliability defined in (17.2) with G a known function, satisfying regularity conditions given by Slud and Vonta (2004, 2005). Here the unknown parameters, with true values (β^0, λ^0), are (β, λ), where $\beta \in R^d$ and $\Lambda(t) \equiv \int_0^t \lambda(s)\,ds$ is a cumulative-hazard function, i.e. $\lambda(s) \geq 0$, $\Lambda(\infty) = \infty$. For notational simplicity, we assume that the variables $Z \in R^d$ have discrete and finite-valued distribution, $\pi_z \equiv P(Z = z)$, but a compactly supported distribution yields the same set of theoretical results. The data are randomly right-censored, i.e., there is an underlying positive random variable C conditionally independent of T given $Z = z$ with $R_z(y) \equiv P(C \geq y \mid Z = z)$.

The observable data for a sample of n independent individuals are $T_i \equiv \min(T_i^0, C_i)$, $\Delta_i \equiv I_{[T_i^0 \leq C_i]}$, $Z_i, i = 1, \ldots, n)$ where T_i^0 is the ith survival time and C_i the ith censoring time. However, following Slud and Vonta (2002) we also impose a nontrivial technical restriction on the (correctly specified) distribution of the data through the function

$$q_z(t) \equiv P(T \geq t \mid Z = z) = R_z(t) \exp\left(-G\left(e^{z^{tr}\beta^0}\Lambda^0(t)\right)\right) \qquad (17.17)$$

in the form

$$q_z(t) \equiv 0 \quad \text{for} \quad t > \tau_0, q_z(\tau_0) > 0. \qquad (17.18)$$

The import of this restriction is that for some fixed time τ_0 beyond which the individuals in a study do have a positive probability of surviving uncensored, the data are automatically right-censored at τ_0.

Theorem 6 *Under regularity conditions (Slud and Vonta 2005), for the case of censored data and semiparametric frailty or transformation models defined in (17.2), the least favourable information bound about β_0 is given as*

$$I_\beta^0 = \sum_z \pi_z \int_0^{\tau_0} z q_z e^{z^{tr}\beta^0} G'\left(e^{z^{tr}\beta^0}\Lambda^0\right)\left(1 + (xG''(x)/G'(x))_{x=e^{z^{tr}\beta^0}\Lambda^0}\right)$$

$$\cdot \left(z + e^{z^{tr}\beta^0}\left(z\Lambda^0 + \nabla_\beta\Lambda_{\beta^0}\right)\frac{G''}{G'}\Big|_{x=e^{z^{tr}\beta^0}\Lambda^0} + \nabla_\beta\log\lambda_{\beta^0}\right)^{tr}\lambda^0\,dt, \qquad (17.19)$$

where the least favourable direction $\gamma = \nabla_\beta\lambda_\beta(t)|_{\beta=\beta_0}$ is defined implicitly through an ordinary differential equations system with initial and terminal conditions given in Slud and Vonta (2005).

In Slud and Vonta (2005) we obtain information bounds for the frailty or semiparametric models in a much more explicit form than those previously based on Sturm–Liouville problem solutions as in Klaassen (1993) or Bickel et al.

(1993). The information bounds are written in a new form, given in (17.16), which allows tractable calculations in frailty and transformation settings. Therefore, we derive information bounds in a case where essentially no previous computations of such bounds have been available.

As examples of the resulting information formula (17.19), we provide numerical bounds for several cases of the two-sample right-censored Clayton–Cuzick (1986) frailty model. Table 17.3 exhibits the quantities $A_{\beta^0, \lambda^0}(1, 1)$ and I_β^0, for the case where the covariate $Z_i = 0$, 1 is the group label for subject i, with approximately half of all n subjects allocated to each of the two groups; where the model (17.2) holds with $G(x) = b^{-1} \log(1 + bx)$, where the group-1 over group-0 log hazard-ratio parameters is $\beta = \beta_1 = \log 2$ with $\lambda^0 \equiv 1$ and where the censoring distributions $R_z(t) = P(C \geq t \,|\, Z = z)$ are Uniform$[0, \tau^{(z)}]$ subject to additional, administrative right-censoring at τ_0, i.e., for $z = 0, 1, R_z(t) = \max(0, 1 - t/\tau^{(z)}) I_{[t \leq \tau_0]}$. (Note that the cases of very large $\tau^{(z)}$ values in Table 17.3 correspond to data uncensored before τ_0). The I_β^0 values were calculated with accuracy approximately 0.0001. The case $b = 0.0001$ corresponds closely to the relatively easily calculated and well-established values for Cox model, which is the limiting case of the Clayton–Cuzick model at $b = 0+$.

The fifth column contains the upper-left entries of the information matrix and the sixth column is the information bound I_β^0.

Remark 2 In Slud and Vonta (2005) we propose modified profile likelihood estimators of the parameter of interest β which are proved to be semiparametric efficient under appropriate regularity conditions. We also discuss the construction of these estimators.

17.3.2 Interval Censored and Truncated Data

When the observation of a process is not continuous in time but instead is being done through a specific time window, say $[Z_1, Z_2]$ so that some individuals could totally be excluded, we have truncation. If moreover, the survival time X of an individual/item is not completely known but rather we only have the information that it falls within an observed interval $[L(X), R(X)]$, we have interval censoring.

Let X be a random variable with density f and survival function conditional on $Z = z$ defined in (17.2). Let v^k be the Lebesgue measure on R^k and recall that $z \in R^d$ and let $dP_Z = \phi(z) dv_d$ for some σ-finite measure v_d possibly equal to v^d. Without loss of generality we consider the right-truncation case. The problem could be formulated as follows. Our observations are $Q_1, \ldots, Q_n$, i.i.d. random vectors, $Q = (L(X), R(X), L(Z_2), Z)$, with density $p(u, v, w, z)$ with respect to a measure v^* (for details see Huber-Carol et al. 2006, 2009) given as

Table 17.3 Information bound calculations for two-sample Clayton–Cuzick frailty model

b	$\tau^{(0)}$	$\tau^{(1)}$	τ_0	$A_{\beta^0,\lambda^0}(1,1)$	I_β^0
0.0001	1.e8	1.e8	20	0.4999	0.2253
0.5	1.e8	1.e8	20	0.2500	0.1160
1	1.e8	1.e8	20	0.1667	0.0770
2	1.e8	1.e8	20	0.1000	0.0458
3	1.e8	1.e8	20	0.0714	0.0326
4	1.e8	1.e8	20	0.0556	0.0253
0.0001	2	4	3.96	0.3773	0.1676
0.5	3	6	5.95	0.2227	0.1054
1	4	8	7.9	0.1564	0.0741
2	5	10	9.95	0.0967	0.0453
3	6	12	11.95	0.0700	0.0325
4	7	14	13.95	0.0548	0.0253

$$p(u,v,w,z) = r(u,v,w) \cdot \frac{\int_u^v f(t|z)\, dt}{\int_0^w f(t|z)\, dt} \phi(z), \qquad (17.20)$$

where v^* is the measure on R^{3+d} which is defined for continuous nonnegative functions $\psi(s) = \psi(u,v,w,z)$ by the relation

$$\iiint \psi(s)\, dv^* = \iiint \psi(u,v,w,z)\, d\left(v^3 \otimes v_d\right)(u,v,w,z)$$
$$+ \iiint \psi(u,v,v,z)\, d\left(v^2 \otimes v_d\right)(u,v,z).$$

From model (17.2) p is equal to

$$r(u,v,w) \cdot \frac{\int_u^v e^{-G\left(e^{\beta^{\mathrm{tr}}z}\Lambda(t)\right)} G'\left(e^{\beta^{\mathrm{tr}}z}\Lambda(t)\right) e^{\beta^{\mathrm{tr}}z}\lambda(t)\, dt}{\int_0^w e^{-G\left(e^{\beta^{\mathrm{tr}}z}\Lambda(t)\right)} G'\left(e^{\beta^{\mathrm{tr}}z}\Lambda(t)\right) e^{\beta^{\mathrm{tr}}z}\lambda(t)\, dt} \cdot \phi(z) \qquad (17.21)$$
$$\equiv r(u,v,w) \cdot \varphi(u,v,w,z|\beta,\lambda) \cdot \phi(z),$$

thus defining function φ as the ratio of integrals in formula (17.21). Here $0 \le u < v < w \le b$, r is the known density with respect to v^{**}, the marginal of v^* integrated over z, of the known joint law of censoring and truncation. Function φ is the likelihood of each unobserved survival conditional on the censoring, truncation and covariate.

The joint law of censoring and truncation with density r with respect to v^{**} has two components, one denoted by r_3 which is absolutely continuous with respect to the Lebesgue measure on R^3 (corresponding to the case where $u < v < w$) and a second one, denoted by r_2, which is absolutely continuous with respect to the Lebesgue measure on R^2 (corresponding to the case where $u < v = w$). For details and an example of such a law r (see Huber-Carol et al. 2009).

We are interested in the efficient estimation of the parameter of interest β in the presence of the unknown cumulative hazard function Λ or equivalently in the presence of the hazard intensity function λ, where obviously $\lambda(t) \geq 0$. We assume that the function $\Lambda(t)$ is finite for finite times t and that $\Lambda(\infty) = \infty$. The data-space $\mathbf{D} = R \times R \times R \times R^d$ consists of vectors $s = (u, v, w, z)$. We denote the true parameters by $\left(\beta^0, \lambda^0\right)$ and we define the probability law for the true model by $d\mu_0(s) \equiv p_0(s)dv^*(s)$, where $p_0(s)$ denotes the density p taken at the true point $\left(\beta^0, \lambda^0\right)$.

Taking the densities $f_Q(s|\beta, \lambda)$ with respect to μ_0, the true law of the observations Q, as in Slud and Vonta (2005), we get that $f_Q(s|\beta, \lambda)$ is equal to

$$\frac{\int_u^v e^{-G}G'\big|_{e^{\beta^{\mathrm{tr}}z}\Lambda(t)}e^{\beta^{\mathrm{tr}}z}\lambda(t)dt}{\int_0^w e^{-G}G'\big|_{e^{\beta^{\mathrm{tr}}z}\Lambda(t)}e^{\beta^{\mathrm{tr}}z}\lambda(t)dt} \cdot \frac{\int_0^w e^{-G}G'\big|_{e^{\left(\beta^0\right)^{\mathrm{tr}}z}\Lambda_0(t)}e^{\left(\beta^0\right)^{\mathrm{tr}}z}\lambda_0(t)dt}{\int_u^v e^{-G}G'\big|_{e^{\left(\beta^0\right)^{\mathrm{tr}}z}\Lambda_0(t)}e^{\left(\beta^0\right)^{\mathrm{tr}}z}\lambda_0(t)dt}. \tag{17.22}$$

We will prove efficiency by following the methodology given in Slud and Vonta (2005) and by providing conditions under which assumptions $\mathbf{A_0 - A_8}$, which ensure efficiency, of that paper are satisfied. We propose to estimate β by a modified profile likelihood estimator, defined in Slud and Vonta (2005). This is work currently in progress.

17.4 Discussion

In this paper we focus on frailty or more generally on transformation models useful in survival analysis or reliability whenever there is heterogeneity in the population that has not been explained by simpler models and the inclusion of a random effect could improve the fit of the data. We deal with topics related to consistency and efficient estimation of parameters involved in these models as well as model selection under various censoring schemes. The models considered are parametric as well as semiparametric. We review our past work and refer to ongoing research.

Acknowledgments We would like to thank the anonymous referee whose comments greatly improved the presentation and clarity of the paper.

References

Aalen O (1998) Heterogeneity in survival analysis. Stat Med 7:1121–1137

Akaike H (1974) A new look at the statistical model identification. IEEE Trans Auto Cont 19(6):716–723

Bickel P, Klaassen C, Ritov Y, Wellner J (1993) Efficient and adaptive inference in semiparametric models. Johns Hopkins Univ Press, Baltimore

Cheng SC, Wei LJ, Ying Z (1995) Analysis of transformation models with censored data. Biometrika 82:835–845

Clayton D, Cuzick J (1985) Multivariate generalizations of the proportional hazards model (with discussion). J Roy Stat Soc A 148:82–117

Clayton D, Cuzick J (1986) The semiparametric Pareto model for regression analysis of survival times. Papers on semiparametric models, MS-R8614, Centrum voor Wiskunde en Informatica, Amsterdam, pp 19–31

Cox DR (1972) Regression models and life tables. J Roy Stat Soc B 34:187–202

Fleming TR, Harrington DP (1991) Counting processes and survival analysis. Wiley, New York

Hougaard P (1984) Life table methods for heterogeneous populations: distributions describing the heterogeneity. Biometrika 71:75–83

Hougaard P (1986) Survival models for heterogeneous populations derived from stable distributions. Biometrika 73:387–396

Huber-Carol C, Vonta F (2004) Frailty models for arbitrarily censored and truncated data. Lifetime Data Anal 10:369–388

Huber-Carol C, Solev V, Vonta F (2006) Estimation of density for arbitrarily censored and truncated data. In: Nikulin MS, Commenges D, Huber-Carol C (eds) Probability, statistics and modelling in public health. Springer, New York, pp 246–265

Huber-Carol C, Solev V, Vonta F (2009) Interval censored and truncated data: rate of convergence of NPMLE of the density. J Stat Plan Infer 139:1734–1749

Klaassen C (1993) Efficient estimation in the Clayton–Cuzick model for survival data. Preprint, University of Amsterdam

Kosorok M, Lee B, Fine J (2004) Robust inference for univariate proportional hazards frailty regression models. Ann Stat 32(4):1448–1491

Lee S, Klein JP (1988) Bivariate models with a random environmental factor. Ind J Prod Reliab Qual Cont 13:1–18

Lindley DV, Singpurwalla NA (1986) Multivariate distributions for the life lengths of components of a system sharing a common environment. J Appl Prob 23:418–431

Little RJA, Rubin DB (1987) Statistical analysis with missing data. Wiley, New York

McGilchrist CA, Aisbett CW (1991) Regression with frailty in survival analysis. Biometrics 47:461–466

Parner E (1998) Asymptotic theory for the correlated gamma-frailty model. Ann Stat 26:183–214

Shibata R (1980) Asymptotically efficient selection of the order of the model for estimating parameters of linear process. Ann Stat 8:147–164

Slud EV, Vonta F (2002) Nonparametric likelihood and consistency of NPMLE's in the transformation model. TR/17/02 Dept of Math Stat, University of Cyprus

Slud EV, Vonta F (2004) Consistency of the NPML estimator in the right-censored transformation model. Scand J Stat 31:21–41

Slud EV, Vonta F (2005) Efficient semiparametric estimators via modified profile likelihood. J Stat Plan Infer 129:339–367

Vaupel JW, Manton KG, Stallard E (1979) The impact of heterogeneity in individual frailty on the dynamics of mortality. Demography 16:439–454

Vonta F (1996) Efficient estimation in a non-proportional hazards model in survival analysis. Scand J Stat 23:49–61

Vonta F (2004) Efficient estimation in regression frailty or transformation models based on an algorithm. Austral New Zeal J Stat 47(4):503–514

Vonta F, Karagrigoriou A (2007) Variable selection strategies in survival models with multiple imputations. Lifetime Data Anal 13(3):295–315

Chapter 18
Goodness-of-Fit Tests for Reliability Modeling

Alex Karagrigoriou

Abstract In this work we provide goodness-of-fit (GoF) tests which are based on measures of divergence and can be used for assessing the appropriateness of both discrete and continuous distributions. The main focus is on continuous distributions like the exponential, Weibull, Inverse Gaussian, Gamma and lognormal which frequently appear in engineering and reliability. An extensive simulation study shows that the new tests maintain good stability in level and high power across a wider range of distributions and sample sizes than other tests.

Keywords Generalized goodness-of-fit tests · Measures of divergence · Lifetime distribution · Multinomial populations

18.1 Introduction

The χ^2 test has passed into common use since its introduction by Pearson (1900). It has, however, long been recognized that in certain cases it is inadequate as a test for goodness-of-fit (GoF) and in particular in the case when the deviations of the observations from the hypothesis tested are consecutively positive (or negative); for the χ^2 test, by taking into account the square of the difference between the observed and expected values, renders it impossible to pay attention to the sign of this difference. It is because of this inadequacy that Neyman (1937) introduced the "smooth" test for GoF.

A. Karagrigoriou (✉)
55 Lomvardou Street, 11474 Athens, Greece
e-mail: alex@ucy.ac.cy

A. Lisnianski and I. Frenkel (eds.), *Recent Advances in System Reliability*,
Springer Series in Reliability Engineering, DOI: 10.1007/978-1-4471-2207-4_18,
© Springer-Verlag London Limited 2012

The smooth Neyman test relies on normalized Legendre polynomials on [0, 1]. The Neyman smooth tests were constructed to be asymptotically locally uniformly most powerful, symmetric, unbiased and of specified size against specified smooth alternatives (Neyman 1937). For the order k smooth alternative, the resulting test statistic is based on the summation of the squares of the first k Legendre polynomials. Then, under the null hypothesis, the test statistic follows the central chi-square distribution with k degrees of freedom.

The smooth tests were found to provide useful diagnostic and inferential tools but, over the years, have been taken over by the so-called omnibus tests of the Cramer-von Mises family. Indeed, another way to view the gof problem is through the Cramer-von Mises family of tests. The general Cramer-von Mises family of test statistics focuses on the squared differences given by

$$Q = \int\limits_{-\infty}^{\infty} [F_n(y) - F_0(y)]^2 \Psi(y) dF_0(y)$$

where $F_n(y)$ is the empirical distribution function, $F_0(y)$ is the hypothesized distribution and $\Psi(y)$ is an appropriate scaling function. For different forms of the function $\Psi(y)$ we obtain the Cramer-von Mises test statistic (Cramer 1928; von Mises 1931), the Anderson–Darling test statistic (Anderson and Darling 1954) and through a modification, the Watson test (Watson 1961). All these test statistics advocate the use of tables or formulas for critical values which depend on the parameters of the underlying distribution.

All the above tests are related to and based on measures of disparity or divergence between the true but unknown and the hypothesized distributions. Measures of divergence between two probability distributions have a very long history. One could consider as pioneers in this field the famous Mathematicians and Statisticians of the twentieth century, Pearson (1900), Mahalanobis (1936), Levy (1925, p. 199–200) and Kolmogorov (1933). The most popular measures of divergence are considered to be the Kullback–Leibler (KL) measure of divergence and the measures associated with the Pearson's chi-square test and the likelihood ratio test. The tests based on these measures can among other areas, be used in reliability applications. In many reliability applications, certain classes of life distributions have been introduced to describe several aging criteria. Such examples include the residual lifetime that represents the remaining life of an aging system and the renewal lifetime of a device replaced upon failure. Among the most well-known families of life distributions are the classes of increasing failure rate (IFR), decreasing failure rate (DFR), increasing failure rate average (IFRA), new better than used (NBU), new better than used in expectation (NBUE) and harmonic new better than used in expectation (HNBUE). For the residual lifetime, if the mean residual life (i.e., the expected value of the remaining life) is a decreasing function then we encounter an important and practical class of life distributions, the decreasing mean residual lifetime distribution (DMRL). In most of these problems the null distribution is the exponential distribution and the alternative is a member of one of the above classes other than the exponential.

In this work we focus on testing problems in reliability, like the ones discussed earlier. In Sect. 18.2 we discuss various measures of divergence and in Sect. 18.3 the associated tests of fit. In Sect. 18.4 we provide simulations for continuous distributions in order to explore the capabilities of the proposed tests for various continuous distributions.

18.2 Divergence Measures

The most famous family of measures is the ϕ-divergence between two functions f and g, known also as Csiszar's measure of information which was introduced and investigated independently by Csiszar (1963) and Ali and Silvey (1966) and is given by

$$D_\phi(f, g) = \int_{\mathcal{X}} g(\mathbf{x}) \phi\left(\frac{f(\mathbf{x})}{g(\mathbf{x})}\right) d\mu(\mathbf{x}), \quad \phi \in \Phi \tag{18.1}$$

where Φ is the class of all convex functions $\phi(x)$, $x \geq 0$, such that at $x = 1$, $\phi(1) = 0$, at $x = 0$, $0\phi(0/0) = 0$ and $0\phi(u/0) = \lim_{u \to \infty} \phi(u)/u$. For various functions for ϕ the measure takes different forms. Note that the well-known KL divergence measure (Kullback and Leibler 1951) is obtained for $\phi(x) = x \log(x) - x + 1 \in \Phi$. If $\phi(u) = 1/2(1 - u)^2$,

$$\phi(u) = \Phi_{2,\lambda}(u) = \frac{1}{\lambda(\lambda + 1)} \left(u^{\lambda+1} - u - \lambda(u - 1)\right), \quad \lambda \neq 0, -1, \tag{18.2}$$

or $\phi(u) = (1 - \sqrt{u})^2$, Csiszar's measure yields the Pearson's chi-squared divergence, the Cressie and Read power divergence (Cressie and Read 1984) and the Matusita's divergence (Matusita 1967), respectively. More examples can be found in Arndt (2001), Pardo (2006) and Vajda (1989, 1995).

A unified analysis has been provided by Read and Cressie (1988) who introduced for both the continuous and the discrete case a family of measures of divergence known as power divergence family of statistics that depends on a parameter λ and is used for GoF tests for multinomial distributions. The Cressie and Read family includes among others the well-known Pearson's χ^2 divergence measure and for multinomial models the log likelihood ratio statistic.

Csiszar's family of measures was recently generalized by Mattheou and Karagrigoriou (2010) to the Φ-family of measures between two functions f and g which is given by

$$D_X^a(g, f) = \int g^{1+a}(\mathbf{x}) \Phi\left(\frac{f(\mathbf{x})}{g(\mathbf{x})}\right) d\mu(\mathbf{x}), \quad \Phi \in \Phi^*, \quad a > 0 \tag{18.3}$$

where μ represents the Lebesgue measure and Φ^* is the class of all convex functions Φ on $[0, \infty)$ such that $\Phi(1) = \Phi'(1) = 0$ and $\Phi''(1) \neq 0$. We also assume the conventions $0\Phi(0/0) = 0$ and $0\Phi(u/0) = \lim_{u\to\infty} \Phi(u)/u$, $u > 0$.

Expression (18.3) covers not only the continuous case but also a discrete setting where the measure μ is a counting measure. Indeed, for the discrete case, the divergence in (18.3) is meaningful for probability mass functions f and g whose support is a subset of the support S_μ, finite or countable, of the counting measure μ that satisfies $\mu(x) = 1$ for $x \in S_\mu$ and 0 otherwise.

The discrete version of the Φ-family of divergence measures is presented in the definition below.

Definition 1   For two discrete distributions $P = (p_1, \ldots, p_m)'$ and $Q = (q_1, \ldots, q_m)'$ with sample space $\Omega = \{x : p(x)q(x) > 0\}$, where $p(x)$ and $q(x)$ are the probability mass functions of the two distributions, the discrete version of the Φ-family of divergence measures with a general function Φ as in (18.3) and $a > 0$ is given by

$$d_a = \sum_{i=1}^{m} q_i^{1+a} \Phi\left(\frac{p_i}{q_i}\right). \tag{18.4}$$

For Φ having the special form

$$\Phi_1(u) = u^{1+a} - \left(1 + \frac{1}{a}\right)u^a + \frac{1}{a} \tag{18.5}$$

we obtain the BHHJ measure of Basu et al. (1998) which was proposed for the development of a minimum divergence estimating method for robust parameter estimation. Observe that for Φ having the special form $\Phi(u) = \phi(u)$ and $a = 0$ we obtain the Csiszar's ϕ-divergence family of measures while for $a = 0$ and for $\Phi(u) = \Phi_{2,\lambda}(u)$ as in (18.2), it reduces to the Cressie and Read power divergence measure. Other important special cases are the ones for which the function $\Phi(u)$ takes the form $\Phi_2(u) = (1 + \lambda)\Phi_{2,\lambda}(u)$ and

$$\Phi_{1,a}(u) = \frac{1}{1+a}\Phi_1(u) = \frac{1}{1+a}\left(u^{1+a} - \left(1 + \frac{1}{a}\right)u^a + \frac{1}{a}\right). \tag{18.6}$$

It is easy to see that for $a \to 0$ the measure $\Phi_2(\cdot)$ reduces to the KL measure. Note that from the statistical point of view, the Cressie and Read (CR) family is very important since as a by-product, a statistic, called power divergence statistic that emerged for GoF purposes for multinomial distributions (Read and Cressie 1988) has received widespread attention.

18.3 Tests of Fit Based on Divergence Measures

It is important to point out that any type of data can be handled by the above tests of fit. To introduce the basic ideas, we discuss first the problem of testing a simple GoF hypothesis. More specifically, assume that $X_1, \ldots, X_n$ are i.i.d. random variables with common distribution function (d.f.) F. Given some specified d.f. F_0, the classical GoF problem is concerned with testing $H_0 : F = F_0$. The above problem is frequently treated by partitioning the range of data into m disjoint intervals and by testing the hypothesis

$$H_0 : \mathbf{p} = \mathbf{p}_0 \tag{18.7}$$

about the vector of parameters of a multinomial distribution.

Let $P = \{E_i\}_{i=1,\ldots,m}$ be a partition of the real line $\mathbb{R}$ in m intervals. Let $\mathbf{p} = (p_1, \ldots, p_m)'$ and $\mathbf{p}_0 = (p_{10}, \ldots, p_{m0})'$ be the true and the hypothesized probabilities of the intervals $E_i, i = 1, \ldots, m$, respectively, in such a way that $p_i = P_F(E_i)$, $i = 1, \ldots, m$ and $p_{i0} = P_{F_0}(E_i) = \int_{E_i} dF_0$, $i = 1, \ldots, m$.

Let $Y_1, \ldots, Y_n$ be a random sample from F, let $n_i = \sum_{j=1}^{n} I_{E_i}(Y_j)$, where

$$I_{E_i}(Y_j) = \begin{cases} 1, & \text{if } Y_j \in E_i \\ 0, & \text{otherwise} \end{cases},$$

$\widehat{\mathbf{p}} = (\hat{p}_1, \hat{p}_2, \ldots, \hat{p}_m)'$ with $\hat{p}_i = n_i/n$, $i = 1, \ldots, m$ be the maximum likelihood estimator (MLE) of p_i, the true probability of the E_i interval and $\sum_{i=1}^{m} n_i = n$. For testing the simple null hypothesis $H_0 : \mathbf{p} = \mathbf{p}_0$, the most commonly used test statistics are Pearson's or chi-squared test statistic, given by

$$X^2 = \sum_{i=1}^{m} \frac{(n_i - np_{i0})^2}{np_{i0}} \tag{18.8}$$

and the likelihood ratio test statistic given by

$$G^2 = 2 \sum_{i=1}^{m} n_i \log\left(\frac{n_i}{np_{i0}}\right). \tag{18.9}$$

Both these test statistics are particular cases of the family of power-divergence test statistics (CR test) which has been introduced by Cressie and Read (1984) and is given by

$$I_n^\lambda(\widehat{\mathbf{p}}, \mathbf{p}_0) = \frac{2n}{\lambda(\lambda+1)} \sum_{i=1}^{m} \hat{p}_i \left(\left(\frac{\hat{p}_i}{p_{i0}}\right)^\lambda - 1\right) = 2n \sum_{i=1}^{m} p_{i0} \Phi_{2,\lambda}\left(\frac{\hat{p}_i}{p_{i0}}\right), \quad \lambda \neq -1, 0$$

$$\tag{18.10}$$

where $-\infty < \lambda < \infty$. Particular values of λ in (18.10) correspond to well-known test statistics: Chi-squared test statistic χ^2 ($\lambda = 1$), likelihood ratio test statistic G^2 ($\lambda \to 0$), Freeman–Tukey test statistic (Freeman and Tukey 1959)($\lambda = -1/2$), minimum discrimination information statistic (Gokhale and Kullback 1978)($\lambda \to -1$), modified chi-squared test statistic ($\lambda = -2$) and Cressie–Read test statistic ($\lambda = 2/3$).

Although the power-divergence test statistics yield to an important family of GoF tests, it is possible to consider the more general Csiszar's family of ϕ-divergence test statistics for testing (18.7) which contains (18.10) as a particular case, is based on the discrete form of (18.1) and is defined by

$$I_n^\phi(\widehat{\mathbf{p}}, \mathbf{p}_0) = \frac{2n}{\phi''(1)} \sum_{i=1}^m p_{i0} \phi\left(\frac{\hat{p}_i}{p_{i0}}\right) \tag{18.11}$$

with $\phi(x)$ a twice continuously differentiable function for $x > 0$ such that $\phi''(1) \neq 0$.

Csiszar's family of measures was recently generalized by Mattheou and Karagrigoriou (2010) to the Φ-family of measures given by (18.4). A general test of fit based on this Φ-family of measures was recently proposed by Mattheou and Karagrigoriou (2010):

$$I_n^\Phi(\widehat{\mathbf{p}}, \mathbf{p}_0) = \frac{2nd_a}{\Phi''(1)}, \quad d_a = \sum_{i=1}^m p_{i0}^{1+a} \Phi\left(\frac{\hat{p}_i}{p_{i0}}\right), \quad \Phi \in \Phi^*. \tag{18.12}$$

Cressie and Read (1984) obtained the asymptotic distribution of the power-divergence test statistics $I_n^\Phi(\widehat{\mathbf{p}}, \mathbf{p}_0)$ for $\Phi(u) = \Phi_{2,a}(u)$, Zografos et al. (1990) extended the result to the family $I_n^\Phi(\widehat{\mathbf{p}}, \mathbf{p}_0)$ for $a = 0$ and $\Phi = \phi \in \Phi^*$ and Mattheou and Karagrigoriou (2010) extended the result to cover any function $\Phi \in \Phi^*$:

Theorem 1 *Under the null hypothesis $H_0 : \mathbf{p} = \mathbf{p}_0 = (p_{10}, \ldots, p_{m0})'$, the asymptotic distribution of the Φ-divergence test statistic, $I_n^\Phi(\widehat{\mathbf{p}}, \mathbf{p}_0)$ divided by a constant c, is a chi-square with $m - 1$ degrees of freedom:*

$$\frac{1}{c} I_n^\Phi(\widehat{\mathbf{p}}, \mathbf{p}_0) \xrightarrow[n \to \infty]{L} \chi_{m-1}^2,$$

where $c = 0.5(\min_i p_{i0}^a + \max_i p_{i0}^a)$.

The following two theorems for the CR test statistic and the ϕ-family of test statistics are special cases of the above theorem for $c = 1$, $a = 0$ and for the appropriate forms of the function $\Phi(\cdot)$.

Theorem 2 *Under the null hypothesis $H_0 : \mathbf{p} = \mathbf{p}_0 = (p_{10}, \ldots, p_{m0})'$, the asymptotic distribution of the divergence test statistic $I_n^\lambda(\widehat{\mathbf{p}}, \mathbf{p}_0)$ with $\Phi = \Phi_{2,\lambda}$ given in (18.10), is chi-square with $m - 1$ degrees of freedom:*

$$I_n^\lambda(\widehat{\mathbf{p}}, \mathbf{p}_0) = I_n^{\Phi_{2,\lambda}}(\widehat{\mathbf{p}}, \mathbf{p}_0) \xrightarrow[n \to \infty]{L} \chi_{m-1}^2.$$

Theorem 3 *Under the null hypothesis $H_0 : \mathbf{p} = \mathbf{p}_0 = (p_{10}, \ldots, p_{m0})'$, the asymptotic distribution of the ϕ-divergence test statistic, $I_n^\phi(\widehat{\mathbf{p}}, \mathbf{p}_0)$ given in (18.11), is chi-square with $m - 1$ degrees of freedom:*

$$I_n^\phi(\widehat{\mathbf{p}}, \mathbf{p}_0) \xrightarrow[n \to \infty]{L} \chi_{m-1}^2.$$

Consider the hypothesis

$$H_0 : p_i = p_{i0} \text{ vs. } H_1 : p_i = p_{ib}, \ i = 1, \ldots, m.$$

Suppose the null hypothesis indicates that $p_i = p_{i0}$, $i = 1, 2, \ldots, m$ when in fact it is $p_i = p_{ib}$, $\forall i$. As it is well known, if p_{i0} and p_{ib} are fixed then as n tends to infinity, the power of the test tends to 1. In order to examine the situation when the power is not close to 1, we must make it continually harder for the test as n increases. This can be done by allowing the alternative hypothesis steadily closer to the null hypothesis. As a result, we define a sequence of alternative hypotheses as follows

$$H_{1,n} : p_i = p_{in} = p_{i0} + d_i/\sqrt{n}, \ \forall i, \tag{18.13}$$

which is known as Pitman transition alternative or Pitman (local) alternative or local contiguous alternative to the null hypothesis $H_0 : p_i = p_{i0}$ (Cochran 1952; Lehmann 1959). In vector notation the null hypothesis and the local contiguous alternative hypotheses take the form

$$H_0 : \mathbf{p} = \mathbf{p}_0 \text{ vs. } H_{1,n} : \mathbf{p} = \mathbf{p}_n = \mathbf{p}_0 + \mathbf{d}/\sqrt{n},$$

where $\mathbf{p}_n = (p_{1n}, \ldots, p_{mn})'$ and $\mathbf{d} = (d_1, \ldots, d_m)'$ is a fixed vector such that $\sum_{i=1}^m d_i = 0$. Observe that as n tends to infinity the local contiguous alternative converges to the null hypothesis at the rate $O(n^{-1/2})$.

The following theorem by Mattheou and Karagrigoriou (2010) provides the asymptotic distribution of the Φ-divergence test, under contiguous alternatives.

Theorem 4 *Under the alternative hypothesis given in (18.13), the asymptotic distribution of the Φ-divergence test statistic, $I_n^\Phi(\widehat{\mathbf{p}}, \mathbf{p}_0)$ divided by a constant c, is a non-central chi-square with $m - 1$ degrees of freedom:*

$$\frac{1}{c} I_n^\Phi(\widehat{\mathbf{p}}, \mathbf{p}_0) \xrightarrow[n \to \infty]{L} \chi_{m-1,\delta}^2,$$

where $c = 0.5(\min_i p_{i0}^a + \max_i p_{i0}^a)$ and noncentrality parameter $\delta = \sum_{i=1}^m \frac{d_i^2}{p_{i0}}$.

An alternative testing procedure which is used exclusively for the MRL class of distributions is based on the following measure of departure (Ahmad 1992):

$$\Delta_F = \int\limits_0^\infty \{2x\bar{F}(x) - v(x)\}dF(x),$$

where $v(x) = \int_x^\infty \bar{F}(u)du$ and $\bar{F}(x) = 1 - F(x)$. The problem of exponentiality has also been investigated by Ebrahimi et al. (1992), who proposed a test of fit using the KL divergence and its relation to the entropy measure. Although the above tests have been used for the exponential distribution, other distributions can also been used. The problem of testing the hypothesis that the underlying distribution belongs to the family of logistic distributions has been addressed and examined by Aguirre and Nikulin (1994). Also, very recently, the problem of normality and uniformity was addressed by Vexler and Gurevich (2010) who proposed an empirical likelihood methodology that standardizes the entropy-based tests. Other related GoF tests have been investigated among others, by Bagdonavicius and Nikulin (2002), Huber-Carol and Vonta (2004), Marsh (2006), Menendez et al. (1997) and Zhang (2002).

We close this section with the problem of composite null hypothesis. Although the simple null hypothesis discussed earlier appears frequently in practice, it is common to test the composite null hypothesis that the unknown distribution belongs to a parametric family $\{F_\theta\}_{\theta\in\Theta}$, where Θ is an open subset in R^k. In this case we can again consider a partition of the original sample space with the probabilities of the elements of the partition depending on the unknown k-dimensional parameter θ. Then, the hypothesis can be tested by the hypotheses

$$H_0 : p_i = p_i(\theta_0) \text{ vs. } H_1 : p_i = p_i(\theta),$$

where θ_0 the true value of the parameter under the null model. Pearson encountered this problem in the well-known chi-square test statistic and suggested the use of a consistent estimator for the unknown parameter. He further claimed that the asymptotic distribution of the resulting test statistic, under the null hypothesis, remains a chi-square random variable with $m-1$ degrees of freedom. Later, Fisher (1924) established that the correct distribution does not have $m - 1$ degrees of freedom. In general, the following theorem holds:

Theorem 5 *Under the composite null hypothesis, the asymptotic distribution of the Φ-divergence test statistic, $I_n^\Phi\left(\widehat{\mathbf{p}}, \mathbf{p}_0(\widehat{\theta})\right)$ divided by a constant c, is a chi-square with $m - k - 1$ degrees of freedom:*

$$\frac{1}{c}I_n^\Phi\left(\widehat{\mathbf{p}}, \mathbf{p}_0(\widehat{\theta})\right) \xrightarrow[n\to\infty]{L} \chi^2_{m-k-1},$$

where $c = 0.5(\min_i p_{i0}^a(\widehat{\theta}) + \max_i p_{i0}^a(\widehat{\theta}))$ and $\widehat{\theta}$ a consistent estimator of θ.

18.4 Applications and Simulations

It is important to point out that any type of data can be viewed as multinomial data by dividing the range of data into m categories. In that sense data related to biomedicine, engineering and reliability that usually come from continuous distributions can be transformed into multinomial data and tests of fit based on the above measures can be applied. Some of the most popular of such continuous distributions are the exponential, lognormal, Gamma, Inverse Gaussian, Weibull, Pareto and Positive Stable distributions. For instance, the family of the two-parameter inverse Gaussian distribution (IG2) is one of the basic models for describing positively skewed data which arise in a variety of fields of applied research as cardiology, hydrology, demography, linguistics, employment service, etc. Such examples include the repair times of an airborne communication transceiver (Chhikara and Folks 1977) and quality characteristics (Sim 2003). Recently, Huberman et al. (1998) have demonstrated the appropriateness of the IG2 family for studying the internet traffic and in particular the number of visited pages per user within an internet site. Most applications of IG2 are justified on the fact that the IG2 is the distribution of the first passage time in Brownian motion with positive drift. Furthermore, distributions like the Weibull, the Positive Stable and the Pareto are frequently encountered in survival modeling. The main problem of determining the appropriate distribution is extremely important for reducing the possibility of erroneous inference. In addition, the existence of censoring schemes in survival modeling makes the determination of the proper distribution an extremely challenging problem. Finally, distributions like the exponential, the Gamma, the lognormal and others are very common in lifetime problems.

In reliability theory one encounters the exponential and Weibull distributions and subsequently the IFR and DFR distributions. Such models are considered suitable in the understanding of aging (IFR) or of objects whose reliability properties improve over time (DFR). An important aspect then is to check the validity of a specific model assumption. Some research has been done in this regard. For example, Gail and Ware (1979) studied grouped censored survival data by comparing with a known survival distribution, while Akritas (1988) constructed a Pearson-type GoF measure for one-sample data that allows for random censorship. Here, in order to assess the applicability and performance of the proposed tests we focus on the comparison of the proposed method with classical GoF statistics under the condition of no censoring. More specifically we test

$$H_0 : F(t) = 1 - \exp(-t) \text{ vs. } H_1 : F(t) = 1 - \exp(t^\gamma). \quad (18.14)$$

Observe that the alternative is the Weibull distribution with shape parameter γ. Note that equivalently, we can test the corresponding hazard or survival functions. The comparison is based on the power and the achieved size of the test. Extending Akritas (1988) alternatives we choose $\gamma = 1/(1 + b/\sqrt{n})$, with $b = -4(1)4$, $n = 20$, 50 and 120 and $m = 3$ number of intervals/categories. Observe that for $b = 0$ the data come from the exponential null model so that the values appearing

Table 18.1 Achieved power at the 5% for contiguous Weibull alternatives with index b, $m = 3$ (symmetric) and $n = 20$

b	KL-test	χ^2-test	Mat-test	CR-test	Φ-test
-4	99.99	99.99	99.99	99.99	99.99
-3	94.60	94.64	92.95	94.66	90.81
-2	52.04	52.12	48.23	52.73	46.32
-1	15.01	14.24	13.74	15.14	14.95
0	5.98	4.45	5.51	5.29	7.78
1	10.73	7.26	10.23	8.76	14.43
2	22.87	15.70	22.35	19.62	28.35
3	39.15	28.05	38.59	34.79	45.19
4	56.15	42.92	55.75	51.17	62.46

in the tables refer to the size of the test. For each combination of alternatives we simulate 10,000 samples. Among the members of the Φ-family we choose the one associated with the function Φ_1 given in (18.5) with index $\alpha = 0.05$. The competing tests are the KL-test, the χ^2 test (χ^2-test), the Matusita test (Mat-test) and the Cressie and Read test (CR-test). For the intervals we are using a symmetric trinomial distribution with $p_1 = p_3 = 0.20$ and $p_2 = 0.60$ (Tables 18.1, 18.2, 18.3) as well as an equiprobable trinomial model with $p_1 = 1/3$, $i = 1, 2, 3$ (Tables 18.4, 18.5, 18.6).

In addition, we compare the standard exponential model with parameter equal to 1 against exponential models with parameter $\gamma = 1 + b/\sqrt{n} \neq 1$, namely

$$H_0 : F(t) = 1 - \exp(-t) \text{ vs. } H_1 : F(t) = 1 - \exp(-t\gamma). \qquad (18.15)$$

We pick again the same choices for the parameter b but present the result only for the equiprobable trinomial model (Tables 18.7, 18.8, 18.9). Recall that the null model is obtained for $b = 0$.

The results clearly show that the Φ test achieves a very good level which is always close to the nominal 5% level. In addition, the power of the test is extremely good in all cases examined. Observe that for the case of the hypothesis (18.14) the Φ-test has a behavior very similar to the Matusita test for both symmetric and equiprobable splitting. At the same time the other three tests have a comparable performance although the KL-test is slightly better for the equiprobable case.

We close this section by applying the proposed test to the data set on mileages for 19 military personnel carriers that failed in service (Grubbs 1971) the sample mean of which is found to be equal to 997.95. The data have been analyzed by Ebrahami et al. (1992) who found that the exponential distribution cannot be rejected at the 10% level. Our methodology confirms this result. Indeed, by applying Theorem 1 for the equiprobable trinomial model and with Φ as in (18.5), we see that $c^{-1} \cdot I_n^{\Phi}(\hat{\mathbf{p}}, \mathbf{p}_0) = 0.4585$, which is smaller than the critical point of $\chi^2_{3-1;0.10} = 4.605$. Thus, the exponential distribution cannot be rejected. Note that

Table 18.2 Achieved power at the 5% for contiguous Weibull alternatives with index b, $m = 3$ (symmetric) and $n = 50$

b	KL-test	χ^2-test	Mat-test	CR-test	Φ-test
−4	100.00	99.99	100.00	99.99	100.00
−3	92.39	89.48	92.91	89.48	93.98
−2	47.06	39.93	50.65	39.93	53.20
−1	13.17	10.32	16.03	10.22	17.16
0	4.86	5.25	6.25	4.83	6.25
1	10.23	13.42	10.65	11.89	10.45
2	24.01	29.55	24.12	27.74	23.88
3	43.53	49.65	43.53	47.70	43.36
4	62.05	67.28	62.05	65.99	61.96

Table 18.3 Achieved power at the 5% for contiguous Weibull alternatives with index b, $m = 3$ (symmetric) and $n = 120$

b	KL-test	χ^2-test	Mat-test	CR-test	Φ-test
−4	99.24	98.95	99.25	99.02	99.27
−3	83.25	80.28	83.48	80.74	84.03
−2	41.72	37.15	42.34	37.96	43.41
−1	13.14	10.68	13.76	11.13	14.27
0	5.19	4.94	5.33	5.12	5.23
1	10.47	11.80	10.17	11.86	9.65
2	26.63	29.62	25.90	29.62	24.79
3	49.54	52.89	48.79	52.89	47.62

Table 18.4 Achieved power at the 5% for contiguous Weibull alternatives with index b, $m = 3$ (equiprobable) and $n = 20$

b	KL-test	χ^2-test	Mat-test	CR-test	Φ-test
−4	100.00	100.00	100.00	100.00	100.00
−3	93.94	92.55	94.54	92.55	94.54
−2	39.19	38.43	40.83	38.43	40.83
−1	11.54	10.60	12.41	10.60	12.41
0	5.94	5.40	6.69	5.40	6.69
1	8.19	7.50	9.44	7.50	9.44
2	13.98	12.66	15.36	12.66	15.36
3	21.55	19.48	23.75	19.48	23.75
4	28.92	26.04	31.36	26.04	31.36

the same conclusion is also reached under the symmetric trinomial model with $p_1 = p_3 = 0.2$ and $p_3 = 0.6$, although the test statistic is much larger: $c^{-1}I_n^{\Phi}(\widehat{\mathbf{p}}, \mathbf{p}_0) = 1.9803$. It should be noted that the equiprobable model is extensively used in the literature, especially in small sample problems. In addition note that in such problems, the number of intervals should not be very large (not larger than 5).

Table 18.5 Achieved power at the 5% for contiguous Weibull alternatives with index b, $m = 3$ (equiprobable) and $n = 50$

b	KL-test	χ^2-test	Mat-test	CR-test	Φ-test
-4	98.26	98.19	98.68	98.19	98.68
-3	69.25	69.80	72.17	69.80	72.17
-2	27.37	28.17	30.16	28.17	30.16
-1	8.89	9.01	10.63	9.01	10.63
0	5.30	5.16	6.37	5.16	6.37
1	7.20	7.22	8.64	7.22	8.64
2	14.22	14.13	16.51	14.13	16.51
3	23.83	23.66	27.04	23.66	27.04
4	34.60	33.75	38.80	33.75	38.80

Table 18.6 Achieved power at the 5% for contiguous Weibull alternatives with index b, $m = 3$ (equiprobable) and $n = 120$

b	KL-test	χ^2-test	Mat-test	CR-test	Φ-test
-4	89.09	88.76	89.09	88.76	89.30
-3	58.20	57.82	58.21	57.82	58.84
-2	25.71	25.22	25.38	25.22	26.10
-1	8.68	8.45	8.57	8.45	8.91
0	5.16	4.91	5.16	4.91	5.55
1	8.47	8.24	8.46	8.24	8.89
2	16.04	15.62	16.17	15.62	17.16
3	28.97	28.15	29.36	28.15	30.46
4	43.20	42.42	43.54	42.42	44.99

Table 18.7 Achieved power at the 5% for contiguous exponential alternatives with index b, $m = 3$ (equiprobable) and $n = 20$

b	KL-test	χ^2-test	Mat-test	CR-test	Φ-test
-4	99.99	100.00	99.99	100.00	99.99
-3	87.41	89.67	87.57	89.67	87.57
-2	43.70	45.35	44.53	45.35	44.53
-1	13.39	12.90	14.45	12.90	14.45
0	5.84	5.16	6.64	5.16	6.64
1	10.49	9.53	11.80	9.53	11.80
2	23.53	20.83	25.63	20.83	25.63
3	40.72	36.81	43.92	36.81	43.92
4	60.00	55.54	63.07	55.54	63.07

Remark Although the Monte Carlo experiments contacted in this work (see (18.14) and (18.15)) refer to the classification problem, one can easily address the formal GoF test by considering general alternatives. In such a case, for the determination of proper and legitimate alternatives one may choose to consider

Table 18.8 Achieved power at the 5% for contiguous exponential alternatives with index b, $m = 3$ (equiprobable) and $n = 50$

b	KL-test	χ^2-test	Mat-test	CR-test	Φ-test
−4	92.27	97.80	97.48	97.80	97.48
−3	73.83	76.20	75.71	76.20	75.71
−2	34.66	36.44	37.25	36.44	37.25
−1	10.74	11.13	12.51	11.13	12.51
0	5.17	5.22	6.10	5.22	6.10
1	9.79	9.86	11.67	9.86	11.67
2	22.79	22.27	25.62	22.27	25.62
3	44.06	42.96	48.61	42.96	48.61
4	65.02	63.42	69.17	63.42	69.17

Table 18.9 Achieved power at the 5% for contiguous exponential alternatives with index b, $m = 3$ (equiprobable) and $n = 120$

b	KL-test	χ^2-test	Mat-test	CR-test	Φ-test
−4	93.49	93.47	93.07	93.47	93.20
−3	69.04	68.72	68.09	68.72	68.39
−2	33.64	33.20	33.06	33.20	33.62
−1	10.94	10.69	10.60	10.69	11.22
0	5.23	5.07	5.23	5.07	5.68
1	9.87	9.52	10.02	9.52	10.52
2	25.13	24.30	25.59	24.30	26.82
3	49.38	48.37	49.99	48.37	51.57
4	71.38	70.61	71.77	70.61	73.22

distributions with the same mean, the same skewness or the same kurtosis as the null distribution (see for instance Ebrahami et al. 1992 and Koutrouvelis et al. 2010).

18.5 Conclusions

The aim of this work is the investigation of generalized tests of fit for lifetime distributions which are based on the Φ-divergence class of measures. In particular, we present various test statistics associated with the above testing problem and calculate the size and the power by simulating samples from decreasing and IFR distributions which often appear in engineering or aging systems and reliability theory. For comparative purposes we are using well-known tests like the χ^2 test, the KL test, the Matusita test and the Cressie and Read test.

The results show that the Φ-test as well as the Matusita test perform better than the KL test in most cases and also have the advantage of distinguishing between null and alternative hypothesis when they are very close.

References

Aguirre N, Nikulin M (1994) Chi-squared goodness-of-fit test for the family of logistic distributions. Kybernetika 30:214–222

Ahmad IA (1992) A new test for mean residual life times. Biometrika 79:416–419

Akritas MG (1988) Pearson-type goodness-of-fit tests: the univariate case. J Amer Stat Ass 83:222–230

Ali SM, Silvey SD (1966) A general class of coefficients of divergence of one distribution from another. J R Stat Soc B 28:131–142

Anderson TW, Darling DA (1954) A test of goodness of fit. J Amer Stat Ass 49:765–769

Arndt C (2001) Information Measures. Springer, Berlin

Bagdonavicius V, Nikulin MS (2002) Goodness-of-fit tests for accelerated life models. In: Huber-Carol C, Balakrishnan N, Nikulin MS, Mesbah M (eds) Goodness-of-fit tests and model validity. Birkhauser, Boston, pp 281–297

Basu A, Harris IR, Hjort NL, Jones MC (1998) Robust and efficient estimation by minimizing a density power divergence. Biometrika 85:549–559

Chhikara RS, Folks JL (1977) The inverse Gaussian distribution as a lifetime model. Technometrics 19:461–468

Cochran WG (1952) The χ^2 test of goodness of fit. Ann Math Stat 23:315–345

Cramer H (1928) On the composition of elementary errors. Skand Aktuarietids 11:13–74, 141–180

Cressie N, Read TRC (1984) Multinomial goodness-of-fit tests. J R Stat Soc B 5:440–454

Csiszar I (1963) Eine Informationstheoretische Ungleichung und ihre Anwendung auf den Bewis der Ergodizitat on Markhoffschen Ketten. Publ Math Inst Hung Acad Sc 8:84–108

Ebrahami N, Habibullah M, Soofi ES (1992) Testing exponentiality based on Kullback–Leibler information. J R Stat Soc B 54:739–748

Fisher RA (1924) The conditions under which χ^2 measures the discrepancy between observation and hypothesis. J R Stat Soc B 87:442–450

Freeman MF, Tukey JW (1959) Transformations related to the angular and the squared root. Ann Math Stat 27:601–611

Gail MH, Ware JH (1979) Comparing observed life table data with a known survival curve in the presence of random censorship. Biometrics 35:385–391

Gokhale DV, Kullback S (1978) The information in contingency tables. Marcel Dekker, New York

Grubbs FE (1971) Approximate fiducial bounds on reliability for the two-parameter negative exponential distribution. Technometrics 13:873–876

Huber-Carol C, Vonta F (2004) Frailty models for arbitrarily censored and truncated data. Lifetime Data Anal 10:369–388

Huberman BA, Pirolli PLT, Pitkow JE, Lukose RM (1998) Strong regularities in World Wide Web surfing. Science 280:95–97

Kolmogorov AN (1933) Sulla determinazione empirica di una legge di distribuzione. Giornale dell'Istituto Italiano degli Attuari 4:83–91

Koutrouvelis IA, Canavos GC, Kallioras AG (2010) Cumulant plots for assessing the Gamma distribution. Commun Stat Theory Meth 39:626–641

Kullback S, Leibler R (1951) On information and sufficiency. Ann Math Stat 22:79–86

Lehmann EL (1959) Testing statistical hypothesis. Wiley, New York

Levy P (1925) Calcul des Probabilites. Gauthiers-Villars, Paris

Mahalanobis BC (1936) On the generalized distance in statistics. Proc. Nation. Acad. Sc. (India) 2:49–55

Marsh P (2006) Data driven likelihood ratio tests for goodness-to-fit with estimated parameters. Discussion Papers in Economics, Department of Economics and Related Studies, The University of York, 2006/20

Mattheou K, Karagrigoriou A (2010) A new family of divergence measures for tests of fit. Aust NZ J Stat 52:187–200

Matusita K (1967) On the notion of affinity of several distributions and some of its applications. Ann Inst Stat Math 19:181–192

Menendez ML, Pardo JA, Pardo L, Pardo MC (1997) Asymptotic approximations for the distributions of the (h, φ)-divergence goodness-of-fit statistics: applications to Renyi's statistic. Kybernetes 26:442–452

Neyman J (1937) 'Smooth' test for goodness of fit. Skand Aktuarietidskr 20:150–199

Pardo L (2006) Statistical inference based on divergence measures. Chapman and Hall/CRC, Boca Raton

Pearson K (1900) On the criterion that a given system of deviations from the probable in the case of a correlated system of variables is such that it can be reasonable supposed to have arisen from random sampling. Philosophy Magazine 50:157–172

Read TRC, Cressie N (1988) Goodness-of-fit statistics for discrete multivariate data. Springer, New York

Sim CH (2003) Inverse Gaussian control charts for monitoring process variability. Commun Stat Simul Comput 32:223–239

Vajda I (1989) Theory of statistical inference and information. Kluwer, Dorfrecht

Vajda I (1995) Information-theoretic methods in Statistics. In: Research Report. Acad Sc Czech Rep Inst Inf Theory Automat, Prague

Vexler A, Gurevich G (2010) Empirical likelihood ratios applied to goodness-of-fit tests based on sample entropy. Comput Stat Data Anal 54:531–545

von Mises R (1931) Wahrscheinlichkeitsrechnung und ihre anwendung in der statistik und theoretischen physik. Deuticke, Leipzig

Watson GS (1961) Goodness-of-fit tests on the circle. Biometrika 48:109–114

Zhang J (2002) Powerful goodness-of-fit tests based on likelihood ratio. J R Stat Soc B 64: 281–294

Zografos K, Ferentinos K, Papaioannou T (1990) Φ-divergence statistics: Sampling properties, multinomial goodness of fit and divergence tests. Commun Stat Theory Meth 19:1785–1802

Chapter 19
On the Markov Three-State Progressive Model

Jacobo de Uña-Álvarez

Abstract In this work we revisit the Markov three-state progressive model. Several characterizations of the Markov condition are discussed, and nonparametric estimators of important targets such as the bivariate distibution of the event times or the transition probabilities are motivated. Three points of special interest are considered: (1) the relative improvements when introducing the Markov condition in the construction of the estimators; (2) bootstrap algorithms to resample under markovianity; and (3) goodness-of-fit testing for the Markov assumption. Simulation studies and some technical derivations are included.

19.1 Introduction

A multi-state model is a model for a continuous-time stochastic process allowing individuals (or units) to move among a finite number of states. In biomedical applications, the states might be based on clinical symptoms (e.g. bleeding episodes), biological markers (e.g. CD4 T-lymphocyte cell counts), some scale of the disease (e.g. stages of cancer or VIH infection), or a non-fatal complication in the course of the illness (e.g. cancer recurrence). In reliability studies, the states may indicate the status of devices of a complex system, such as deterioration levels or types of failures. The simplest form of a multi-state model is the mortality model, with states 'alive' and 'dead' and one possible transition. By splitting the 'alive' state into two transient states which are visited in a successive way

J. de Uña-Álvarez (✉)
Department of Statistics and OR, Facultad de CC Económicas y Empresariales,
Universidad de Vigo, Campus Lagoas-Marcosende, 36310 Vigo, Spain
e-mail: jacobo@uvigo.es

A. Lisnianski and I. Frenkel (eds.), *Recent Advances in System Reliability*,
Springer Series in Reliability Engineering, DOI: 10.1007/978-1-4471-2207-4_19,
© Springer-Verlag London Limited 2012

(with no chance of coming back) we obtain the three-state progressive model: State 1→State 2→State 3. This model is convenient when there exists an intermediate event (e.g. a recurrence, or the deterioration of an electronic device) which may influence the survival prognosis. See Datta et al. (2000), Hougaard (2000), Andersen and Keiding (2002), or Meira-Machado et al. (2009) for an introduction to this area.

A typical situation in which a k-state progressive model is useful is when analyzing recurrent event data, which arise when each individual may go through a well-defined event several times along his history. Then, the inter-event times are referred to as the gap times, and they are of course determined by the times at which the recurrences take place (i.e. the recurrence times). See Cook and Lawless (2007) for an up-to-date account of statistical methods for recurrent event data. In the three-state progressive model, the interest is often focused on a given couple of (successive) gap times (T_1, T_2), and the model may be represented by a pair (Z, T), where $Z = T_1$ is the sojourn time in State 1 (or the transition time from State 1 to State 2), while $T = T_1 + T_2$ is the transition time to State 3 (i.e. the survival time of the process). Of course, we always have $Z \leq T$. By assuming $P(T_2 = 0) = 0$ (which holds in particular under continuity), we have indeed $Z < T$ with probability 1.

In this paper we revisit the three-state progressive model. Specifically, we focus on the Markov three-state progressive model, which is suitable when, conditionally on the present state, the future evolution of the process is independent of the past. Some representations of the Markov condition in this simple model are given, and nonparametric estimation of important targets such as the conditional distribution function (df) of T given Z, the joint df of (Z, T), or the transition probabilities is discussed. Finally we come up with the time-honored Aalen–Johansen estimator (Aalen and Johansen 1978), but on the way we throw some light on three points that have not been addressed in much detail before:

1. How much do we gain by assuming that the process is Markovian?
2. How to incorporate the Markov condition in the bootstraping of a multi-state model?
3. How can the Markov condition be efficiently tested?

Point (1) has been considered by Meira-Machado et al. (2006) in the scope of the illness–death (or disability) progressive model. These authors compared by simulations the Aalen–Johansen estimators for the transition probabilities to some non-Markovian counterparts (which reduce to sampling proportions in the uncensored case; note that in general this is not true for the Aalen–Johansen estimator, for which the number of individuals at risk may be influenced by left-truncation). Here we rather focus on the comparison of Markovian and non-Markovian estimators for the joint df of (Z, T). To the best of our knowledge, there is only a small literature on issue (2), see e.g. Datta and McCormick (1993, 1995), Fuh and Ip (2005), or Bertail and Clémençon (2006), all of which are confined to discrete-time processes (i.e. Markov chains). Finally, there exist some methods for answering question (3), the most popular being the proportional hazards approach, see Sect. 19.3 below; however, there exist other testing approaches which enjoy more flexibility and it is interesting to review them here.

The rest of the paper is organized as follows. In Sect. 19.2 we introduce the required notations for the three-state progressive model and some basic characterizations of markovianity. Nonparametric estimation and bootstrap algorithms are considered, including the possibility of right censoring. In Sect. 19.3 we discuss the problem of testing if a given three-state progressive model is Markov. Some simulation results are provided in Sect. 19.4, while Sect. 19.5 reports a final discussion. All the technical derivations are deferred to the Appendix.

19.2 Nonparametric Estimation

Markov condition claims that, given the present, the future evolution of the process is independent of the past. For a three-state progressive model, there exists a simple representation of this condition. Put $\lambda(t|s)$ for the hazard rate function of T evaluated at time t, conditionally on $Z = s$. Then, the Markov condition is satisfied if and only if $\lambda(t|s)$ is free of s, $t \geq s$, for each s (e.g. Hougaard 2000). As a consequence, we may introduce a function ψ through

$$\psi(t) = \lambda(t|s), \quad t \geq s,$$

which does not depend on the particular s value.

Also, the conditional df of T given $Z = s$ is given by

$$F_{T|Z}(t|s) = 1 - \exp\left[-\int_s^t \lambda(y|s)\mathrm{d}y\right] = 1 - \frac{\overline{S}(t)}{\overline{S}(s)},$$

where $\overline{S}(t) = \exp[-\Psi(t)]$ and $\Psi(t) = \int_0^t \psi(y)\mathrm{d}y$. Since $\overline{S}(0) = 1$, we have $F_{T|Z}(t|0) = 1 - \overline{S}(t)$, i.e. $\overline{S}$ is just the survival function of T for those individuals starting in State 2 at time zero. Thus, under the Markov condition, the joint df of (Z, T) is written as

$$F_{Z,T}(z, t) = \int_0^z F_{T|Z}(t|x)F_Z(\mathrm{d}x) = \int_0^z \left(1 - \frac{\overline{S}(t)}{\overline{S}(x)}\right)F_Z(\mathrm{d}x),$$

where F_Z stands for the df of Z. This expression is a useful characterization of markovianity; in particular, it can be used to introduce goodness-of-fit tests for markovianity (see Sect. 19.3).

From the above expression for $F_{Z,T}$, the following equality is obtained:

$$\Psi(t) = \int_0^t \frac{F_T(\mathrm{d}y)}{M(y)},$$

where F_T is the marginal df of T and $M(t) = P(Z < t \leq T)$; see Lemma 1 in the Appendix for a proof. Hence, $\psi(t)$ represents the proportion of individuals entering State 3 at time t among those being in State 2 just prior to time t. Below, this equality is used to introduce suitable estimators which incorporate the Markov condition. Note that all the above expressions are valid under Markovianity, but they do not hold for a general three-state progressive model.

In practice, the observation of the process is typically influenced by a censoring risk. This means that, because of follow-up limitations, lost cases and so on, rather than (Z, T) one observes $\left(\widetilde{Z}, \widetilde{T}, \Delta_1, \Delta\right)$ where $\widetilde{Z} = \min(Z, C)$, $\widetilde{T} = \min(T, C)$, $\Delta_1 = I(Z \leq C)$, and $\Delta = I(T \leq C)$. Here C denotes the potential censoring time, which we assume to be independent of the process (that is, C and (Z, T) are assumed to be independent). We put G for the df of C. Note that Δ_1 and Δ are the censoring indicators pertaining to Z and T, respectively. Three different situations are possible: $(\Delta_1, \Delta) = (1, 1)$, meaning that both Z and T are uncensored (i.e. the individual is observed to reach State 3); $(\Delta_1, \Delta) = (1, 0)$, meaning that Z is uncensored but T is censored (individual censored while being in State 2); and $(\Delta_1, \Delta) = (0, 0)$, when individual is never observed to leave State 1. The sampling information is represented by $\left(\widetilde{Z}_i, \widetilde{T}_i, \Delta_{1i}, \Delta_i\right)$, $i = 1, \ldots, n$, iid copies of $\left(\widetilde{Z}, \widetilde{T}, \Delta_1, \Delta\right)$.

With these data at hand, the marginal df's of Z and T can be estimated (when needed) from the Kaplan–Meier estimator based on the $\left(\widetilde{Z}_i, \Delta_{1i}\right)$'s and the $\left(\widetilde{T}_i, \Delta_i\right)$'s, respectively, while G can be estimated by the Kaplan–Meier estimator based on the $\left(\widetilde{T}_i, 1 - \Delta_i\right)$'s. Estimation of other targets such as the joint df of (Z, T) is not so easy. This problem is related to the estimation of the bivariate df of censored gap times, which has received much attention in the literature; see de Uña-Álvarez and Amorim (2011) and the references therein. All these papers deal with estimation procedures which do not depend on the Markov assumption. Here, we focus on nonparametric estimators which make use of the Markov condition.

For estimation purposes, note that (from the independence between C and (Z, T))

$$\Psi(t) = \int_0^t \frac{F_T(\mathrm{d}y)}{M(y)} = \int_0^t \frac{H^1(\mathrm{d}y)}{\widetilde{M}(y)},$$

where we put $H^1(y) = P(\widetilde{T} \leq y, \Delta = 1)$ and $\widetilde{M}(y) = P(\widetilde{Z} < y \leq \widetilde{T})$; hence, $\Psi(t)$ may be estimated by

$$\Psi_n(t) = \int_0^t \frac{H_n^1(\mathrm{d}y)}{\widetilde{M}_n(y)},$$

where $H_n^1(y) = \frac{1}{n} \sum_{i=1}^n I(\widetilde{T}_i \leq y, \Delta_i = 1)$ and $\widetilde{M}_n(y) = \frac{1}{n} \sum_{i=1}^n I(\widetilde{Z}_i < y \leq \widetilde{T}_i)$.

The resulting estimator for $\overline{S}(t)$ is

$$\overline{S}_n(t) = \prod_{\widetilde{T}_i \leq t} \left[1 - \frac{\Delta_i}{n\widetilde{M}_n(\widetilde{T}_i)} \right],$$

which has the same structure of the product-limit estimator under independent left-truncation, see e.g. Woodroofe (1985). The function $\overline{S}(t)$ determines the transition probabilities from State 2, these are $p_{22}(s,t)$ and $p_{23}(s,t) = 1 - p_{22}(s,t)$, where

$$p_{22}(s,t) = P(T > t | Z \leq s < T) = P(T > t | Z = s) = \frac{\overline{S}(t)}{\overline{S}(s)}.$$

Hence, $\overline{S}_n(t)$ may be plugged-in to obtain the estimator

$$\widehat{p}_{22}(s,t) = \frac{\overline{S}_n(t)}{\overline{S}_n(s)} = \prod_{s < \widetilde{T}_i \leq t} \left[1 - \frac{\Delta_i}{n\widetilde{M}_n(\widetilde{T}_i)} \right],$$

which is just the Aalen–Johansen estimator (Aalen and Johansen 1978). From our derivations, it becomes clear that the conditional df of T given $Z = s$ and the joint df of (Z,T) may be estimated through

$$\widehat{F}_{T|Z}(t|s) = 1 - \frac{\overline{S}_n(t)}{\overline{S}_n(s)} = 1 - \prod_{s < \widetilde{T}_i \leq t} \left[1 - \frac{\Delta_i}{n\widetilde{M}_n(\widetilde{T}_i)} \right]$$

and

$$\widehat{F}_{Z,T}(z,t) = \int_0^z \left(1 - \frac{\overline{S}_n(t)}{\overline{S}_n(x)} \right) \widehat{F}_Z(dx)$$

respectively, where $\widehat{F}_Z$ stands for the Kaplan–Meier estimator of F_Z. All these nonparametric estimators are Markovian, i.e. they are only valid under the Markov assumption.

For some applications it is necessary to perform bootstrap resampling from the $(\widetilde{Z}_i, \widetilde{T}_i, \Delta_{1i}, \Delta_i)$'s under the Markov condition. One application of such a bootstrap would be the approximation of the null distribution of a test statistic for testing the Markov condition. Another application is the estimation of standard error of complicated estimators. An algorithm for the bootstrapping of the Markov three-state progressive model (Z,T) would be as follows.

Step 1. Draw Z_i^* from the Kaplan–Meier estimator of F_Z, $\widehat{F}_Z$.

Step 2. Given Z_i^*, draw T_i^* from $\widehat{F}_{T|Z}(t|Z_i^*)$.

Step 3. Draw independently C_i^* from the Kaplan–Meier estimator of G, say $\widehat{G}$.

Step 4. Compute $\widetilde{Z}_i^* = \min(Z_i^*, C_i^*)$, $\widetilde{T}_i^* = \min(T_i^*, C_i^*)$, $\Delta_{1i}^* = I(Z_i^* \leq C_i^*)$, and $\Delta_i^* = I(T_i^* \leq C_i^*)$.

Repeat Steps 1–4 until a bootstrap resample of n data is formed. The aim of the bootstrap algorithm is to reproduce the Markov condition, and this explains Step 2. Note that a different scheme is obtained if we replace Step 2 by $T_i^* \sim \widehat{P}(T > \cdot \,| T \geq Z_i^*)$; in such a case, we are imposing relationship $F_{T|Z}(t|s) = 1 - S_T(t)/S_T(s)$ (here $S_T = 1 - F_T$ stands for the survival function of T), which is true only when T is distributed as $T|Z = 0$.

19.3 Testing the Markov Condition

There is some literature on testing that a three-state progressive model is Markov. Jones and Crowley (1992) proposed a class of test statistics under the proportional hazards assumption $\lambda(t|s) = \lambda_0(t)e^{\beta s}$, and studied their asymptotic properties. These authors recommend the Cox partial likelihood score (log-rank) test as the best one inside the proposed family. Chang et al. (2001) introduced a test based on the differences of Nelson–Aalen estimators for data stratified by the first gap time. Simulations reported by these authors suggest that their test is preferable to the log-rank test when hazards are not proportional. This is not surprising, since the proportional hazards assumption may mislead the inference when it is violated. More evidence on this was recently provided in Rodríguez-Girondo and de Uña-Álvarez (2010). These authors proposed a new method for testing Markovianity based on a local Kendall's Tau τ_t, which measures the association between future and past given the present $\{Z < t \leq T\}$. The inspection of the p-values of τ_t along time t was then used to propose a significance trace as a graphical device to assess Markovianity. In this section we provide some mathematical derivations to explain why the test statistic in Chang et al. (2001) works. At the same time, we suggest a different test statistic based on a comparison between the Markovian and non-markovian nonparametric estimators of $F_{Z,T}$.

Under the Markov assumption it holds that

$$E\left[\frac{I(z < Z, T \leq t)}{M_z(T)}\right] = E\left[\frac{I(z < T \leq t)}{M(T)}\right], \quad z < t,$$

where $M_z(y) = P(z < Z < y \leq T)$. Note that, with this notation, $M(y) = M_0(y)$. A formal proof of this equality is given in the Appendix, Lemma 2. This allows for the construction of a test statistic for the Markov condition. Introduce the function

$$A(t, z) = E\left[\frac{I(z < Z, T \leq t)}{M_z(T)}\right],$$

so equality above is just $A(t, z) = A(t, 0) - A(z, 0)$. Introduce the counterpart of $A(t, z)$ under censoring,

$$\widetilde{A}(t, z) = E\left[\frac{I\left(z < \widetilde{Z}, \widetilde{T} \le t\right)\Delta}{\widetilde{M}_z(\widetilde{T})}\right],$$

where $\widetilde{M}_z(y) = P\left(z < \widetilde{Z} < y \le \widetilde{T}\right) = M_z(y)P(C \ge y)$. By conditioning on (Z, T) it is easily seen that $\widetilde{A}(t, z) = A(t, z)$, which in turn leads to the natural estimator

$$A_n(t, z) = \frac{1}{n}\sum_{i=1}^{n}\frac{I\left(z < \widetilde{Z}_i, \widetilde{T}_i \le t\right)\Delta_i}{\widetilde{M}_{z,n}(\widetilde{T}_i)},$$

where

$$\widetilde{M}_{z,n}(y) = \frac{1}{n}\sum_{i=1}^{n}I\left(z < \widetilde{Z}_i < y \le \widetilde{T}_i\right).$$

The test statistic in Chang et al. (2001) is based on the empirical process $W_n^A(z, t) = \sqrt{n}[A_n(t, z) - A_n(t, 0) + A_n(z, 0)]$. More specifically, the Markov assumption is rejected when $\sup\left|W_n^A(z, t)\right|$ is large, where the sup is taken over special subsets of the plane.

Alternatively, one may consider the process

$$W_n^F(z, t) = \sqrt{n}\left[\widehat{F}_{Z,T}(z, t) - \widehat{F}_{Z,T}^0(z, t)\right],$$

where $\widehat{F}_{Z,T}^0$ stands for a non-Markovian estimator of $F_{Z,T}$. For example, $\widehat{F}_{Z,T}^0(z, t)$ could be defined as an average of the indicators $I\left(\widetilde{Z}_i \le z, \widetilde{T}_i \le t\right)$ randomly weighted by the Kaplan–Meier estimator of F_T (e.g. de Uña-Álvarez and Amorim 2011), which is an extension of the ordinary empirical df to the censored setup. To the best of our knowledge, investigation of W_n^F is missing so far, but it is a promising way of constructing an omnibus test for Markovianity. We are presently investigating the theoretical properties of the empirical process W_n^F, as well as the performance of the above-mentioned bootstrap in the approximation of its null distribution; results will be provided elsewhere.

There exists an interesting connection between the Markov condition for (Z, T) and the quasi-independence assumption in a left-truncated scenario. For seeing this, let U and V be the time up to events E_U and E_V, respectively. Assume that some individuals experience E_U before E_V (i.e. $U \le V$), and that others do not. Moreover, assume that (U, V) is observed only when $U \le V$, so the variable V is left-truncated by U. For example, event E_V could be 'death' and event E_U could be 'relapse'; the mentioned situation occurs when individuals dying without having a relapse are discarded. Still, one is interested in estimating the distribution of V for the total population, and hence some correction for left-truncation is needed. Such a suitable correction is given by the Lynden–Bell product-limit estimator, see e.g. Woodroofe (1985), provided that (U, V) satisfies the quasi-independence assumption

$$H_0^Q : P(U \leq u, V > v | U \leq V) = cP(\overline{U} \leq u)P(\overline{V} > v), \quad u \leq v,$$

for some random pair $(\overline{U}, \overline{V})$ and some constant c (Tsai 1990). Checking this assumption and testing for the markovianity of the conditional process (U, V) given $U \leq V$ (which is well-represented by a three-state progressive model) is the very same issue (see Lemma 3 in the Appendix for a proof). We only mention that some methods for testing for H_0^Q have been proposed (Emura and Wang 2010 and the references therein).

19.4 Simulation Study

In this section we perform a simulation study to evaluate the relative performance of the Markovian and non-Markovian estimators of $F_{Z,T}$: $\widehat{F}_{Z,T}$ and $\widehat{F}_{Z,T}^0$. For simplicity, we only consider the uncensored case, where a random sample (Z_i, T_i), $i = 1, \ldots, n$ of (Z, T) is available. The Markovian estimator is then given by

$$\widehat{F}_{Z,T}(z, t) = \frac{1}{n} \sum_{i=1}^{n} I(Z_i \leq z) \left(1 - \frac{\overline{S}_n(t)}{\overline{S}_n(Z_i)} \right),$$

where $\overline{S}_n(t) = \prod_{T_i \leq t} \left[1 - \frac{1}{nM_n(T_i)} \right]$ and $M_n(y) = \frac{1}{n} \sum_{i=1}^{n} I(Z_i < y \leq T_i)$.

The non-Markovian estimator is just the ordinary empirical df of the (Z_i, T_i)'s, namely

$$\widehat{F}_{Z,T}^0(z, t) = \frac{1}{n} \sum_{i=1}^{n} I(Z_i \leq z, T_i \leq t).$$

Two simulated scenarios are considered. The first scenario fits the Markov condition, while the second violates the markovianity. Of course, it is expected that $\widehat{F}_{Z,T}(z, t)$ performs better than $\widehat{F}_{Z,T}^0(z, t)$ in the first scenario, while the contrary could be expected in the second scenario due to the systematic bias of $\widehat{F}_{Z,T}(z, t)$. Both scenarios correspond to the specification

$$\log(T - Z) = f(Z) + \varepsilon,$$

where the error term ε is independent of Z. The resulting conditional hazard of T given $Z = s$ is given by

$$\lambda(t|s) = \lambda_0((t - s)e^{-f(s)})e^{-f(s)},$$

Table 19.1 Bias, standard deviation, and MSE of the Markovian and non-markovian estimators for Scenario 1

n	(z, t)	Bias($\hat{F}$)	Bias($\hat{F}^0$)	sd($\hat{F}$)	sd($\hat{F}^0$)	MSE($\hat{F}$)	MSE($\hat{F}^0$)
50	(0.75, 1)	−0.000294	−0.000109	0.06481	0.06772	0.00420	0.00459
	(0.5, 1)	−0.000148	−0.000169	0.05617	0.06267	0.00315	0.00393
	(0.25, 1)	0.000115	−0.000179	0.04259	0.04932	0.00181	0.00243
	(0.5, 1.25)	0.000040	0.000137	0.05807	0.06557	0.00337	0.00430
	(0.5, 3)	0.000174	0.000238	0.06793	0.07032	0.00461	0.00494
250	(0.75, 1)	0.000322	0.000157	0.02877	0.03013	0.00083	0.00091
	(0.5, 1)	0.000245	0.000234	0.02460	0.02763	0.00061	0.00076
	(0.25, 1)	0.000262	0.000399	0.01879	0.02246	0.00035	0.00050
	(0.5, 1.25)	0.000428	0.000254	0.02541	0.02924	0.00065	0.00086
	(0.5, 3)	0.000433	0.000507	0.03008	0.03124	0.00090	0.00098

Averages along 10,000 Monte Carlo trials

Table 19.2 Bias, standard deviation, and MSE of the Markovian and non-markovian estimators for Scenario 2

n	(z, t)	Bias($\hat{F}$)	Bias($\hat{F}^0$)	sd($\hat{F}$)	sd($\hat{F}^0$)	MSE($\hat{F}$)	MSE($\hat{F}^0$)
50	(0.75, 1)	−0.003687	0.001147	0.06091	0.06295	0.00372	0.00396
	(0.5, 1)	−0.011916	0.000471	0.05277	0.05893	0.00293	0.00347
	(0.25, 1)	−0.013322	0.000256	0.03917	0.04791	0.00171	0.00230
	(0.5, 1.25)	−0.020459	0.000646	0.05509	0.06266	0.00345	0.00393
	(0.5, 3)	−0.031609	0.000701	0.06652	0.07009	0.00542	0.00491
250	(0.75, 1)	−0.005286	−0.000194	0.02712	0.02815	0.00076	0.00079
	(0.5, 1)	−0.013532	−0.000160	0.02356	0.02621	0.00074	0.00069
	(0.25, 1)	−0.014265	−0.000187	0.01758	0.02149	0.00051	0.00046
	(0.5, 1.25)	−0.022455	−0.000318	0.02457	0.02799	0.00111	0.00078
	(0.5, 3)	−0.034161	−0.000925	0.02965	0.03122	0.00205	0.00098

Averages along 10,000 Monte Carlo trials

where the baseline hazard λ_0 is the hazard function of $W = \exp(\varepsilon)$. The Z is uniformly distributed on the interval $[0, 1]$.

Scenario 1 (Markov). $\lambda_0(.) \equiv 1$ (i.e. $W \sim \text{Exp}(1)$) and $f(.) \equiv 0$
Scenario 2 (non-Markov). $\lambda_0 \equiv 1$ (i.e. $W \sim \text{Exp}(1)$) and $f(s) = s$

Note that in Scenario 1 the second gap time $T_2 = T - Z$ is distributed as $\text{Exp}(1)$, being independent of the first gap time $T_1 = Z$; while in Scenario 2 the conditional distribution of T_2 given $T_1 = s$ is $\text{Exp}(e^{-s})$.

In Table 19.1 (for Scenario 1) and Table 19.2 (Scenario 2) we report the bias, the standard deviation (sd), and the mean square error (MSE) of the two estimators $\hat{F}_{Z,T}(z, t)$ and $\hat{F}^0_{Z,T}(z, t)$ for several pairs (z, t). The values $F_{Z,T}(z, t)$ being estimated are given separately in Table 19.3. The Markovian estimator has a smaller MSE in Scenario 1, and its bias is negligible (recall that the ordinary empirical df is perfeclty unbiased). In Scenario 2 (which violates the Markov condition) it is

Table 19.3 Values of $F_{Z,T}(z,t)$ for the two simulated scenarios

(z, t)	Scenario 1	Scenario 2
(0.75, 1)	0.339079	0.267825
(0.5, 1)	0.261349	0.221973
(0.25, 1)	0.145513	0.134438
(0.5, 1.25)	0.314138	0.270280
(0.5, 3)	0.467702	0.438371

seen that the bias of the Markovian estimator increases with the length of the interval (z, t), and that for $n = 250$ the systematic bias of $\widehat{F}_{Z,T}(z,t)$ results in an MSE larger than that of the ordinary empirical df. In particular, for $(z,t) = (0.5, 3)$ the MSE of the Markovian estimator is more than twice the MSE of the non-Markovian estimator

19.5 Discussion

In this work the Markov three-state progressive model has been revisited. Different characterizations of the model have been discussed. In particular, a closed-form expression for the bivariate df of the two successive event times has been provided. This expression and the several characterizations have been used to discuss possible goodness-of-fit test statistics for the Markov condition. Besides, a method for bootstrapping a continuous-time Markov three-state model has been introduced. Finally, a simulation study has been conducted to illustrate the relative benefits of including the Markov condition in the construction of estimators.

One of the main motivations for testing the Markov assumption is the fact that the time-honored Aalen–Johansen estimator of a transition matrix may be inconsistent when Markovianity is violated. Interestingly, it has been quoted that particular transition probabilities such as the occupation probabilities are consistently estimated by the Aalen–Johansen estimator even when the Markov condition fails (Aalen et al. 2001; Datta and Satten 2001). Besides, there exist nonparametric estimators of transition probabilities or gap times distributions not relying on the Markov condition [Meira-Machado et al. (2006), de Uña-Álvarez and Meira-Machado (2008) and references therein]. Non-Markovian estimators of other targets such as the state entry and exit time distributions and the state waiting time distributions have been also introduced (Datta and Ferguson 2011). Still, since by using the Markov condition one may construct estimators with improved variance, in practice it is interesting to formally test for this simplifying assumption.

Another interesting point is whether it is possible to extend the existing testing methods for the Markov condition to more complex multi-state models. The answer to this question greatly depends on the model complexity. For the illness–death progressive model, in which individuals may directly go from State 1 to State 3 without visiting the intermediate State 2, characterizations of Markovianity similar to those discussed here can be given. Indeed, testing methods can be applied exactly as discussed by using the subsample of individuals passing through State 2. We cannot give such a statement for other, more involved multi-state models, since each particular model needs to be addressed in a special way. This is without a doubt an exciting research area for the future.

Acknowledgments This work is supported by the Grants MTM2008-03129 of the Spanish Ministerio de Ciencia e Innovación and 10PXIB300068PR of the Xunta de Galicia. Financial support from the INBIOMED project (DXPCTSUG, Ref. 2009/063) is also acknowledged.

Appendix

In this Appendix we give the technical derivations of three results that have been referred in Sect. 19.2 and Sect. 19.3.

Lemma 1 Assume that (Z, T) is Markov. Then, we have for each t.

$$\Psi(t) = \int_0^t \frac{F_T(\mathrm{d}y)}{M(y)}.$$

Proof From the bivariate df of (Z, T) we get that the marginal df of T satisfies

$$F_T(t) = F_{Z,T}(t, t) = F_Z(t) - \overline{S}(t) \int_0^t \frac{F_Z(\mathrm{d}y)}{\overline{S}(y)}.$$

Since $\overline{S}'(t) = -\psi(t)\overline{S}(t)$ we have

$$F_T(\mathrm{d}t) = \psi(t)\overline{S}(t) \int_0^t \frac{F_Z(\mathrm{d}y)}{\overline{S}(y)} \mathrm{d}t.$$

We also have

$$M(y) = \iint_{\{u < y \le v\}} F_{Z,T}(\mathrm{d}u, \mathrm{d}v) = \iint_{\{u < y \le v\}} \psi(v) \frac{\overline{S}(v)}{\overline{S}(u)} F_Z(\mathrm{d}u) \mathrm{d}v$$

$$= \int_y^\infty \psi(v)\overline{S}(v)\mathrm{d}v \int_0^y \frac{F_Z(\mathrm{d}u)}{\overline{S}(u)}.$$

Hence,

$$\int_0^t \frac{F_T(\mathrm{d}y)}{M(y)} = \int_0^t \frac{\psi(y)\overline{S}(y)\mathrm{d}y}{\int_y^\infty \psi(v)\overline{S}(v)\mathrm{d}v} = -\log \overline{S}(t),$$

where for the last equality we have used $\overline{S}(t) = \int_t^\infty \psi\overline{S}$. Since $\Psi(t) = -\log \overline{S}(t)$, this completes the proof.

Lemma 2 Assume that (Z, T) is Markov. Then, for each $z < t$ we have $A(t, z) = A(t, 0) - A(z, 0)$. Conversely, if (Z, T) is not Markov, there exists a pair (t, z) for which $A(t, z) \neq A(t, 0) - A(z, 0)$.

Proof Write

$$
A(t, z) = \iint_{\{z < x < y \leq t\}} \frac{F_{Z,T}(dx, dy)}{M_z(y)} = \iint_{\{z < x < y \leq t\}} \frac{\psi(y)}{M_z(y)} \frac{\overline{S}(y)}{\overline{S}(x)} F_Z(dx) dy
$$

$$
= \int_z^t \frac{\psi(y) \overline{S}(y)}{M_z(y)} \left[\int_z^y \frac{F_Z(dx)}{\overline{S}(x)} \right] dy.
$$

Now,

$$
M_z(y) = \iint_{\{z < u < y \leq v\}} F_{Z,T}(du, dv) = \iint_{\{z < u < y \leq v\}} \psi(v) \frac{\overline{S}(v)}{\overline{S}(u)} F_Z(du) dv
$$

$$
= \int_y^\infty \psi(v) \overline{S}(v) dv \int_z^y \frac{F_Z(du)}{\overline{S}(u)}.
$$

Summarizing, we have

$$
A(t, z) = \int_z^t \frac{\psi(y) \overline{S}(y)}{\int_y^\infty \psi(v) \overline{S}(v) dv} dy.
$$

From this, equation $A(t, z) = A(t, 0) - A(z, 0)$ is immediately satisfied. The second assertion can be checked by tracing the above steps back.

Lemma 3 H_0^Q holds if and only if (U, V) conditionally on $U \leq V$ is Markov.

Proof Assume that $(U, V) | U \leq V$ is Markov. Then, the conditional df of (U, V) given $U \leq V$, say $F_{U,V|U \leq V}$, satisfies for $u \leq v$

$$
F_{U,V|U \leq V}(u, v) = \int_0^u \left(1 - \frac{\overline{S}(v)}{\overline{S}(x)} \right) F_{U|U \leq V}(dx),
$$

where $\overline{S}(t) = \exp[-\int_0^t \psi]$ stands for the survival function of $V | U = 0, U \leq V$, and $U | U \leq V \sim F_{U|U \leq V}$. Then,

$$
P(U \leq u, V > v | U \leq V) = \iint_{\{x \leq u, y > v\}} F_{U,V|U \leq V}(dx, dy)
$$

$$
= \iint_{\{x \leq u, y > v\}} \psi(y) \frac{\overline{S}(y)}{\overline{S}(x)} F_{U|U \leq V}(dx) dy
$$

$$
= \int_0^u \frac{F_{U|U \leq V}(dx)}{\overline{S}(x)} \int_v^\infty \psi(y) \overline{S}(y) dy,
$$

and hence H_0^Q is satisfied. The complementary assertion is proved by tracing the above steps back.

References

Aalen O, Johansen S (1978) An empirical transition matrix for nonhomogeneous Markov chains based on censored observations. Scand J Stat 5:141–150

Aalen O, Borgan O, Fekjaer H (2001) Covariate adjustment of event histories estimated from Markov chains: the additive approach. Biometrics 57(4):993–1001

Andersen PK, Keiding N (2002) Multi-state models for event history analysis. Stat Methods Med Res 11(2):911–915

Bertail P, Clémençon S (2006) Regenerative block bootstrap for Markov chains. Bernoulli 12(4):689–712

Chang IS, Chuang YC, Hsiung CA (2001) Goodness-of-fit tests for semi-Markov and Markov survival models with one intermediate state. Scand J Stat 28(3):505–525

Cook RJ, Lawless JF (2007) The statistical analysis of recurrent events. Springer, New York

Datta S, Ferguson AN (2011) Nonparametric estimation of marginal temporal functionals in a multistate model. In: Lisnianski A, Frenkel I (eds) Recent advances in system reliability: signatures, multi-state systems and statistical inference. Springer, London (to appear)

Datta S, McCormick W (1993) Regeneration-based bootstrap for Markov chains. Canad J Statist 21(2):181–193

Datta S, McCormick WP (1995) Some continuous edgeworth expansions for Markov chains with applications to bootstrap. J Multivar Anal 52(1):83–106

Datta S, Satten GA (2001) Validity of the Aalen–Johansen estimators of stage occupation probabilities and Nelson Aalen integrated transition hazards for non-Markov models. Stat Probab Lett 55(4):403–411

Datta S, Satten GA, Datta S (2000) Nonparametric estimation for the three-stage irreversible illness-death model. Biometrics 56(3):841–847

de Uña-Álvarez J, Amorim AP (2011) A semiparametric estimator of the bivariate distribution function for censored gap times. Biometrical J 53(1):113–127

de Uña-Álvarez J, Meira-Machado L (2008) A simple estimator of the bivariate distribution function for censored gap times. Stat Probab Lett 78(15):2440–2445

Emura T, Wang W (2010) Testing quasi-independence for truncation data. J Multivar Anal 101(1):223–239

Fuh CD, Ip EH (2005) Bootstrap and Bayesian bootstrap clones for censored Markov chains. J Stat Plan Inf 128(1):459–474

Hougaard P (2000) Analysis of multivariate survival data. Springer, New York

Jones MP, Crowley J (1992) Nonparametric tests of the Markov model for survival data. Biometrika 79(3):513–522

Meira-Machado L, de Uña-Álvarez J, Cadarso-Suárez C (2006) Nonparametric estimation of transition probabilities in a non-Markov illness-death model. Lifetime Data Anal 12(3):325–344

Meira-Machado L, de Uña-Álvarez J, Cadarso-Suárez C, Andersen PK (2009) Multistate models for the analysis of time-to-event data. Stat Methods Med Res 18:195–222

Rodríguez-Girondo M, de Uña-Álvarez J (2010) Testing markovianity in the three-state progressive model via future-past association. Report 10/01. Discussion papers in statistics and OR, University of Vigo

Tsai WY (1990) Testing the assumption of independence of truncation time and failure time. Biometrika 77(1):169–177

Woodroofe M (1985) Estimating a distribution function with truncated data. Ann Stat 13(1):163–177

Chapter 20
Multi-State Semi-Markov Model of the Operation Reliability

Franciszek Grabski

Abstract Semi-Markov processes theory delivers some methods that allow us to construct models of the operation process, especially in the reliability aspect. Semi-Markov model of the many steps operation is presented in the chapter. The considered operation consists of many stages which follow in turn. The duration of each stage is supposed to be the non-negative random variable. On each step the operation may be perturbed by faults. Time to event causing an operation perturbation on each stage is a random variable having the exponential distribution. The repair time of an object and continuation of operation process on each stage is assumed to be non-negative random variable. Time to a total failure is a non-negative random variable with an exponential distribution. Two cases are investigated: (a) the final failure finishes operation process and (b) it is possible a renewal of the object after the total failure. A renewal time in this case, is a non-negative random variable. Some results of the semi-Markov processes theory enable to obtain reliability parameters and characteristics of the operation process. A multi-state reliability function, the corresponding expectations for a model without repair and the 1-level and 2-level limiting availability coefficient for the model with repair are calculated.

Keywords Operation · Reliability · Multi-state model · Semi-Markov process

F. Grabski (✉)
Department of Mathematics and Physics, Polish Naval University,
Śmidowicza 69, 81-103 Gdynia, Poland
e-mail: franciszekgr@onet.eu

A. Lisnianski and I. Frenkel (eds.), *Recent Advances in System Reliability*,
Springer Series in Reliability Engineering, DOI: 10.1007/978-1-4471-2207-4_20,
© Springer-Verlag London Limited 2012

20.1 Introduction

The operation processes like transport processes, production processes and many other consist of some stages which are realised in turn. In each stage there are possible failures that lead to perturbation or to a failure of the operation. The main goal of this paper is to construct a stochastic model of an operation process and it's analysis. Some concepts of a semi-Markov process theory (Feller 1964; Grabski 2002; Korolyuk and Turbin 1976, 1982; Limnios and Oprisian 2001; Silvestrov 1980) are applied to construct a model of the operation process. Markov and semi-Markov processes for modelling multi-state systems are applied in many different reliability problems (Kolowrocki 2004; Lisnianski and Frenkel 2009; Lisnianski and Levitin 2003). To construct a multi-state semi-Markov model we define the renewal kernel. Characteristics of the semi-Markov process are used as reliability measures of the considered multi-state system.

20.2 Description and Assumptions

Suppose that an operation consists of r stages which follow in turn. We assume that duration of an ith stage of the operation, $i, i = 1, \ldots, r$ is a non-negative random variable ξ_i, $i = 1, \ldots, r$ with distribution given by CDF

$$F_i(t) = P(\xi_i \leq t), \quad i = 1, \ldots, r. \tag{20.1}$$

On each step the operation may be perturbed by faults. We assume that on each stage no more than one failure causing perturbation of the operation may occur. Time to event causing an operation perturbation on ith stage is a non-negative random variable v_i, $i = 1, \ldots, r$ having the exponential distribution with CDF

$$P(v_i \leq t) = 1 - e^{-\alpha_i t}, \quad i = 1, \ldots, r. \tag{20.2}$$

The repair time of an object and continuation of operation process on ith stage is non-negative random variable ζ_i, $i = 1, \ldots, r$ with CDF

$$P(\zeta_i \leq t) = H_i(t), \quad i = 1, \ldots, r. \tag{20.3}$$

In many cases we can assume

$$\zeta_i = [\xi_i - v_i]_+, \quad i = 1, \ldots, r. \tag{20.4}$$

If the object has not been repaired in ith stage, the operation process is continued until moment of a total failure. Time to a total failure (a length of a time interval from an instant of a dangerous event to a moment of the failure) is a non-negative random variable η_i, $i = 1, \ldots, r$ having the exponential distribution with CDF

$$P(\eta_i \leq t) = 1 - e^{-\lambda_i t}, \quad i = 1, \ldots, r. \tag{20.5}$$

We investigate two cases:

1. The renewal of the object is impossible and the total failure finishes operation process.
2. It is possible to renew the object after total failure and a renewal (repair) time ϑ is a non-negative random variable with CDF

$$K(x) = P(\vartheta \leq t) \tag{20.6}$$

with a finite second moment $E(\vartheta^2)$. After renewal the operation starts from the state 1. In this case the operation process is cyclic.

We also assume that random variables ξ_i, η_i, v_i, ϑ, ζ_i, $i = 1, \ldots, r$ are mutually independent and the operation process is generated by their independent copies.

20.3 Construction of semi-Markov Model

To construct a reliability model of an operation, we have to start from definition of the states.

- i, $(i = 1, \ldots, r)$—operation process stage,
- $i + r$, $(i = 1, \ldots, r)$—fault in the ith operation stage,
- $2r + 1$—total failure in the operation stage.

Under above assumptions, stochastic process describing the overall operation in reliability aspect is a semi-Markov process $\{X(t) : t \geq 0\}$ with a set of the states $S = \{1, 2, \ldots, 2r, 2r + 1\}$. A flow graph of the process transitions, for $r = 3$, is shown in Fig. 20.1.

To obtain a semi-Markov model we have to define all non-negative elements of the semi-Markov kernel

$$Q(t) = [Q_{ij}(t) : i, j \in S], \tag{20.7}$$

where $Q_{ij}(t) = P(X(\tau_{n+1}) = j, \tau_{n+1} - \tau_n \leq t | X(\tau_n) = i)$ denotes transition probability from state i to state j during the time no greater than t.

First, we define transition probabilities from state i to state $j = i + 1, i = 1, \ldots, r - 1$ at the time no greater than t. From assumptions it follows that

$$Q_{i,i+1}(t) = P(\xi_i \leq t, \ v_i > \xi_i) = \int_0^t e^{-\alpha_i x} f_i(x) dx. \tag{20.8}$$

In similar way we obtain

Fig. 20.1 A flow graph
of the process transitions
for $r = 3$

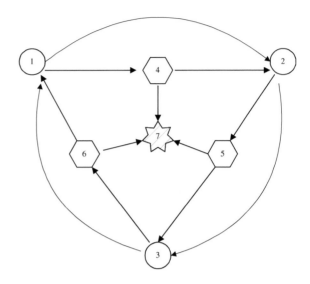

$$Q_{r1}(t) = \int_0^t e^{-\alpha_r x} f_r(x) \mathrm{d}x. \tag{20.9}$$

For $i = 1, \ldots, r$ we have

$$Q_{i,i+r}(t) = P(v_i \leq t, \ v_i < \xi_i) = \int_0^t \alpha_i e^{-\alpha_i u} [1 - F_i(u)] \mathrm{d}u. \tag{20.10}$$

For $i = 1, \ldots, r - 1$ we get

$$Q_{i+r,i+1}(t) = P(\zeta_i \leq t, \eta_i > \zeta_i) = \int_0^t e^{-\lambda_i x} h_i(x) \mathrm{d}x. \tag{20.11}$$

Similarly

$$Q_{2r,1}(t) = P(\zeta_i \leq t, \eta_i > \zeta_i) = \int_0^t e^{-\lambda_r x} h_r(x) \mathrm{d}x. \tag{20.12}$$

For $i = 1, \ldots, r$ we have

$$Q_{i+r,\,2r+1}(t) = P(\eta_i \leq t, \zeta_i > \eta) = \int_0^t \lambda_i e^{\lambda_i x}[1 - H_i(x)]dx. \tag{20.13}$$

For the model without repair we suppose

$$Q_{2r+1,\,2r+1}(t) = U(t), \; t \geq 0, \tag{20.14}$$

where $U(t)$ is any CDF of a non-negative random variable. For the model with repair we have

$$Q_{2r+1,\,1}(t) = K(t), \; t \geq 0. \tag{20.15}$$

20.4 Model for $r = 3$ Without Repair

A kernel of the semi-Markov process is a matrix

$$Q(t) = \begin{bmatrix} 0 & Q_{12}(t) & 0 & Q_{14}(t) & 0 & 0 & 0 \\ 0 & 0 & Q_{23}(t) & 0 & Q_{25}(t) & 0 & 0 \\ Q_{31}(t) & 0 & 0 & 0 & 0 & Q_{36}(t) & 0 \\ 0 & Q_{42}(t) & 0 & 0 & 0 & 0 & Q_{47}(t) \\ 0 & 0 & Q_{53}(t) & 0 & 0 & 0 & Q_{57}(t) \\ Q_{61}(t) & 0 & 0 & 0 & 0 & 0 & Q_{67}(t) \\ 0 & 0 & 0 & 0 & 0 & 0 & Q_{77}(t) \end{bmatrix}, \tag{20.16}$$

where its elements are defined by (7)–(13). A function

$$G_i(t) = P(\tau_{n+1} - \tau_n \leq t | X(\tau_n) = i) = \sum_{j \in S} Q_{ij}(t) \tag{20.17}$$

is CDF of a random variable T_i denoting the time spent in state i, when we do not know the successor state. The random variable T_i is called waiting time in state i. From (20.8), (20.9) and (20.17) for $i = 1, 2, 3$ we obtain

$$G_i(t) = \int_0^t e^{-\alpha_i x} f_i(x) dx + \int_0^t \alpha_i e^{-\alpha_i x}[1 - F_i(x)]dx.$$

Integration by parts leads us to a formula

$$G_i(t) = 1 - e^{-\alpha_i t}[1 - F_i(t)], \quad i = 1, 2, 3 \tag{20.18}$$

which are CDF of random variables

$$T_i = \min(\xi_i, \nu_i), \quad i = 1, 2, 3. \tag{20.19}$$

We calculate expected values of those random variables as

$$E(T_i) = \int_0^\infty [1 - G_i(t)]dt = \int_0^\infty e^{\alpha_i t}[1 - F_i(t)]dt, \quad i = 1,2,3$$

In a similar way, for the states 4, 5 and 6 we have

$$G_{i+3}(t) = 1 - e^{-\lambda_i t}[1 - H_i(t)], \quad i = 1,2,3 \tag{20.20}$$

A transition probability matrix of an embedded Markov chain is

$$P = \begin{bmatrix} 0 & p_{12} & 0 & p_{14} & 0 & 0 & 0 \\ 0 & 0 & p_{23} & 0 & p_{25} & 0 & 0 \\ p_{31} & 0 & 0 & 0 & 0 & p_{36} & 0 \\ 0 & p_{42} & 0 & 0 & 0 & 0 & p_{47} \\ 0 & 0 & p_{53} & 0 & 0 & 0 & p_{57} \\ p_{61} & 0 & 0 & 0 & 0 & 0 & p_{67} \\ 0 & 0 & 0 & 0 & 0 & 0 & 1 \end{bmatrix} \tag{20.21}$$

where

$$p_{ij} = \lim_{t \to \infty} Q_{ij}(t). \tag{20.22}$$

20.5 Reliability Characteristics and Parameters

Some characteristics and parameters of a semi-Markov process allow us to obtain the reliability characteristics and parameters of the operation process. A random variable

$$\Theta_A = \inf\{t \geq 0 : X(t) \in A\} \tag{20.23}$$

denotes an instant of a first passage to a subset A. The function

$$\Phi_{iA}(t) = P(\Theta_A \leq t | X(0) = i), \quad t \geq 0, i \in A' \tag{20.24}$$

is CDF of a random variable Θ_{iA}, which denotes the first passage time from state $i \in A'$ to subset A. From theorem presented in Grabski (2002) it follows that the Laplace-Stieltjes transforms of those functions satisfy a system of linear equations which have a following matrix form

$$(I - q_{A'}(s))\varphi_{A'}(s) = b(s), \tag{20.25}$$

where $I = [\delta_{ij} : i,j \in A']$, $q_{A'}(s) = [\tilde{q}_{ij}(s) : i,j \in A']$,

$$\varphi_{A'}(s) = [\tilde{\varphi}_{iA}(s) : i \in A']^T, \quad b(s) = \left[\sum_{j \in A} \tilde{q}_{ij}(s) : i \in A'\right]^T$$

$$\tilde{q}_{ij}(s) = \int_0^\infty e^{-st} dQ_{ij}(t), \quad \tilde{\varphi}_{iA}(s) = \int_0^\infty e^{-st} d\Phi_{iA}(t).$$

One of the method which allows to obtain an expectation of a random variable Θ_{iA}, $i \in A'$ consists in solving the system of linear equations, which have a following matrix form

$$(I - P_{A'})\overline{\Theta}_{A'} = \overline{T}_{A'}, \tag{20.26}$$

where $P_{A'} = [p_{ij} : i, j \in A']$, $\overline{\Theta}_{A'} = [E(\Theta_{iA}) : i \in A']^T$, $\overline{T}_{A'} = [E(T_i) : i \in A']^T$.

Second moments $E(\Theta_{iA}^2)$, $i \in A'$ are solutions of the linear system equations, which in matrix form are

$$(I - P_{A'})\overline{\Theta}_{A'}^2 = B_{A'} \tag{20.27}$$

where

$$P_{A'} = [p_{ij} : i, j \in A'], \quad \overline{\Theta}_{A'} = [E(\Theta_{iA}^2) : i \in A']^T$$
$$B_{A'} = [b_{iA} : i \in A']^T, \quad b_{iA} = E(T_i^2) + 2\sum_{k \in A'} p_{ik} E(T_{ik}) E(\Theta_{kA}).$$

For 3−stage model of the operation process, taking $i = 1$ and $A = \{4, 5, 6, 7\}$, we obtain a random variable Θ_{1A} the value of which denotes time of a completely reliable operation (without any perturbations and failures). For $i = 1$ and $A = \{7\}$ the value of the random variable Θ_{17} denotes time of the partially reliable operation (with perturbation but without total failure). The function

$$R_{[1]}(t) = P(\Theta_{1A} > t) = 1 - \Phi_{1A}(t), \quad t \geq 0 \tag{20.28}$$

we call a 1-level reliability function. The value of this function at a moment t denotes probability that the object is completely "up" at the instant t in the operation process. The function

$$R_{[2]}(t) = P(\Theta_{17} > t) = 1 - \Phi_{17}(t), \quad t \geq 0 \tag{20.29}$$

we call a 2-level reliability function. The value of this function at the moment t is probability that the object is capable to make limited tasks. The Laplace transforms of those functions are

$$\tilde{R}_{[1]}(s) = \frac{1 - \tilde{\varphi}_{1A}(s)}{s}, \tag{20.30}$$

$$\tilde{R}_{[2]}(s) = \frac{1 - \tilde{\varphi}_{17}(s)}{s}. \qquad (20.31)$$

The Eq. (20.25) for $A' = \{1, 2, 3\}$ and $A = \{4, 5, 6, 7\}$ takes a form:

$$\begin{bmatrix} 1 & -\tilde{q}_{12}(s) & 0 \\ 0 & 1 & -\tilde{q}_{23}(s) \\ -\tilde{q}_{31}(s) & 0 & 1 \end{bmatrix} \cdot \begin{bmatrix} \tilde{\varphi}_{1A}(s) \\ \tilde{\varphi}_{2A}(s) \\ \tilde{\varphi}_{3A}(s) \end{bmatrix} = \begin{bmatrix} \tilde{q}_{14}(s) \\ \tilde{q}_{25}(s) \\ \tilde{q}_{36}(s) \end{bmatrix}. \qquad (20.32)$$

The solution is the matrix

$$\varphi_{A'}(s) = \begin{bmatrix} \tilde{\varphi}_{1A}(s) \\ \tilde{\varphi}_{2A}(s) \\ \tilde{\varphi}_{3A}(s) \end{bmatrix}, \qquad (20.33)$$

where

$$\begin{aligned}
\tilde{\varphi}_{1A}(s) &= \frac{\tilde{q}_{14}(s) + \tilde{q}_{12}(s)\tilde{q}_{25}(s) + \tilde{q}_{12}(s)\tilde{q}_{23}(s)\tilde{q}_{36}(s)}{1 - \tilde{q}_{12}(s)\tilde{q}_{23}(s)\tilde{q}_{31}(s)}, \\
\tilde{\varphi}_{2A}(s) &= \frac{\tilde{q}_{25}(s) + \tilde{q}_{23}(s)\tilde{q}_{36}(s) + \tilde{q}_{23}(s)\tilde{q}_{31}(s)\tilde{q}_{14}(s)}{1 - \tilde{q}_{12}(s)\tilde{q}_{23}(s)\tilde{q}_{31}(s)}. \\
\tilde{\varphi}_{3A}(s) &= \frac{\tilde{q}_{36}(s) + \tilde{q}_{31}(s)\tilde{q}_{14}(s) + \tilde{q}_{31}(s)\tilde{q}_{12}(s)\tilde{q}_{25}(s)}{1 - \tilde{q}_{12}(s)\tilde{q}_{23}(s)\tilde{q}_{31}(s)}.
\end{aligned} \qquad (20.34)$$

Hence, using formula (35), we obtain the Laplace transform of the 1-level reliability function

$$\tilde{R}_{[1]}(s) = \frac{1 - \tilde{q}_{14}(s) - \tilde{q}_{12}(s)\tilde{q}_{25}(s) - \tilde{q}_{12}(s)\tilde{q}_{23}(s)\tilde{q}_{36}(s)}{s\left(1 - \tilde{q}_{12}(s)\tilde{q}_{23}(s)\tilde{q}_{31}(s)\right)}. \qquad (20.35)$$

The same way we can obtain a Laplace transform $\tilde{R}_{[2]}(s)$.

20.5.1 Example 1

We investigate two cases assuming that duration of ith stage of the operation has distribution with CDF:

$$F_i(t) = 1 - e^{-\kappa_i t}, \quad t \geq 0, \quad i = 1, 2, 3. \qquad (20.36)$$

$$F_i(t) = \begin{cases} 0, & \text{for} \quad t \leq c_i, \\ 1, & \text{for} \quad t > c_i. \end{cases} \qquad (20.37)$$

We also suppose that duration of an ith stage of operation is a non-negative random variable ξ_i having distribution given by CDF

Fig. 20.2 The 1- and 2-level reliability functions $R_{[1]}(t)$, $R_{[2]}(t)$, $t \geq 0$

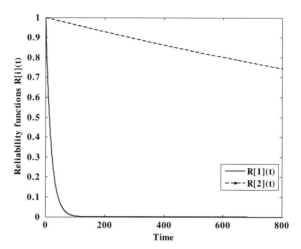

$$H_i(t) = 1 - e^{-\kappa_i t}, \quad t \geq 0, \quad i = 1, 2, 3. \tag{20.38}$$

For

$$\lambda_1 = 0.01, \quad \lambda_2 = 0.01, \quad \lambda_3 = 0.01,$$
$$\alpha_1 = 0.08, \quad \alpha_2 = 0.06, \quad \alpha_3 = 0.04,$$
$$\kappa_1 = 1.86, \quad \kappa_2 = 2.94, \quad \kappa_3 = 0.94$$

the 1- and 2-level reliability functions we obtain as the inverse Laplace transforms, which are shown in Fig. 20.2.

$$R_{[1]}(t) = 0.9891\, e^{-0.0544\, t} + 2e^{-2.8699\, t}[0.0054 \cos(1.3207\, t) - 0.0022 \sin(1.3207\, t)],$$
$$R_{[2]}(t) = 0.0001e^{-3.0097\, t} + 0.00008e^{-1.95\, t} - 0.00042e^{-0.9899\, t} + e^{-0.00037 0008\, t}$$
$$+ e^{-2.86994\, t}[-0.00011 \cos(1.32\, t) + 0.00016 \sin(1.32\, t)].$$

The corresponding expectations we obtain from (20.26)

$$E(\Theta_{17}) = 2703.48, \quad E(\Theta_{1A}) = 18.16.$$

As a solution of Eq. (20.27) we get second moments. Applying well known equations we obtain standard deviations:

$$\sigma(\Theta_{17}) = 2702.46, \quad \sigma(\Theta_{1A}) = 18.35.$$

Assuming that duration of ith stage of the operation is deterministic (CDF is given by (20.40)) and supposing

$$c_i = \frac{1}{\kappa_i}, \quad c_j = \frac{1}{\kappa_j}, \quad j = 3 + i, \quad i = 1, 2, 3,$$

from Eq. (20.26), we get

$$E(\Theta_{17}) = 2752.55, \quad E(\Theta_{1A}) = 18.50.$$

From (20.27) we obtain second moments. Finally the standard deviations are

$$\sigma(\Theta_{17}) = 2751.73, \quad \sigma(\Theta_{1A}) = 18.66.$$

Let us notice that the mean time to a fatal failure in case of the process with a deterministic duration of the operation stages is greater than for the process with exponentially distributed random duration of them.

20.6 Model for $r = 3$ with Repair

For this case semi-Markov kernel is

$$Q(t) = \begin{bmatrix} 0 & Q_{12}(t) & 0 & Q_{14}(t) & 0 & 0 & 0 \\ 0 & 0 & Q_{23}(t) & 0 & Q_{25}(t) & 0 & 0 \\ Q_{31}(t) & 0 & 0 & 0 & 0 & Q_{36}(t) & 0 \\ 0 & Q_{42}(t) & 0 & 0 & 0 & 0 & Q_{47}(t) \\ 0 & 0 & Q_{53}(t) & 0 & 0 & 0 & Q_{57}(t) \\ Q_{61}(t) & 0 & 0 & 0 & 0 & 0 & Q_{67}(t) \\ Q_{71}(t) & 0 & 0 & 0 & 0 & 0 & 0 \end{bmatrix}, \quad (20.39)$$

where its elements (apart from $Q_{71}(t)$) are defined by (20.7)–(20.13). For element from last row we have

$$Q_{71}(t) = K(t). \tag{20.40}$$

An important characteristic is a conditional probability

$$P_{ij}(t) = P(X(t) = i \mid X(0) = i), \quad i, j \in S, \quad t \geq 0, \tag{20.41}$$

which is called *the interval transition probability from state i to state j* is an important characteristic. Feller (1964) has introduced equations that allow to calculate these interval transition probability based on semi-Markov kernel. In many cases the interval transitions probabilities $P_{ij}(t)$, $t \geq 0$ and the states probabilities

$$P_j(t) = P(X(t) = j), \ t \geq 0, \quad j \in S \tag{20.42}$$

approach constant values for large t.
     Let

$$P_j = \lim_{t \to \infty} P_j(t) = \lim_{t \to \infty} P(X(t) = j), \quad j \in S, \tag{20.43}$$

$$P_{ij} = \lim_{t \to \infty} P_{ij}(t) = \lim_{t \to \infty} P(X(t) = j | X(0) = i), \quad j \in S. \tag{20.44}$$

For our case, from theorem presented in Korolyuk and Turbin (1976) it follows that there exist limits of $P_{ij}(t)$, $i, j \in S$ and $P_j(t)$, $j \in S$ for $t \to \infty$ and

$$P_{ij} = \lim_{t \to \infty} P_{ij}(t) = P_j = \lim_{t \to \infty} P_j(t) = \frac{\pi_j E(T_j)}{\sum_{i \in S} \pi_i E(T_i)}, \tag{20.45}$$

where π_i, $i \in S$ form the stationary distribution of the embedded Markov chain $\{X(\tau_n) : n = 0, 1, 2, \ldots\}$.

Let us recall that the stationary distribution of $\{X(\tau_n) : n = 0, 1, 2, \ldots\}$ is the unique solution of the linear system of equations

$$\sum_{i \in S} \pi_i p_{ij} = \pi_j, \quad j \in S, \quad \sum_{j \in S} \pi_j = 1, \tag{20.46}$$

where

$$p_{ij} = \lim_{t \to \infty} Q_{ij}(t).$$

20.6.1 Example 2

We accept all of assumptions from Example 1. Moreover we assume that the expectation of repair time is $E(\vartheta) = 50$ and the second moment is $E(\vartheta^2) = 2900$ From (20.45) and (20.46) we get limiting probabilities of states $P_1 = 0.2608$, $P_2 = 0.1686$, $P_3 = 0.5161$, $P_4 = 0.0112$, $P_5 = 0.0034$, $P_6 = 0.0217$, $P_7 = 0.0182$.

The 1-level and 2-level limiting coefficient of availability are given by equations

$$K_{[1]} = P_1 + P_2 + P_3, \quad K_{[2]} = P_4 + P_5 + P_6.$$

Finally we obtain $K_{[1]} = 0.9455$, $K_{[2]} = 0.0363$.
Limiting probability of the object total failure is $P_7 = 0.0182$.

20.7 Conclusions

The multi-state semi-Markov models of the operation process allow us to calculate many reliability measures. To construct a multi-state semi-Markov model it is necessary to define the renewal kernel or another equivalent mathematical objects. Characteristics of the semi-Markov process are applied as reliability measures of the considered multi-state system. The 1- and 2-level reliability functions, the corresponding expectations, second moments and the standard deviations, for a model without repair, are obtained. For the model with repair, the states limiting probabilities and 1- and 2-level limiting coefficients availability are calculated. From presented example it follows that the mean time to a fatal failure in case of the process with the deterministic duration of the operation stages is greater than that one for the process with the exponentially distributed random duration of the operation stages.

References

Feller W (1964) On semi-Markov processes. Proc Nat Acad Sci USA 51(4):653–659

Grabski F (2002) Semi-Markov models of reliability and operation. Warszawa, IBS PAN (in Polish)

Kolowrocki K (2004) Reliability of large systems. Elsevier, NY

Korolyuk VS, Turbin AF (1976) Semi-Markov processes and their applications. Naukova Dumka, Kiev (in Russian)

Korolyuk VS, Turbin AF (1982) Markov renewal processes in systems reliability problems. Naukova Dumka, Kiev (in Russian)

Limnios N, Oprisian G (2001) Semi-Markov procesess and reliability. Birkhauser, Boston

Lisnianski A, Frenkel I (2009) Non-homogeneous Markov reward model for aging multi-state system under minimal repair. Int J Perform Eng 4(5):303–312

Lisnianski A and Levitin G (2003) Multi-state system reliability: assessment, optimization and applications. World Scientific, Sigapore

Silvestrov DC (1980) Semi-Markov processes with a discrete state space. Sovetskoe Radio, Moskow (in Russian)

Chapter 21
Reliability of Continuous-State Systems in View of Heavy-Tailed Distributed Performance Features

Emil Bashkansky and Tamar Gadrich

Abstract The paper examines the performance distribution for continuous-state systems that may be essentially different from Gaussian. Notable progress in understanding the nature of rare events and the implications for heavy-tailed distributions has been achieved in the last decade. The insights, however, even though they generated robust interest in such distributions, are not reflected strongly enough in reliability engineering practice. The paper presents four simple mechanisms of heavy-tail formation in reliability engineering contexts, their importance and the phenomena where they may appear. It also discusses the implications of this knowledge for optimization of quality controller performance.

Keywords Continuous-State reliability · Performance measures · Heavy-tail formation

21.1 Introduction

Beginning in the mid 1970s researchers started developing multi-state and continuous-state reliability models designed to address the issue of performance degradation (Lisnianski and Levitin 2003; Lisnianski et al. 2010). Many real systems undergo continuous performance change during their lifetime, moving

E. Bashkansky (✉) · T. Gadrich
Department of Industrial Engineering and Management,
Ort Braude College of Engineering, P.O. Box 78, Karmiel, Israel
e-mail: ebashkan@braude.ac.il

T. Gadrich
e-mail: tamarg@braude.ac.il

A. Lisnianski and I. Frenkel (eds.), *Recent Advances in System Reliability*,
Springer Series in Reliability Engineering, DOI: 10.1007/978-1-4471-2207-4_21,
© Springer-Verlag London Limited 2012

from perfect operation to complete or partial failure. This phenomenon can be described as the random process $X(t)$, which denotes the performance state as a function of time t that takes any value in a prescribed range (e.g., in binary state reliability, the values are $\{0, 1\}$ where 0 means failure and 1 means success). Every system follows its own curve on the X-t plane starting at the $X(0)$ point, so for a fixed t we can consider the distribution of X and for a defined threshold level of performance—the lifetime distribution. Understanding the kinds of distribution is critical when building appropriate reliability models (Zuo et al. 1999) and determining optimal warranty time. The familiar reliability measures for continuous-state systems (CSSs) are, for example: (1) expected value of state at time t; (2) variance of state number at time t; and (3) probability that the performance measure is greater than some given threshold at time t (Yang and Kapur 1997).

Low-level system reliability can be caused either by the probability of exceeding tolerable deviation limits, or by wrongly predicted system performance in such out-of-tolerance areas. Managers and engineers, schooled in the central limit theorem concept, may have difficultly in seeing system performance distribution *tails* as something other than Gaussian. Moreover, any deviation from the Gaussian law is sometimes interpreted as an unsuccessful choice of basic variables, which can be eliminated by suitable nonlinear transformation.

The anomalistic, abnormal behavior of some systems on distribution tails, resulting in what is known as *heavy tails*, is not, however, something new or insufficiently explored. On the contrary, the term heavy tails is widely used to describe a broad spectrum of phenomena in economics, geophysics, finance, material crush analysis, computer science, linguistics, healthcare, sociology and other scientific domains where a Gaussian approach is inappropriate and leads to underestimation of the real risk. Even Pareto distribution, widely utilized by reliability engineers, is essentially different from any other distribution used in reliability theory. Always of interest to the wider public, heavy tails became especially popular following the publication of Nassim Taleb's bestseller (2010) "The Black Swan" and the financial crashes of the 2000s. Multivariate models of financial systems using the famous Li's copula (Li 2000) have proved nonrobust and failed to predict the 2008 economic meltdown due to a dramatic impact of heavy tails in the multivariate case. The ideas presented in the Taleb's book have become a paradigm for many researchers engaged in risks analysis (Kenett and Tapiero 2009).

Nevertheless, the empirical usage of heavy-tailed distributions for the description of various phenomena and evaluation of various performance measures does not help us comprehend the mechanism of their formation. Understanding the mechanisms of these non-reliable/non-quality formations is necessary both for their prediction and for preventive action.

Nowadays researchers believe that heavy tails are usually formed only in nonlinear systems that have a great number of degrees of freedom and few restrictions. Only in such systems can small changes in one place bring about great changes in other places and in the entire system as a whole. The authors of the present paper have encountered a great number of abnormal, heavy-tailed distributions while investigating the performance of different systems. As a result of trying to understand the mechanisms behind them, the authors became convinced of the existence of a spectrum of relatively straightforward mechanisms causing heavy-tailed distributions in some simple models used in reliability engineering. The aim of the present paper is to familiarize engineers with four relatively simple and quite commonly used situations/mechanisms in which the appearance of heavy tails results emerges from studying the nature of different phenomena. Avoiding unjustified generalizations, we nevertheless believe that recognizing these mechanisms leads to certain insights that may, sometimes, allow better understanding of the studied phenomena and also enable us to find a method for predicting and/or preventing the damage caused by poor performance. One such phenomenon, appearing in the field of controller performance feature optimization, is analyzed in more detail in Sect. 21.4.

21.2 Heavy-, Long- and Power-Tailed Distributions

In this section we briefly survey the various definitions that are customarily used when dealing with heavy-, long- or power-tailed distributions.

Definition 1 A distribution is *heavy-tailed* (HTD) if its tail, i.e., the complement to the cumulative distribution function of the random variable X, denoted by $F^C(x) = P(X > x)$, decays slower than exponentially. This means that for all $\alpha > 0$,

$$e^{\alpha x} F^C(x) \to \infty, \quad \text{as } x \to \infty. \tag{21.1}$$

Definition 2 A distribution is *long-tailed* (LTD) if and only if

$$\lim_{x \to \infty} \frac{F^C(x+y)}{F^C(x)} = 1, \quad y \geq 0. \tag{21.2}$$

Definition 3 A non-negative random variable X is said to have *power-tailed* distribution (PTD) if

$$F^C(x) = P(X \geq x) \sim cx^{-\alpha}, \quad \text{as } x \to \infty, \quad c > 0, \quad \alpha > 0. \tag{21.3}$$

Note that all power-tailed distributions are long and heavy-tailed but the inverse is not true (Shortle et al. 2004). Also, all long-tailed distributions are heavy-tailed but the inverse is not true.

Remark As a result of Definition 3, it follows that if $1 < \alpha < 2$, power-tailed distribution has a finite mean and an infinite variance, and if $\alpha \leq 1$, it has both an infinite variance and infinite mean. In order to check if given data follow a power-tailed distribution, we employ a log–log complementary distribution (CD) plot, i.e., we check if the following exists:

$$\frac{\frac{d}{dx}\log P(X > x)}{\frac{d}{dx}\log x} = -\alpha, \quad x > \theta \tag{21.4}$$

for some θ. In practice, we produce the CD plot and look for approximately linear behavior at the tail of the distribution.

21.3 Four Mechanisms of Heavy-Tail Formation

21.3.1 Pareto Distribution

Perhaps the most well-known result associated with HTD in reliability is the popular Pareto principle, as initially applied by Joseph M. Juran (Godfrey and Kenett 2007). Juran's (1950) brief and well-known formulation (also known as the 80/20 rule, or the law of the vital few and trivial many) states that for many events 80/20 means that roughly 80% of the effects come from 20% of the causes. This is the basis for Pareto charts used in exploratory data analysis to find the dominant causes of reliability problems. Sometimes these charts are based on failure costs or severity analysis in order to set work priorities for attacking the few vital problems and as a method of separating important issues from the many trivial issues. Most reliability engineers need to be working on the top few items. Periodic reviews of the Pareto distribution are important for accountability (who has solved what problems) and to define what new targets have come up that require immediate attention. Pareto distributions explain why some work orders always get maintenance priority while other tasks are relegated to the category of "whenever we get time to solve the problem."

The Pareto principle is an illustration of the *power-tailed* behavior of the *Pareto distribution*, originally used to describe the allocation of wealth among individuals.

The Pareto distribution, which is hyperbolic over its entire range (therefore, its asymptotic shape is also hyperbolic), has the following density function:

$$f_X(x) = \frac{\alpha x_{min}^{\alpha}}{x^{\alpha+1}}, \quad x > x_{min}, \quad \alpha > 0, \quad x_{min} > 0, \tag{21.5}$$

and its complement to cumulative distribution function is given by:

$$F^C(x) = P(X > x) = \left(\frac{x_{min}}{x}\right)^{\alpha}, \quad x > x_{min}, \quad \alpha > 0, \quad x_{min} > 0. \tag{21.6}$$

It is easy to check that this distribution follows all three definitions given in Sect. 21.2, i.e., it is a power-tailed distribution, so it is also long and heavy distributed.

The finiteness of the first two moments of the Pareto distribution depends on the shape of parameter α. The moments are given by:

$$E(X) = \begin{cases} \frac{\alpha x_{\min}}{\alpha-1}, & \alpha > 1, \\ \infty, & \alpha \le 1, \end{cases} \quad VAR(X) = \begin{cases} \frac{\alpha x_{\min}^2}{(\alpha-1)^2(\alpha-2)}, & \alpha > 2, \\ \infty, & \alpha \le 2. \end{cases}$$

For the Pareto principle, explained above, implying that for many events, roughly $W\%$ of the effects comes from $P\%$ of the causes, it follows from Newman (2005) that the following relation exists:

$$W = P^{(\alpha-2)/(\alpha-1)} \Leftrightarrow \alpha = \frac{\ln W - 2\ln P}{\ln(W/P)}, \tag{21.7}$$

if the effects are distributed according to the Pareto distribution with parameter α. The pure Pareto 80/20 principle implies $\alpha = 2.160964$, i.e., finite mean and variance.

The authors of the present paper assumed that the Pareto distribution is so popular because of its special properties: scale invariance and self-similarity.

Scale invariance This is the only distribution that is the same whatever scale one uses to look at it, i.e., if the scale or units by which we measure X by a factor γ changes, the shape of the distribution remains unchanged, except for an overall normalization constant. Let X be a Pareto-distributed random variable with shape parameter α and $Y = \gamma X$ be a new random variable, then:

$$f_Y(y) = f_X(x) \cdot \frac{dx}{dy}\Big|_{x=\frac{y}{\gamma}} = \frac{\alpha \cdot y_{\min}^\alpha}{y^{\alpha+1}}, \quad (y_{\min} = \gamma x_{\min}), \tag{21.8}$$

i.e., Y is distributed according to the Pareto distribution with the same shape parameter α.

Self-similarity The conditional probability distribution of a Pareto-distributed random variable, given the event that it is greater than or equal to a particular number $y_{\min} = \gamma x_{\min}$ ($\gamma > 1$), is a Pareto distribution with the same Pareto shape parameter α but with minimum $y_{\min}$ instead of x_m. In other words, the following exists:

$$P(X > x|X > y_{\min}) = \frac{P(X > x)}{P(X > y_{\min})} = \frac{\left(\frac{x_{\min}}{x}\right)^\alpha}{\left(\frac{x_{\min}}{y_{\min}}\right)^\alpha} = \left(\frac{y_{\min}}{x}\right)^\alpha, \quad x > y_{\min}. \tag{21.9}$$

A self-similar object is exactly or approximately similar to a part of itself. A self-similar phenomenon behaves the same when viewed under different degrees of magnification. Many objects and processes in the real world are statistically self-similar: parts of such objects or processes show the same statistical properties at many scales. Arrangement of leaves on a plant stem (phyllotaxis), fractals, logarithmical spiral, coastlines, network traffic, optimal energy transfer via mediation (Bashkansky and Netzer 2006) and many others are examples of self-similar structures. Sometimes this phenomenon can lead to undesirable results as, for example, degradation of network performance in long-tailed traffic (Park et al. 1997). Often these multi-level structures are described by an exponential function or its discrete analog-geometric progression. Under some natural and obvious assumptions, the equilibrium distribution of parameters characterizing such a structure is the Pareto distribution (Champernowne 1953).

As regards reliability engineering (RE), we often utilize self-similar multi-state hierarchical structures in cases such as classifying performance decrease or failure by its occurrence, severity (according to damage caused by failure) or detectability (or test capability) of hierarchic levels. Such classification, used in FMECA and risk analysis, is usually made using a geometric sequence scale. Consequently, the distribution of performance decrease or failure by occurence/severity/detectability levels seems to approximately follow Pareto distribution.

One more mechanism that generates Pareto distribution relevant to RE problems will be discussed in Sect. 21.3.2.

21.3.2 Lognormal Distribution and Multiplicative Generative Models

If a random variable Y has a normal distribution with mean μ and standard deviation σ, then a random variable $X = \exp(Y)$ has a *lognormal* distribution with the following density function:

$$f_X(x) = \frac{1}{\sqrt{2\pi}\sigma x} e^{-(\ln x - \mu)^2/2\sigma^2}, \quad x > 0, \tag{21.10}$$

and its cumulative distribution function is given by:

$$F_X(x) = P\left(e^Y \leq x\right) = \Phi\left(\frac{\ln x - \mu}{\sigma}\right). \tag{21.11}$$

It can be shown that a lognormal distribution is a long-tailed, but not power-tailed distribution.

The lognormal distribution also has a finite mean and variance, given by:

$$E(X) = \exp\left[\mu + \frac{1}{2}\sigma^2\right],$$
$$\text{VAR}(X) = \exp\left[2\mu + 2\sigma^2\right] - \exp\left[2\mu + \sigma^2\right].$$

Lognormal distributions, widely used in reliability and maintainability engineering, are most commonly supposed to be generated by a multiplicative process. Let us start with an entity of size X_0 (the performance level at time 0 that can be measured in continuous-state reliability models). At each step j, the entity may grow (as rings on a tree trunk, sediment on water heaters, sediment on electrodes) or decrease (as the temperature of water heated every morning at the same time or as battery capacity after recharging or fuel economy after tank refilling), according to a random multiplicative factor F_j, so that $X_j = F_j \cdot X_{j-1} (j = 1, 2, \ldots)$ and $\ln X_j = \ln X_0 + \sum_{k=1}^{j} \ln F_k$.

Assuming the random variables $\ln F_k$ are i.i.d. (stochastically self-similar!) with finite mean and variance, the central limit theorem (CLT) says that for sufficiently large j, $\ln X_j$ converges approximately to a normal distribution, and hence X_j asymptotically approaches a lognormal distribution.

This mechanism gives insight into why the lognormal has been a successful model for describing survival time of many failure mechanisms based on degradation processes. Let X_j be the amount of degradation for a particular failure process taken at successive discrete instances of time as the process moves toward failure. Assume $\left(\frac{\Delta X}{X}\right)_j = \frac{X_i - X_{j-1}}{X_{j-1}} = \varepsilon_j$ where ε_j are small, independent random perturbations or "shocks" to the system that move the failure process along. In other words, the increase in the amount of degradation from one instant to the next is a small random multiple of the total amount of degradation already present, i.e., $X_j = (1 + \varepsilon_j) \cdot X_{j-1}$. Since failure occurs when the amount of degradation reaches a critical point, time to failure for this type of process will be modeled successfully by a lognormal. This is what is meant by multiplicative performance degradation. Examples of failure mechanisms that might be successfully modeled by the lognormal distribution based on the multiplicative degradation model are (Nist/Sematech 2003):

1. Chemical reactions leading to shrinking of battery capacity;
2. Aging as result of fragmentation;
3. Crack growth or propagation;
4. Decrease in boiler effectiveness caused by sediment on heating element;
5. Corrosion and so on…

The difference between lognormal and Pareto distributions is not as great as might seem at first glance (Mitzenmacher 2004). Sometimes, the available empirical data are insufficient to allow us to distinguish between power law and lognormal tails. A small change in the lognormal generative process yields a generative process with a power law distribution. If there exists a lower bound ε_0, so that $X_j = \max(\varepsilon_0, F_j \cdot X_{j-1})$, then X_j converges to a power law distribution and the lognormal model is easily transformed into a power law model. Both distributions are integrated in the so-called double Pareto-lognormal distribution, but its description is beyond the scope of this article and we refer readers to Mitzenmacher (2004) for more details.

21.3.3 Inverse (Reciprocal) of Quantity Distribution

Reciprocal (inverse) quantities characterizing performance of continuous-state systems (CSS) are used in many areas of engineering in general and in reliability engineering in particular. Moreover, sometimes the quantity and its reciprocal are used to characterize the same feature as, for example, in:

- Resistance—conductivity,
- Frequency—period,
- Failure rate—MTBF,
- Pixel size—dots per inch (DPI),
- Fuel consumption—fuel economy.

Let us consider the last example in the above list in terms of the performance measure of a car engine.

Fuel consumption is the amount of fuel used per unit distance; for example, liters per 100 km (l/100 km). This is the measure generally used across Europe and Canada. In this case the lower the value, the more economical a vehicle is. A lower number means better.

Fuel economy is the distance traveled per unit volume of fuel used; for example, miles per gallon (mpg). This measure is popular in the USA and the UK. It gives an answer to the question "How far can I go on this tank before I have to fill up again?" In this case, the higher the value, the more fuel efficient a vehicle is (the more distance it can travel with a certain volume of fuel). A higher number means better.

Converting from mpg to l/100 km (or vice versa) requires us to use the reciprocal function. In the computer and microelectronics industry people are more comfortable with the idea that "smaller is better", while in the automotive industry they prefer the "larger is better" approach. Among other applications of inverse values, note *surface-to-volume ratios* (or size reciprocal) characterizing reactivity (it is also a performance measure!) of nanotechnology entities.

One of the most important engineering features widely used to characterize performance quality in electronics, communications, statistical reliability control and so on is the *signal-to-noise ratio*, abbreviated as S/N. This is one of the main parameters used in reliability experiment design and many other fields of reliability and quality engineering. It is important that the S/N ratio have an inverse relation to the stochastic noise signal. The higher the ratio, the less obtrusive the background noise is.

Suppose some quantity Y has a distribution $f_Y(y)$ that passes through zero, thus having both positive and negative values (as noise in the S/N ratio). Suppose further that the quantity we are really interested in is the reciprocal $X = 1/Y$, which will have the distribution:

$$f_X(x) = \left|\frac{dy}{dx}\right| \cdot f_Y(y) = \frac{f_Y(y)}{x^2}. \tag{21.12}$$

The large values of x, those in the tail of the distribution, correspond to the small values of Y close to zero and thus the large x tail is given by $f(x) \sim x^{-2}$, where the constant of proportionality is close to $f(y = 0)$. More generally, any quantity $X = Y^{-\gamma}$ for some γ will have a power-law tail on its distribution $f(x) \sim x^{-\alpha}$ with $\alpha = 1 + \frac{1}{\gamma}$. For example, if $f_Y(y) \sim N(0, \sigma^2)$ and $X = 1/Y$, then:

$$f_X(x) = \frac{e^{-\frac{1}{2 \cdot \sigma^2 \cdot x^2}}}{x^2 \cdot \sigma \sqrt{2\pi}} \sim x^{-2}, \quad \text{as } x \to \infty. \tag{21.13}$$

The denominator also plays a cardinal role in the so-called *ratio distribution* constructed as the distribution of the ratio $Z = X/Y$ of two random variables X and Y having two other known distributions. The S/N ratio for both noise and signals having a random nature is the best example. Note that the well-known heavy-tailed *Cauchy distribution* is an example of such a ratio distribution, which occurs when X and Y are independent standard normal distributed random variables. For almost every distribution $f_Y(y)$ that passes through zero, distribution of the ratio $Z = X/Y$ will be heavy-tailed because it contains a significant Cauchy term. It is extremely important to take this simple fact into account when defining a threshold for an inspected performance feature measured indirectly by its reciprocal or with the help of a ratio.

21.3.4 Quadratic Loss Distribution

A loss function represents the loss associated with a parameter deviation causing a decrease in the desired performance as a function of such deviation. The use of a quadratic loss function is common, for example, when using Taguchi methods. If the target value of the parameter is T, then a quadratic loss function is $L(x) = c \cdot (x - T)^2$ for some constant c; often the value of the constant makes no difference to a decision, and can then be ignored by setting it equal to 1. Sometimes $T = 0$ (contamination level, delay time, defect proportion), so: $L(x) \sim x^2$. Powering always makes tailed distribution "heavier". For example, it was proved that if X is a geometric random variable with parameter p ($0 < p < 1$), then for any $r > 1$, the distribution of x^r is heavy-tailed (Su and Tang 2003).

Let us consider a controller performing periodic, sequentially sampled testing at successive moments, one following another (usually by a fixed period of time—$\Delta\tau$), as a tool for checking whether a manufacturing process is in a statistically stable state (Montgomery 2008). If something happens, which is followed by a change in the controlled process, the event will not necessarily be immediately detected in the first subsequent test. More likely, it will be detected only after several tests, called *run length* (RL). RL is a random variable and follows the geometric distribution:

$$P(\text{RL} = x) = \beta^{x-1} \cdot (1 - \beta), \quad x = 1, 2, 3, \ldots, \tag{21.14}$$

where β is the probability of missing the change in every individual test, with expectation $E(\text{RL}) = (1 - \beta)^{-1}$ called the *average run length* (ARL). Multiplying RL by time period $\Delta\tau$ yields the delay time τ between the appearance of the disturbance and its detection. This time is called *time to signal* (TS) and its expectation, called *average time to signal* (ATS), is usually used to characterize an important performance feature of control tools—their response time.

If the damage caused by the delay follows a quadratic loss function $Loss \sim \tau^r$, (such a scenario often happens when the damage propagates to a wider area than the immediate vicinity of the damage, as, for example, an oil leak), or, more generally, follows a power-loss function, $Loss \sim \tau^r$, the distribution of loss in accordance to the effect mentioned in the first paragraph of this section will be heavy-tailed. This important fact reflects the typical for "Black Swan" concept (Taleb 2010) combination of a rare event (long delay) with large impact (quadratic damage). One challenge in handling "Black Swan" is to determine optimal decisions that will reduce risk. This issue will be discussed in Sect. 21.4.

21.4 The Example of a Performance Optimization Problem with a Heavy-tailed Distributed Characteristic

Awareness of the possibility of any characteristic being heavy-tailed distributed, as well as understanding the mechanism of its formation, can help us formulate CSS reliability problems more effectively. Both actions are likely to lead to solutions significantly different from traditional ones.

As an example, let us consider the setting up of a control device to monitor a process by sequential samples using a fixed sampling interval (FSI), a fairly usual practice. Such control is traditionally characterized by a false-alarm rate, α(depends on control limits only), and average time to signal (ATS) about a change in a process (Montgomery 2008). The last feature characterizes the delay in process change detection as described in Sect. 21.3. It depends on the process change value defining the probability of exposing this change in a single sample $1 - \beta$ and the average number of time periods/samples (ARL) we have before the monitoring signals an out-of-control situation. As noted in Sect. 3.4, this number is random and follows a geometric distribution.

Suppose we investigate the properties of a more adaptive and more intelligent approach, when the sampling interval between each pair of samples is not fixed but rather depends on what is observed in the first sample. The idea is that the time interval until the next sample is lengthened if there is no indication of significant change in the process and shortened if a sample shows some indication of a significant change in the process (Reynolds et al. 1988). We set the modified control to have the same average sampling rate and false-alarm rate.

When solving such an optimal control problem, the question is: what should the optimization criterion be? If the damage caused by the delay follows a quadratic law, that is proportional to the square of the delay, its mean may be extremely large *due to the heavy nature of its tail* (Sect. 21.3.4). Thus, a small change in a process, characterized by prolonged delay in detection, may lead to very significant damage (note that a similar effect is known in queuing theory as the "bus waiting time paradox").

Consequently, the correct reliability engineering problem that we should be attempting to resolve does not revolve around minimizing the ATS, but focuses on how to minimize expected "heavy-tailed distributed" loss. The change in the definition of the performance measure resulting from our awareness of the mechanism behind heavy-tail formation strongly influences the type of the solution we seek. For instance, when the goal is to minimize the ATS, the solution is of the "bang–bang" type with fixed warning limits (Reynolds et al. 1988). However, in the case of a heavy distributed quadratic delay loss performance index, the solution is continuous and much more complicated.

21.5 Conclusion

The paper deals with performance distribution for CSSs. It is shown that some-times these distributions may be essentially different from Gaussian and relate to heavy-tailed distributions indicated by the use of a log–log plot, for example. Every reliability engineer must be aware of the possibility of their presence. In such cases traditional statistical approaches to performance determination should be revised and reanalyzed in the light of recent developments in the field of heavy-tailed distributions formation. We focus on the four simplest mechanisms that may cause heavy-tailed distribution formation. Understanding these mechanisms leads to certain insights that may, in turn, direct us to better understanding of studied phenomena and help to predict or/and prevent the low reliability level when a heavy tail is discovered. One such phenomenon, appearing in the field of adaptive controller performance optimization, was discussed.

References

Bashkansky E, Netzer N (2006) The role of mediation in collisions and related analogs. Am J Phys 74(12):1083–1087
Champernowne D (1953) A model of income distribution. Econ J 63:318–351
Godfrey AB, Kenett RS (2007) Joseph M. Juran, a perspective on past contributions and future impact. Qual Reliab Eng Int 23(6):653–663
Juran JM (1950) Pareto, Lorenz, Cournot, Bernoulli. Juran and Others. Ind Qual Cont 17(4):25
Kenett RS, Tapiero C (2009) Quality, risk and the Taleb quadrants. Risk and Decision Anal 1(4):231–246
Li DX (2000) On default correlation: a copula function approach. J Fixed Income 9(4):43–54
Lisnianski A, Levitin G (2003) Multi-state system reliability: assessment, optimization and applications. Springer, London
Lisnianski A, Frenkel I, Ding Y (2010) Multi-state system reliability analysis and optimization for engineers and industrial managers. Springer, London
Mitzenmacher M (2004) A brief history of generative models for power law and lognormal distributions. Internet Math 1(2):226–251
Montgomery DC (2008) Introduction to statistical quality control. Wiley, NY
Newman MEJ (2005) Power laws, Pareto distributions and Zipf's law. Contemp Phys 46(5):323–351
Nist/Sematech (2003) e-Handbook of statistical methods. U.S. Commerce Department's Technology Administration. http://www.itl.nist.gov/div898/handbook/

Park K, Kim G, Crovella ME (1997) On the effect of traffic self-similarity on network performance. In: Proceedings of the SPIE international conference on performance and control of network systems, Dallas TX, pp 296–310, 3–5 November 1997

Reynolds MR, Amin RW Jr, Arnold JC, Nachlas JA (1988) X-charts with variable sampling intervals. Technometrics 30(2):181–192

Shortle JF, Brill PH, Fischer MJ, Gross D, Masi DMB (2004) An algorithm to compute the waiting time distribution for the M/G/1 queue. INFORMS J Comp 16(2):152–161

Su C, Tang Q (2003) Characterizations on heavy-tailed distributions by means of hazard rate. Acta Math Applicatae Sinica (English Series) 19(1):135–142

Taleb N (2010) The black swan: the impact of the highly improbable. Random Home Publishing Group, NY

Yang K, Kapur KC (1997) Customer driven reliability: integration of QFD and robust design. In: Proceedings of annual reliability and maintainability IEEE symposium, Philadelphia, PA, pp 339–345

Zuo MJ, Renyan J, Yam RCM (1999) Approaches for reliability modelling of continuous-state devices. IEEE Trans Reliab 48(1):9–18

Chapter 22
On Optimal Control of Systems on Their Life Time

Vladimir Rykov and Dmitry Efrosinin

Abstract Absorbing controllable Markov processes as models of some technical and/or biological objects behavior at their lifetime are proposed. A problem of optimal control of these objects during their lifetime for two-level control policies with respect to two objective functionals as control goals are considered; these goals are: the maximization of mean lifetime and the mean reward during it. The closed formulas for the main characteristic of the process and objective functional are given and some numerical examples of optimal policies calculation for both functionals are done.

Keywords Degradation models · Absorbing controllable Markov processes · Optimal control

22.1 Introduction and Motivation

During the previous years intensive attention to aging and degradation models for technical and biological objects has been drawn. Some wearing and aging models were the focus of many investigators in the framework of shock models. An excellent review and contribution of the earlier papers devoted to the topic can be

V. Rykov (✉)
Department of Applied Mathematics and Computer Modeling,
Russian State University of Oil and Gas, Leninsky prospect, 65,
119991 Moscow, Russia
e-mail: vladimir_rykov@mail.ru

D. Efrosinin
Institute for Stochastic, Johannes Kepler University of Linz,
Linz, Austria
e-mail: dmitry.efrosinin@jku.at

A. Lisnianski and I. Frenkel (eds.), *Recent Advances in System Reliability*,
Springer Series in Reliability Engineering, DOI: 10.1007/978-1-4471-2207-4_22,
© Springer-Verlag London Limited 2012

found in Esary et al. (1973). In Singpurwalla (2006a) the hazard potential notion as random life resource has been introduced and considered. Then this notion was investigated in different directions in Singpurwalla (2006b).

Any technical system and biological object possesses some initial (generally speaking, random) life resource, and during its life it goes through different states, in which its life resource is consumed at different rates. At the same time, due to self-regulation possibilities some partial renovations are usually possible. Therefore, because there are no infinitely long leaving objects and any repair is possible only from the state of partial failure, the modeling of the degradation process during the life period of an object is a most interesting topic. From the mathematical point of view the degradation during an object's life period can be described by the Birth and Death (B & D) type process with absorbtion at some state. Moreover, most biological and technical systems allow to control their resources during their life-time. Some additional financial or life resources are needed for that. Therefore the problem and its optimal consumption becomes very important.

The aging and degradation models suppose the study of the systems with gradual failures for which multi-state reliability models have been elaborated (for the history and bibliography see, Lisniansky and Levitin 2003). In some of our previous papers (Rykov and Dimitrov 2002; Dimitrov et al. 2004; Rykov et al. 2004) and the bibliography therein) the model for a complex hierarchical system has been proposed and the methods for its steady state and time-dependent characteristics investigation have been done. Controllable fault tolerance reliability systems have been considered in Rykov and Efrosinin (2004) and Rykov and Buldaeva (2004). Generalized birth and death processes as degradation models have been proposed in Rykov (2006a, b). In Rykov and Efrosinin (2007) some degradation models with a random initial life resource and process of its consumption have been proposed, and characteristics of lifetime were calculated in terms of these models.

In this paper we consider the problem of optimal control of system resources in the framework of birth and death type process of their consumption. The paper is organized as follows. In the next section a birth and death type process of resources consumption in the framework of a general degradation model is presented and the problem of their optimal control is formulated. The problem solution in the third section is done jointly with formulas for the main charac-teristic calculation. Some numerical examples in the fourth section are presented. The paper is concluded by a discussion on further problems.

22.2 Controllable Markov Degradation Model

22.2.1 Assumptions

Most up-to-date complex technical systems also as biological objects with sufficiently high organization during their life period pass over different states of evolution and existence. From the reliability point of view these states can be divided into three groups: the states of normal functioning, the dangerous

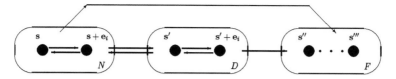

Fig. 22.1 The structure of the degradation process

(degradation) states and the failure states, see Fig. 22.1, where the states are denoted by the letter s, normal, degradation and failure states are joined into the subsets N, D and F, respectively, and possible transitions are shown with arrows.

In the simplest case it is supposed that

- the nature of the degradation process admits an ordering of the system states in the sense of its age $E = \{1, 2, \ldots, n, n+1 = F\}$ with only one failure state $F = n+1$;
- the aging is described by the transition from state i to state $i+1$ (gradual failures) with intensity λ_i;
- the self-regulation process is described by the transition from state i to state $i-1$ with intensity μ_i;
- some "accident" failures are also possible, and they transfer the process from any state i to the failure one with intensity $v \ll \lambda_i$;
- the inner control (preventive repair of the system) allows to transfer the process from some state k to another state $l < k$ with intensity μ_{kl}.

22.2.2 The Process and Control Policy

To model the system behavior consider an absorbing process of birth and death type with finite states space $E = \{1, \ldots, n, n+1 = F\}$ and absorption at state F and denote by $X = \{X(t)\}_{t \geq 0}$ the state of the system at time t.

The model is assumed to involve a two-level threshold control policy f_{kl} with levels k and l, $(l < k)$. According to this policy one has to choose the states $k = \overline{1, n}$ for the beginning of the preventive repair and $l = \overline{0, k-1}$ for the starting state after this repair, respectively, i.e. the deep of the repair is equal to $k - l$, $1 \leq k - l \leq k$. It implies that the preventive recovery can be complete or partial. When the decision about preventive repair is taken the process can pass from state k to state l with given intensity μ_{kl}.

Obviously, the process $\{X(t)\}_{t \geq 0}$ is a Markov absorbing process with state space E and transition intensities

$$
\alpha_{ij} = \begin{cases}
\lambda_i, & \text{for } j = i+1, \quad i = 0, \ldots, n \ (n+1 = F), \\
\mu_i, & \text{for } j = i-1, \quad i = 1, \ldots, n, \\
v, & \text{for } j = n+1 = F \quad i = 0, \ldots, n-1, \\
\mu_{kl}, & \text{for } i = k, \quad j = l.
\end{cases}
$$

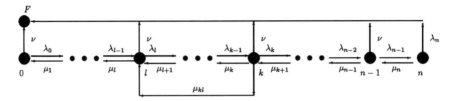

Fig. 22.2 The transition diagram for controlable degradation model under two-level policy f_{kl}

The transition diagram is presented in Fig. 22.2.

Denote by T the lifetime of an object (time to absorbtion)

$$T = \inf\{t : X(t) = F\} \tag{22.1}$$

and by $\pi_i(t)$—the state probabilities of the process X at epoch t of its lifetime T,

$$\pi_i(t) = P\{X(t) = i, \ t < T\}, \quad i \in E. \tag{22.2}$$

22.2.3 Cost Structure and Risk Functional

The following notations are used throughout the paper:

c_i—the reward per unit of time the system being in state i
c_{kl}—the fixed repair cost in fault stage k with return to stage l
S_n—jump point s of the process X.

In these notations the reward functional takes the form

$$L(T) = \int_0^T \sum_{0 \leq i \leq n} c_i 1_{\{X(u)=i\}} du - c_{kl} \sum_{S_n \leq T} 1_{\{X(S_n-0)=k, X(S_n+0)=l\}}. \tag{22.3}$$

and as the objective functional g_{kl} of the system we consider the mean reward during the lifetime $g_{kl} = E^{f_{kl}}[L(T)]$.

In special cases when $c_i = 1, \ c_{kl} = 0$ this functional gives mean lifetime of the system T

$$r_{kl} = E^{f_{kl}}[T] = E^{f_{kl}}\left[\int_0^T \sum_{0 \leq i \leq n} 1_{\{X(u)=i\}} du \right].$$

In this paper we are focusing on the maximization of the objective functions $g_{k,l}$ and $r_{k,l}$, i.e. $g^* = \max_{f_{k,l}}\{g_{kl}\}$, $r^* = \max_{f_{k,l}}\{r_{kl}\}$ and on the evaluation of the corresponding optimal control policies f^* and $\hat{f}^*$, where

$$f^* = \arg\max_{f_{kl}}\{g_{kl}\}, \quad \hat{f}^* = \arg\max_{f_{kl}}\{r_{kl}\}.$$

In the following sections, the characteristic measures calculated for the optimal policies f^* and $\hat{f}^*$ will be labeled by the upper index $*$.

22.3 The Problem Solution

For the considered model the state probabilities $\pi_i(t)$ of a system during its lifetime T satisfy the Kolmogorov's system of differential equations

$$\begin{cases} \dot{\pi}_i(t) = \lambda_{i-1}\pi_{i-1}(t)1_{\{i\geq 1\}} - (\lambda_i + \mu_i + v)\pi_i(t) + \mu_{i+1}\pi_{i+1}(t)1_{\{i\leq n-1\}}, \\ \qquad\qquad i = \overline{0,n}\ i \neq l,\ i \neq k, \\ \dot{\pi}_l(t) = \lambda_{l-1}\pi_{l-1}(t) - (\lambda_l + \mu_l + v)\pi_l(t) + \mu_{l+1}\pi_{l+1}(t) + \mu_{kl}\pi_k(t), \\ \dot{\pi}_k(t) = \lambda_{k-1}\pi_{k-1}(t) - (\lambda_k + \mu_k + v + \mu_{kl})\pi_k(t) + \mu_{k+1}\pi_{k+1}(t), \\ \dot{\pi}_{n+1}(t) = \lambda_n\pi_n(t) + v\sum_{0\leq i\leq n}\pi_i(t). \end{cases} \quad (22.4)$$

with initial conditions

$$\pi_0(0) = 1, \quad \pi_i(0) = 0, \quad i \in E, \quad i \neq 0. \quad (22.5)$$

Here and later, derivatives are denoted with upper point, while sign of prime is used for transposition of matrices and vectors.

Theorem 1 *The reward functional for the model has the form*

$$g_{kl} = \sum_{0\leq i\leq n} c_i \int_0^\infty \pi_i(u)du - c_{kl}\mu_{kl}\int_0^\infty \pi_k(u)du. \quad (22.6)$$

Proof Taking into account the form of reward functional (22.3) by simple calculations one has

$$g_{kl} = E^{f_{kl}}[L[X(t), \, t \le T]$$

$$= E^{f_{kl}}\left[\int\limits_0^T \sum\limits_{0 \le i \le n} c_i \mathbb{1}_{\{X(u)=i\}} du - c_{kl} \sum\limits_{S_n \le T} \mathbb{1}_{\{X(S_n-0)=k, X(S_n+0)=l\}}\right]$$

$$= E^{(k,l)}\left[\int\limits_0^\infty \sum\limits_{0 \le i \le n} c_i \mathbb{1}_{\{X(u)=i, \, u \le T\}} du - c_{kl} \sum\limits_{S_n} \mathbb{1}_{\{X(S_n-0)=k, X(S_n+0)=l, S_n \le T\}}\right]$$

$$= \int\limits_0^\infty \left[\sum\limits_{0 \le i \le n} c_i \pi_i(u) - c_{kl}\mu_{kl}\pi_k(u)\right] du$$

$$= \sum\limits_{0 \le i \le n} c_i \int\limits_0^\infty \pi_i(u) du - c_{kl}\mu_{kl} \int\limits_0^\infty \pi_k(u) du.$$

Corollary 1 *In terms of Laplace transforms*

$$\tilde{\pi}_i(s) = \int\limits_0^\infty e^{-st}\pi_i(t) dt$$

of the probabilities $\pi_i(t)$ the reward functional takes the form

$$g_{kl} = \sum\limits_{0 \le i \le n} c_i \tilde{\pi}_i(0) - c_{kl}\mu_{kl}\tilde{\pi}_k(0).$$

Therefore, the problem is reduced to the last expression of maximization with respect to different policies.

The Equations (22.4) with initial conditions (22.5) in terms of Laplace transforms take the form

$$\begin{cases} s\tilde{\pi}_i(s) = \mathbb{1}_{\{i=0\}} + \lambda_{i-1}\tilde{\pi}_{i-1}(s)\mathbb{1}_{\{i \ge 1\}} - (\lambda_i + \mu_i + \nu)\tilde{\pi}_i(s) + \mu_{i+1}\tilde{\pi}_{i+1}(s)\mathbb{1}_{\{i \le n-1\}}, \\ \qquad\qquad\qquad\qquad\qquad\qquad\qquad\qquad i = \overline{0,n} \; i \ne l, \; i \ne k, \\ s\tilde{\pi}_l(s) = \lambda_{l-1}\tilde{\pi}_{l-1}(s) - (\lambda_l + \mu_l + \nu)\tilde{\pi}_l(s) + \mu_{l+1}\tilde{\pi}_{l+1}(s) + \mu_{kl}\tilde{\pi}_k(s), \\ s\tilde{\pi}_k(s) = \lambda_{k-1}\tilde{\pi}_{k-1}(s) - (\lambda_k + \mu_k + \nu + \mu_{kl})\tilde{\pi}_k(s) + \mu_{k+1}\tilde{\pi}_{k+1}(s), \\ s\tilde{\pi}_F(s) = \lambda_n\tilde{\pi}_n(s) + \nu \sum\limits_{0 \le i \le n} \tilde{\pi}_i(s). \end{cases} \qquad (22.7)$$

The form of these equations allows to get their solution in closed form.

Theorem 2 *The Laplace transforms $\tilde{\pi}_i(s)$, $i \in E\backslash\{F\}$ satisfy to the following relations*

$$\tilde{\pi}_i(s) = q_{0i}(s) = \prod_{j=0}^{i} M_j(s), \tag{22.8}$$

where

$$M_n(s) = \rho_n(s),$$

$$M_j(s) = \frac{\rho_j(s)}{1 - \tau_j(s)M_{j+1}(s) - \theta(s)q_{l+1k}(s)1_{\{j=l\}}}, \quad j = \overline{0, n-1},$$

$$q_{l+1k}(s) = \prod_{j=l+1}^{k} M_j(s), \quad q_{jj-1}(s) = 1$$

and

$$\rho_0(s) = \frac{1}{s + \lambda_0 + v}, \quad \rho_j(s) = \frac{\lambda_{j-1}}{s + \lambda_j + \mu_j + v + \mu_{kl}1_{\{j=k\}}}, \quad (j = \overline{1, n-1}),$$

$$\rho_n(s) = \frac{\lambda_{n-1}}{s + \lambda_n + \mu_n + v + \mu_{kl}1_{\{k=n\}}},$$

$$\tau_0(s) = \frac{\mu_1}{s + \lambda_0 + v}, \quad \tau_j(s) = \frac{\mu_{j+1}}{s + \lambda_j + \mu_j + v + \mu_{kl}1_{\{j=k\}}}, \quad (j = \overline{1, n-1}),$$

$$\theta(s) = \frac{\mu_{kl}}{s + \lambda_l + \mu_l1_{\{l \geq 1\}} + v}.$$

Proof The equation for the $\tilde{\pi}_n(s)$ of system (22.7) implies the relation $\pi_n(s) = \rho_n(s)\pi_{n-1}(s)$, where $\rho_n(s)$ is defined as above. By virtue of the backward substitution scheme we find that the values $\tilde{\pi}_i(s)$ are of the form

$$\tilde{\pi}_i(s) = M_i(s)\tilde{\pi}_{i-1}(s), \quad i = \overline{0, n},$$

where the values $M_i(s)$ after some algebra can be recursively calculated as shown in the statement. By implementing the further substitution we get a product from solution (22.8) that completes the proof.

For numerical results presentation it is more convenient to have recursive expressions for calculation functionals g_{kl} and r_{kl}. For these reasons denote by $\tilde{r}_i(s) = E^{f_{kl}}[e^{-sT}|X(0) = i]$ the conditional Laplace transform of the residual life time T given initial state $i \in E\backslash\{F\}$. Since the process stays in state i exponentially distributed time with parameter $a_i = \lambda_i + \mu_i1_{\{i \geq 1\}} + v$ and jumps to state $i - 1$ with probability λ_i/a_i to state $i + 1$ with probability μ_i/a_i and to state F with probability v/a_i where appropriate conditional Laplace transforms are equal, respectively $r_{i-1}(s)$, $r_{i+1}(s)$ and 1, therefore for the conditional Laplace transforms $\tilde{r}_i(s)$, $i \in E\backslash\{F\}$ of the residual lifetimes the following equations hold

$$r_i(s) = \frac{\lambda_i}{a_i}\frac{a_i}{s + a_i}r_{i+1}(s) + \frac{\mu_i}{a_i}\frac{a_i}{s + a_i}r_{i-1}(s) + \frac{v}{a_i}\frac{a_i}{s + a_i},$$

which after some algebra gives

$$(s + \lambda_i + \mu_i 1_{\{i \geq 1\}} + v)\tilde{r}_i(s) = v + \lambda_i \tilde{r}_{i+1}(s) + \mu_i \tilde{r}_{i-1}(s)1_{\{i \geq 1\}},$$
$$i = \overline{0, n-1}\backslash\{k\},$$
$$(s + \lambda_k + \mu_k + \mu_{kl} + v)\tilde{r}_k(s) = v + \lambda_k \tilde{r}_{k+1}(s) + \mu_k \tilde{r}_{k-1}(s) + \mu_{kl}\tilde{r}_l(s),$$
$$(s + \lambda_n + \mu_n)\tilde{r}_n(s) = \lambda_n + \mu_n \tilde{r}_{n-1}(s).$$
\tag{22.9}

Taking into account the structure of these system of equations its solution can be represented in the form given in the following theorem.

Theorem 3 *The Laplace transforms* $\tilde{r}_i(s)$, $i \in E\backslash\{F\}$ *of the conditional residual time* T_i *distributions satisfy the following relations*

$$\tilde{r}_i(s) = \hat{q}_{in}(s) + \sum_{j=i-1}^{n-1} \hat{q}_{ij}(s)L_{j+1}(s), \ i = \overline{0, n},$$
\tag{22.10}

where

$$\hat{M}_0(s) = \hat{\rho}_0(s), \quad L_0(s) = \sigma_0(s),$$

$$\hat{M}_j(s) = \frac{\hat{\rho}_j(s)}{1 - \hat{\tau}_j(s)\hat{M}_{j-1}(s) - \hat{\theta}(s)\hat{q}_{lk-1}(s)1_{\{j=k\}}}, \quad j = \overline{1, n},$$

$$L_j(s) = \frac{\sigma_j(s)1_{\{j<n\}} + \hat{\tau}_j(s)L_{j-1}(s)}{1 - \hat{\tau}_j(s)\hat{M}_{j-1}(s) - \hat{\theta}(s)\hat{q}_{lk-1}(s)1_{\{j=k\}}}, \quad j = \overline{1, n},$$

$$\hat{q}_{lk-1}(s) = \prod_{j=l}^{k-1}\hat{M}_j(s), \quad \hat{q}_{jj-1}(s) = 1,$$

and

$$\hat{\rho}_0(s) = \frac{\lambda_0}{s + \lambda_0 + v}, \quad \hat{\rho}_j(s) = \frac{\lambda_j}{s + \lambda_j + \mu_j + v + \mu_{kl}1_{\{j=k\}}}, \quad j = \overline{1, n-1},$$

$$\hat{\rho}_n(s) = \frac{\lambda_n}{s + \lambda_n + \mu_n + \mu_{kl}1_{\{k=n\}}},$$

$$\hat{\tau}_j(s) = \frac{\mu_j}{s + \lambda_j + \mu_j + v1_{\{k<n\}} + \mu_{kl}1_{\{j=k\}}}, \quad j = \overline{1, n},$$

$$\hat{\theta}(s) = \frac{\mu_{kl}}{s + \lambda_k + \mu_k + v1_{\{k<n\}} + \mu_{kl}},$$

$$\sigma_0(s) = \frac{v}{s + \lambda_0 + v}, \quad \sigma_j(s) = \frac{v}{s + \lambda_j + \mu_j + v + \mu_{kl}1_{\{j=k\}}}, \quad j = \overline{1, n-1}.$$

Proof From system (22.9) one can get the relation $\tilde{r}_0(s) = \hat{\rho}_0(s)\tilde{r}_1(s) + \sigma_0(s)$. Now by applying the forward substitution scheme we get

$$\tilde{r}_i(s) = \hat{M}_i(s)\tilde{r}_{i+1}(s) + L_i(s), \quad i = \overline{0,n},$$

where the involved variables $\hat{M}_i(s)$ and $L_i(s)$ after some algebra satisfy the recurrent relations defined in the statement. Hence one can easily obtain the formula (22.10) that completes the proof.

Corollary 2 *The latter result obviously implies the relation for the reliability function $R(t)$ and mean lifetime $r_{k,l}$ in the form*

$$\tilde{R}(s) = \frac{1}{s}(1 - \tilde{r}_0(s)), \quad r_{kl} = -\frac{d}{ds}\tilde{r}_0(s)\big|_{s=0}.$$

Denote by N the number of the preventive repairs during the life cycle, by $\psi_i(j) = P[N = j/X(0) = i]$ the conditional probability distribution of an object given initial state i, and by $\tilde{\psi}_i(z) = E[z^{N_i}/X(0) = i]$ the corresponding probability generating function. This function satisfies the system of Equations (22.9) for $s = 0$, but the service rate μ_{kl} is replaced by $z\mu_{kl}$.

Corollary 3 *The generating function $\tilde{\psi}_i(z)$, $i = \overline{0,n}$ of the number of the preventive repairs during the life cycle of an object given initial state i satisfies the relation*

$$\tilde{\psi}_i(z) = \hat{q}_{in}(z) + \sum_{j=i-1}^{n-1} \hat{q}_{ij}(z)L_{j+1}(z), \quad i = \overline{0,n}, \tag{22.11}$$

where the variables involved are the same as in Theorem 2 by setting $s = 0$ and defining

$$\hat{M}_k(z) = \frac{\hat{\rho}_k(0)}{1 - \hat{\tau}_k(0)\hat{M}_{k-1}(0) - z\hat{\theta}(0)\hat{q}_{lk-1}(0)},$$

$$L_k(z) = \frac{\sigma_k(0)1_{\{k<n\}} + \hat{\tau}_k(0)L_{k-1}(0)}{1 - \hat{\tau}_k(0)\hat{M}_{k-1}(0) - z\hat{\theta}(0)\hat{q}_{lk-1}(0)}.$$

Differentiation of the function $\tilde{\psi}_0(z)$ over the parameter z by setting $z = 1$ leads to the mean number of the preventive repairs ψ_{kl} during a lifetime.

22.4 Numerical Examples

In this section we propose some numerical examples for evaluation of optimal policies starting from the absolutely new state 0 up to the complete failure state F:

1. to maximize the mean reward $g_{k,l}$ for the given cost structure,
2. to maximize the mean lifetime $r_{k,l}$.

To make the numerical examples more suitable to the real situations it should be assumed that

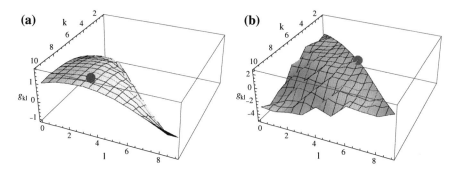

Fig. 22.3 Mean reward g_{kl} with labeled optimal control policy $f^* = (9, 3)$ for Example 1 (**a**) and the same for Example 2 (**b**), where $f^* = (6, 5)$

$$\lambda_i \gg \nu\,(i = \overline{0, n}), \quad \lambda_i > \mu_i\,(i = \overline{1, n}),$$
$$\mu_{ij} \geq \mu_{i+1j}\,(i = \overline{1, n-1}, j = \overline{0, i-1}), \quad \mu_{ij} \leq \mu_{ij+1}, \quad (i = \overline{1, n}, j = \overline{0, i-2}).$$

In the Figs. 22.3 and 22.4 the mean reward g_{kl} with optimal control policy for Example 1 (a) and for Example 2 (b), and appropriate mean lifetime r_{kl} with optimal control policy for Example 1 (a) and for Example 2 (b) are shown.

Further, we fix the values of the following system parameters:

$$n = 10, \; \nu = 0.001, \; c_i = 0.001i, \; i = \overline{0, n},$$
$$\vec{\lambda} = (0.009, 0.010, 0.012, 0.015, 0.026, 0.038, 0.058, 0.100, 0.180, 0.200, 0.230)',$$
$$\vec{\mu} = (0.001, 0.005, 0.006, 0.007, 0.013, 0.019, 0.029, 0.050, 0.090, 0.100)'$$

and, without loss of generality, suppose that

$$\mu_{ij} = \mu_{i-1j-1}, \quad c_{ij} = c_{i-1j-1}, \quad i = \overline{2, n}, j = \overline{1, i-1}.$$

Taking into account the above comments it is enough to present only the values of μ_{k0} and $c_{k0}, k = \overline{1, n}$, in order to define all possible elements μ_{kl} and c_{kl}, $i = \overline{1, n}, j = \overline{0, i-1}$. The following four examples are used to conduct the numerical analysis:

Example 1

$$[\mu_{k0}] = [3.429, 3.000, 2.571, 2.143, 1.714, 1.286, 1.000, 0.958, 0.643, 0.354], \; k = \overline{1, n},$$
$$[c_{k0}] = [0.054, 0.293, 0.432, 0.571, 1.720, 1.849, 3.988, 4.870, 6.760, 8.980], \; k = \overline{1, n}.$$

Example 2

$$[\mu_{k0}] = [1.636, 1.582, 1.527, 1.437, 1.428, 1.364, 1.309, 1.255, 1.200, 1.145], \; k = \overline{1, n},$$
$$[c_{k0}] = [0.364, 0.400, 0.436, 0.473, 0.509, 0.545, 0.582, 0.518, 0.655, 0.791], \; k = \overline{1, n}.$$

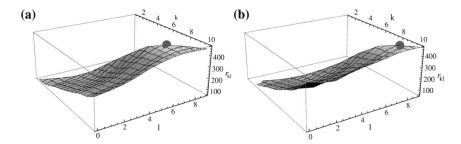

Fig. 22.4 Mean lifetime r_{kl} with labeled optimal control policy $\hat{f}^* = (8,7)$ for Example 1 (**a**) and the same for Example 2 (**b**), where $\hat{f}^* = (9,8)$

Table 22.1 The values of the characteristic measures with and without preventive repair

Example no.	f^*	$\hat{f}^*$	g^*	g	r^*	r	ψ^*	K_1	K_2
1	(6,5)	(9,8)	2.121	0.942	482.809	462.745	8.303	0.103	0.006
2	(9,3)	(8,7)	1.402	0.942	480.801	462.745	4.906	0.699	0.460
3	(10,2)	(9,8)	0.926	0.942	482.809	462.745	8.303	0.103	1.000
4	(10,3)	(10,9)	1.550	0.942	482.552	462.745	9.143	0.700	0.460

Example 3

$$[\mu_{k0}] = [3.429, 3.000, 2.571, 2.143, 1.714, 1.286, 1.000, 0.958, 0.643, 0.354], \; k = \overline{1,n},$$
$$[c_{k0}] = [1.000, 1.000, 1.000, 1.000, 1.000, 1.000, 1.000, 1.000, 1.000, 1.000], \; k = \overline{1,n}.$$

Example 4

$$[\mu_{k0}] = [6.327, 6.116, 5.905, 5.695, 5.484, 5.273, 5.062, 4.851, 4.640, 4.429], \; k = \overline{1,n},$$
$$[c_{k0}] = [0.364, 0.400, 0.436, 0.473, 0.509, 0.545, 0.582, 0.518, 0.655, 0.791], \; k = \overline{1,n}.$$

In Example 1, the earlier failure stages have the high preventive repair intensities and the low repair costs while the later stages—vice versa. Example 2 includes relative equal and enough large repair intensities and low repair costs for all failure stages. In Example 3, the repair intensities are as in Example 1 while the costs are fixed to the value 1. Finally Example 4 considers the case of very high repair intensities and very low repair costs for all the failure stages.

The numerical results are summarized in Table 22.1.

The proposed results confirm the fact that the system with controllable preventive repair can be significantly superior in mean reward and mean lifetime compared to the non-controllable case. Also it can be noticed that the coefficients of homogeneity $K_1 = \frac{\mu_{n0}}{\mu_{10}}$ and $K_2 = \frac{c_{n0}}{c_{10}}$ of the preventive repair intensities and costs

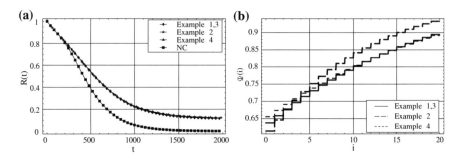

Fig. 22.5 The reliability function $R(t)$ (**a**) and distribution function $\Psi(i)$(**b**)

considerably influence the characteristic measures of the system as well as the optimal threshold levels. The functions of the mean reward g_{kl} and the mean lifetime r_{kl} for varied pairs (k, l) are illustrated in Figs. 22.3 and 22.4 (labeled by "a" and "b"), respectively for Examples 1 and 2.

Next, the diagram in Fig. 22.5 illustrates the reliability function $R(t)$ (labeled by "a") and distribution function $\Psi(i) = \sum_{j=0}^{i} \psi_0(j)$ (labeled by "b") for the number of preventive repairs. We notice that three functions for Examples 1–4 are graphically indistinguishable in the displayed domain. However, the function for Example 2 has a heavier tail. The curve in non-controllable case (NC) lies far below the others. In the right diagram the jump at the point $i = 0$ equals the probability that the complete failure state F reaches earlier than state k, where the preventive repair is possible. We see that the probability $\Psi(0)$ for Example 2 is the smallest one while with increasing i it converges more quickly to the unit value. The functions for Examples 1,3 and 4 have higher values $\Psi(0)$ but they exhibit the heavier tails. Moreover, the numerical examples show that the optimal policy $\hat{f}^* = \arg\max_{f_{kl}}\{\psi_{kl}\}$ leads to the maximum value ψ^*.

22.5 Conclusion

The two-level control policy seems to be quite appropriate for modeling some special maintenance programs in reliability engineering when a degradation process of a unit is observable and there is a possibility to maintain a signal system.

In the particular case of the Markov degradation process the algorithm for the calculation of the reliability characteristics is given. This model can be applied for design and simulation of preventive maintenance programs with observable degradation process.

As the further problems the problem of optimal control of the systems at *random* life cycle (with random life resource) and for more general degradation models (for example generalized Birth & Death process) should be considered.

Generalized Birth & Death processes, which are a special class of Semi-Markov processes are introduced for modeling the degradation processes. The special parameterization of the processes allows to give a more convenient presentation of the results. Special attention is focused on the conditional state probabilities given life cycle, which are the most interesting for the degradation processes.

Acknowledgments This paper was partially supported by the RFFI Grant No. 08-07-90102.

References

Dimitrov B, Green D, Rykov V, Stanchev P (2004) Reliability model for biological objects. In: Antonov V, Huber C, Nikulin M, Polischook V (eds) Longevity, Aging and degradation models. Transactions of the first Russian-French conference (LAD-2004), Saint Petersburg, June 7–9, 2004, pp. 230–240

Esary JD, Marshall AW, Proshan F (1973) Shock models and wear processes. Ann Prob 1(1):627–649

Lisniansky A, Levitin G (2003) Multi-state system reliability assessment optimization and application. World Scientific, Singapore

Rykov V (2006a) Generalized birth and death processes as degradation models. In: Vonta F, Nikulin M, Limnios N, Huber-Carol C (eds) Statistical methods for biomedical and technical systems. Birkhauser, Boston, pp 95–108

Rykov V (2006b) Generalized birth and death processes and their application to aging models. Autom Remote Control 3:103–120 (in Russian)

Rykov V, Buldaeva E (2004) On reliability control of fault tolerance units: regenerative approach. In: Transactions of XXIV international seminar on stability problems for stochastic modes, September 10–17, 2004, Jurmala, Latvia, TTI, Riga, Latvia

Rykov V, Dimitrov B (2002) On multi-state reliability systems. In: Applied stochastic models and information processes. Proceedings of the international seminar "Applied stochastic models and information processes", Dedicated to the 60th Birthday of Vladimir Kalashnikov (8–13 September 2002, Petrozavodsk, Russia), pp 128–135

Rykov V, Efrosinin D (2004) Reliability control of biological systems with failures. In: Antonov V, Huber C, Nikulin M, Polischook V (eds) Longevity, aging and degradation models. Transactions of the first Russian-French conference (LAD-2004), Saint Petersburg, June 7–9, 2004; pp 241–255

Rykov V, Efrosinin D (2007) Risk analysis of controllable degradation model with preventive repaire. In: Bedford T, Walls L, Quigley J, Alkali B, Daneshkman A, Hardman G (eds) Proceedings of the international conference on mathematical methods in reliability (MMR2007), Glasgow, Scotland

Rykov V, Dimitrov B, Green D Jr, Snanchev P (2004). Reliability of complex hierarchical systems with fault tolerance units. In: Communication of fourth international conference on mathematical methods in reliability, methodology and practice, June 21–25, 2004, Santa-Fe, New Mexico

Singpurwalla ND (2006a) The hazard potential: introduction and overview. J Amer Stat Ass 101(476):1705–1717

Singpurwalla ND (2006b) Reliability and risk A Bayesian perspective. Wiley, Chichester